IET ENERGY ENGINEERING SERIES 175

n-Type Crystalline Silicon Photovoltaics

Other volumes in this series:

n-Type Crystalline Silicon Photovoltaics

Technology, applications and economics

Edited by
Delfina Muñoz and Radovan Kopecek

The Institution of Engineering and Technology

Published by The Institution of Engineering and Technology, London, United Kingdom

The Institution of Engineering and Technology is registered as a Charity in England & Wales (no. 211014) and Scotland (no. SC038698).

The Institution of Engineering and Technology
Futures Place
Kings Way, Stevenage
Hertfordshire SG1 2UA, United Kingdom

www.theiet.org

British Library Cataloguing in Publication Data
A catalogue record for this product is available from the British Library

ISBN 978-1-83953-176-7 (hardback)
ISBN 978-1-83953-177-4 (PDF)

Typeset in India by MPS Limited

Cover image: *Image credit: "ISC Konstanz"*

Contents

3   n-type silicon solar cells   69

Sébastien Dubois, Bertrand Paviet-Salomon, Jia Chen, Wilfried Favre, Raphaël Cabal, Armand Bettinelli and Radovan Kopecek

4   n-type silicon modules   **171**
Andreas Schneider, Emilio Muñoz-Cerón, Jorge Rabanal-Arabach,
Eszter Voroshazi, Vincent Barth and Rubén Contreras Lisperguer

About the editors

Dr Delfina Muñoz is a Strategic project manager and senior expert at the Solar Technologies Department of the French Atomic Energy and Alternative Energies Commission in France.

She started the heterojunction solar cells research in her Ph.D. at Universidad Politécnica de Cataluña developing laboratory heterojunction devices. In 2008, she joined CEA-INES and since then she has been improving heterojunction technology from the lab to the fab. She is very active at the European level leading international projects on heterojunction solar cell combining device structure with advanced concepts. She is on the board of ATAMOSTEC in Chile and at the ETIP PV steering committee contributing to roadmap and strategic agendas for PV. Her expertise goes from materials to modules with more than 100 contributions worldwide on PV. She still combines her project activity with the laboratory, directing Ph.D. students and developing heterojunction solar cells for the future photovoltaic technology.

Dr Radovan Kopecek obtain the Dipl. Phys. degree at the University of Stuttgart in 1998. In 2002, he completed his Ph.D. dissertation in the field of c-Si thin film silicon solar cells in Prof. Ernst Bucher´s group at the University of Konstanz. One of the founders of ISC Konstanz, Dr Kopecek is since 2007 the leader of the advanced solar cells department dealing with several European, national research projects and technology transfer in the field of silicon feedstock and solar cell development, focusing on n-type devices. Since 2022, he has been a board member of ESMC, advancing deployment of solar energy, and from 2022, board member of ATAMOSTEC in Chile. He has published numerous key papers.

List of contributors

The first n-type c-Si book for PV technology and economy was written by many n-type believers that support this technology by R&D, writing papers, organizing workshops and being involved in technology transfer of new cell and module concepts. The nPV workshop started in 2011 in Konstanz to connect R&D, industry and investors to bring future n-type technology faster and sustainable into the PV market to reach the lowest LCOEs in future.

Dr Jérémie Aimé received his Ph.D. from Grenoble Alpes University in 2009. He joined industry working for companies in France and Singapore from 2010 to 2019. He joined Commissariat à l'Energie Atomique et aux Energies Alternatives (CEA) in 2019 as the Head of Photovoltaics Systems Laboratory dealing with PV systems and power electronics activities before moving as the Head of Applied Photovoltaics Laboratory dealing with power modules durability, PV systems, and circular economy.

Dr Amran Al-Ashouri is a postdoctoral researcher in the Young Investigator Group Perovskite Tandem Solar Cells at Helmholtz-Zentrum Berlin (HZB). He focuses on the elimination of interfacial non-radiative recombination losses in tandem-relevant perovskite solar cells and received his Ph.D. from the Technical University Berlin for his studies on establishing self-assembled monolayers as high-performance hole-selective layers. He studied physics at the University of Duisburg-Essen, graduating in quantum optics, and dedicated part of his research on understanding how to overcome voltage and fill factor losses in perovskite solar cells. He is the co-founder of Quantum Yield Berlin GmbH (QYB), which provides a simple absolute photoluminescence setup for research acceleration, and is committed to improving climate crisis.

Vincent Barth received his Ph.D. in 2014 from Sorbonne University in the field of organic photovoltaic, working both on the molecule design and synthesis, and their use as a donor material or in hole transport layer in devices. He joined the National Institute of Solar Energy (CEA-INES) in 2017 to work on PV module technology. He works on the development and manufacturing of HJT modules, participating in several EU-funded projects (AMPERE, GOPV, Highlite) as well as industrial projects. His work focuses on cell interconnection and in particular on reducing the use of silver and other critical materials in photovoltaics.

Dr Armand Bettinelli received his Ph.D. in 1987 for his work on cofiring of alumina and tungsten at the Strasbourg University (France). He worked in the industry as a technical manager in the field of High and Low Cofired Ceramics then Plasma Display Panels, fields all using high-level screen-printing. In 2005, he joined CEA-INES where he hold a senior expert position in c-Si Solar Cell Metallization and Interconnection. From 2005, beside the optimization of the screen-printing for PV cells using mesh screens and stencils, Armand introduced and optimized at CEA the Smartwire^{TM} and the Ribbon gluing interconnections for heterojunction cells, more recently the shingling, taking always care of the link between cell metallization and cell interconnection quality.

Dr Mathieu Boccard received the Ph.D. degree in thin-film silicon tandem solar cells from the Photovoltaics and Thin-Film Electronics Laboratory (PV-Lab), Ecole Polytechnique Fédérale de Lausanne (EPFL), Neuchâtel, Switzerland, in 2012. He then joined Arizona State University, Tempe, AZ, USA, for postdoctoral research on high-efficiency crystalline silicon, CdTe-based, and perovskite/silicon tandem solar cells. From 2017 to 2022, he has been leading the silicon heterojunction activities in the PV-Lab, EPFL, working on fundamentals and novel materials for carrier-selective contacts, multijunction devices, and location-dependent solar cell design.

Dr Raphaël Cabal studied semiconductor physics at the Grenoble Institute of Technology, France, where he graduated with an engineer degree in microelectronics and a master in material science in 2007. His Ph.D. thesis (obtained in 2011 at INSA Lyon) focused on solar cell rearside passivation. Since then, he has been working as a project leader at CEA Tech INES on different topics such as cast-mono gettering and defect characterization, but also simplified processing homojunction n-type solar cells and more recently on poly-Si-based passivated contact solar cells.

Dr Jia Chen obtained his Ph.D. from the National University of Singapore. In 2014, he joined IMEC Belgium, focusing on efficiency improvement of n-type silicon solar cells. Since 2018, he is working in Jolywood China as the CTO responsible for their n-type cell and module research and technology, and led the team to build the world's first n-type TOPCon solar cell production line with GW level capacity.

Dr Gianluca Coletti is a program manager at TNO in the Netherlands. He is also adjunct professor at the University of New South Wales in Australia. Gianluca Coletti has a master in Physics from the University of Rome "La Sapienza" in Italy and obtained his PhD at the University of Utrecht in the Netherlands. From 2000 to 2003 worked at the research lab of Eurosolare SpA (Italy) and from 2004 to 2018 worked as a senior research at ECN Solar Energy (the Netherlands). Since 2018, he is a senior scientist and program manager of PV Tandem Technology and Applications at TNO – partner in Solliance. His research focuses on the development and technology transfer of innovative cells and modules technologies. He is one of the world recognized experts in the area of defects in silicon with 20 years of experience in academic and industrial R&D with institutes and companies worldwide. He is chairman of the Silicon Material for Photovoltaic Applications workshop.

Hervé Colin is a research Engineer at the French Institute for Solar Energy (INES), Le Bourget-du-lac, France, in the Photovoltaic System laboratory of the Commissariat à l'Energie Atomique et aux Energies Alternatives (CEA). He is specialized in PV system modeling and simulation, monitoring data treatment and analysis, and performance assessment. He has been involved or in charge of projects dealing with PV plants monitoring (monitoring equipment and data acquisition), performance assessment and fault detection/diagnosis, grid connection rules improvements, simulation of various types of PV systems (ground, floating, agrivoltaic, etc.), methods for production estimations, benchmark analysis of bifacial systems, solar road systems operation.

Dr Rubén Contreras Lisperguer has more than 25 years of experience working in the areas of energy, sustainable development, agriculture, and climate change. From 2002 to 2004, he conducted scientific research on Climate Change in the Antarctic. From 2006 to 2015, he was a Senior Energy and Climate Change Specialist at the Organization of American States (OAS), Washington DC. Subsequently from 2015 to 2017, he was the Regional Program Officer at the International Renewable Energy Agency (IRENA) based in Abu Dhabi, United Arab Emirates. Since 2018, he serves as the Officer leading the Energy Agenda at the United Nations Economic Commission for Latin America and the Caribbean (UN-ECLAC). Mr Contreras Lisperguer is an Engineer in Meteorology, with a Master of Science in Solar PV Technologies, and Master of Engineering in Risk Prevention. He received his Ph.D. in Renewable Energies (cum laude) from the University of Jaen in 2020, where he developed research and proposed a novel theoretical limit for the application of the circular economy in the PV industry.

Dr Sébastien Dubois studied materials' sciences at the National Institute for Applied Sciences (INSA) in Lyon and graduated with an engineering degree in 2004. In 2007, he received his Ph.D. in Physics from Marseille University. After joining CEA-INES in 2007, he conducted and supervised researches in the field of defect engineering and homojunction solar cells. He is currently responsible for the Alternative Cells Laboratory of CEA, with activities on high-temperature passivated contacts, heterojunction cells, and tandem devices. He is the member of the Scientific Committees of several international conferences and teaches the basics of silicon photovoltaics at the University of Nantes (France).

Dr Wilfried Favre, Head of the Heterojunction Solar Cells Laboratory at CEA. He obtained his Ph.D. degree in Physics from the Université d'Orsay (Paris 11) in 2011 for his work on characterization and modeling of heterojunction devices at GeePs Laboratory (CentraleSupelec, Paris). He then joined CEA at INES (2 years post-doc then permanent position) to develop high efficiency heterojunction solar cells at lab and pre-industrial stages (several MW pilot line at CEA). Since 2016, he is working on industrial transfer of the silicon heterojunction technology to production lines notably the Enel Green Power one based in Catania, Sicily. In 2020, he took the lead of the heterojunction solar cells laboratory (>35 persons). He has co-authored more than 50 communications in journals and international conferences, collected several patents, and supervised Ph.D. students and postdoctorates.

Sean Gaal is a senior scientist with Dow, working with the production of silicon, chlorosilanes and other related processes since 2013. His first industrial role was with BHP Billiton in Australia, after which he completed a PhD in Process Metallurgy from the University of New South Wales, followed by a Post Doc at the Norwegian University of Science and Technology. In 2005, he joined SINTEF, researching a range of metallurgical processes, then progressing in 2010 to Norsun where he conducted research and development for the production of CZ ingots. His experience covers a wide range of metallurgical processes, always seeking to apply his knowledge to new problems.

Dr Nouha Gazbour is an expert on Life Cycle Assessment and Photovoltaic Market and a sustainability manager in CEA INES. She has been coordinating the activity of environmental and economic analysis of photovoltaic systems within the CEA at INES since 2015. She has a deep expertise on the photovoltaic solar energy market, its environmental impact and its economic interest. Since 2019, she was heavily involved in the European Innovation FUND which is part of the DG CLIMATE program for the development of innovative low carbon technologies. She has been in charge of several environmental and economic modelling tasks for strategic bilateral industrial projects, and has enjoyed working with many experts in the PV field. She holds an engineering degree from the National Engineering School of Monastir (Tunisia) and a Ph.D. in environmental engineering from the University of Grenoble and the Art et Métiers Institute (ENSAM) in 2019. During her thesis, she developed a tool that allows a systemic integration of eco-design in photovoltaic technologies from the R&D stage through Life Cycle Assessment (LCA). The tool developed is now a key tool to guide technological choices at INES and has been transferred to several other sectors such as microelectrics, batteries, and hydrogen.

Stéphane Guillerez joined the Direction of Energy Programs (DPE) of the Commissariat à l'Energie Atomique et aux Energies Alternatives (CEA) as a program manager in charge of solar photovoltaics since march 2022. Before he was research engineer and occupied research management positions inside the Solar Technology Department at INES. His technical skills are materials and photovoltaic modules.

Harry Guo obtained the bachelor's degree at the University of Liverpool and his major was Electrical Engineering in 2013. In 2014, he completed his master's degree at the University of Hong Kong and major was energy engineering. Since 2015, he joined LONGi Green Energy Technology Co., Ltd. as a technical support manager and product manager, who is mainly responsible for continuous quality improvement and theoretical study in the field of silicon wafer and verification collaboration with global institutes focusing on helping chasing champion solar cells.

Can Han is a researcher in Photovoltaic Materials and Devices (PVMD) group at Delft University of Technology which she started in September 2021. From 2017 to 2021, she finished her Ph.D. project at the PVMD group. Her Ph.D. research topic is contacting schemes for crystalline silicon solar cells, which includes high-mobility transparent conductive oxides and bifacial copper-plating metallization approach development. Her special interest lies in bifacial solar cell design with reduced consumptions of indium and silver. From 2013 to 2017, she was an engineer at Ningbo Institute of Materials Technology & Engineering, CAS, China. The topic was high-performance industrial crystalline silicon solar cells.

Prof Dr Olindo Isabella received the Ph.D. degree (cum laude) from Delft University of Technology for his work on advanced light management for thin-film silicon solar cells. After a visiting research period at the National Institute of Advanced Industrial Science and Technology, Tsukuba (Japan), he started in 2013 his academic career in the Photovoltaic Material and Devices group at Delft University of Technology. His areas of expertise are opto-electrical modelling, implementation of advanced light management techniques in thin-film and wafer-based silicon solar cells as well as related supporting materials, smart PV modules and advanced energy yield modelling of PV systems. He leads the Photovoltaic Materials and Devices group at TU Delft since 2019 and in 2021, he was appointed full professor in PV technologies and applications. Olindo is the Director of the TU Delft PV Technology Centre and is the principal investigator of Solar Urban thematic area of TU Delft Urban Energy Institute and of Urban Energy at the Institute of Advanced Metropolitan Solutions in Amsterdam.

Dr Markus Klenk studied physics at the University of Constance where he also received his Ph.D. on $CuGaSe_2$ thin film photovoltaics in 2001. He started his work in the PV industry at the Sunways AG where he worked until 2010, first in the R&D, later as Head of Quality Assurance for cells & modules. This was followed by positions as Senior Technologist in the "integrated factory" – department of the centrotherm PV AG and the rct solutions GmbH. Since 2015, he is working as a research associate at the ZHAW (Zurich University of Applied Science) in Switzerland, where he continues his PV activities.

Dr Eike Köhnen studied nanotechnology at the Leibniz University Hannover, where he obtained his Master of Science degree in 2017. Afterwards he joined the young investigator group perovskite tandem solar cells at Helmholtz-Zentrum Berlin (HZB). He focused on improving the performance of monolithic perovskite/silicon tandem solar cells by optimizing optical and electrical properties of the perovskite subcell as well as investigating the subcell performance within the device. In 2021, he obtained his Ph.D. by the Technical University Berlin and continued his work on tandem solar cells as a postdoc at HZB.

Dr Lars Korte is a deputy group leader of the Young Investigator Group Perovskite Tandem Solar Cells at Helmholtz-Zentrum Berlin (HZB). He studied physics in Göttingen, Germany, then went to Laboratoire de Chimie Métallurgique des Terres Rares, CNRS, France, as a visiting researcher. He carried out his Ph. D. research at Hahn-Meitner-Institut Berlin, and obtained his Ph.D. in physics from U Marburg in 2006. Lars is permanent staff member at HZB since 2010. As deputy head of HZB's Institute for Silicon Photovoltaics, he led the research on amorphous/silicon heterojunction solar cells. Since 2015, his work is focusing on perovskite/silicon tandem cells. Lars' research interests include the electronic properties of thin films and heterointerfaces; materials and device characterization with photoelectron spectroscopy (XPS, UPS, near-UV) and photoelectrical methods; high efficiency silicon heterojunction solar cells and perovskite-/silicon-based tandem cells. He is author/co-author of more than 150 publications, with an h-index of 39. Lars is an appointed lecturer at Technical University Berlin's faculty Electrical Engineering.

Dr Joris Libal joined ISC Konstanz in 2012 where he works as a R&D project manager, focusing on business development and technology transfer in the areas of high-efficiency n-type solar cells and innovative module technology. He received his diploma in physics from the University of Tübingen and a Ph.D. in the field of n-type crystalline silicon solar cells from the University of Konstanz in 2006. He subsequently conducted postdoctoral research at the Università di Milano-Bicocca in Italy, where he investigated the properties of solar grade silicon. From 2008 until 2012, he worked as a R&D manager for the Italian company Silfab, being responsible for the company's internal and external research and development projects in the field of solar cells and PV modules. He further coordinated the PV module certification process for the company's production sites in Canada and Croatia.

Dr Liu received her bachelor of engineering (Honours, first class) in 2010 and PhD in engineering (photovoltaics) in 2015, both from the Australian National University (ANU). She has since worked as a researcher in the solar PV group at ANU. She was awarded an ACAP Postdoctoral Research Fellowship in 2018. She received the Ulrich Gösele Young Scientist Award in 2022, in recognition of her research contribution to gettering and defect engineering in silicon photovoltaics.

Prof Dr Emilio Muñoz Cerón finished Industrial Engineering in 2008, receiving his Ph.D. in 2014 from the University of Jaén (Spain). He currently works as an associate professor (Senior Lecturer) in the area of engineering projects at the University of Jaén. His main R&D lines are focused on the technical characterization and economic analysis of operating PV projects, specially focused on bifacial technology. He is also involved in the performance analysis of Floating PV systems and in the design of PV solutions such as self-consumption adapted to consumption profiles and infrastructure integration systems (hybrid PV-noise barriers).

Eivind Øvrelid is a senior scientist in SINTEF Materials and Chemistry. Øvrelid has been working with PV technology with main focus on crystal growing research and development since 1999. He is the author of more than 60 technical articles published in peer-reviewed international journals, co-editor of "solar Silicon Processes" ISBN 9780367874780.

Dr Bertrand Paviet-Salomon received the M.Sc. degree in 2009 and the Engineer Diploma degree in theoretical and applied optics from the Institut d'Optique, Paris, France. From 2009 to 2012, he was pursuing the Ph.D. degree with the French National Institute for Solar Energy, Le-Bourget-du-Lac, France, working on laser processes for crystalline silicon solar cells. He received the Ph.D. degree in electronics and photonics from the University of Strasbourg, Strasbourg, France, in 2012. From 2012 to 2014, he was a Postdoctoral Researcher with the Photovoltaics and Thin-Film Electronics Laboratory, École Polytechnique Fédérale de Lausanne, Neuchâtel, Switzerland, working on high-efficiency back-contacted silicon heterojunction solar cells. In 2014, he joined the PV-Center, Centre Suisse d' Électronique et de Microtechnique, Neuchâtel, Switzerland, and since 2021, the group leader of the crystalline silicon solar cells activities there.

Dr Paul Procel received his Ph.D. degree from the University of Calabria, Rende, Italy in 2017 for his research in numerical simulations of advanced silicon solar cells. In 2015, for an internship, he joined the Photovoltaic Materials and Devices Group at Delft University of Technology, where he worked on advanced opto-electrical simulations of interdigitated back contacted solar cells. From 2017, he joined Delft University of Technology as a post-doc researcher focusing on design and development of silicon heterojunction solar cells. His research interests include the opto-electrical properties of materials and interfaces, simulations, and design of solar cells and semiconductor devices.

Prof Dr Jorge Rabanal-Arabach is full time assistant professor at the University of Antofagasta, and researcher at the Center for Energy Development Antofagasta (CDEA). He obtained his B.Eng. degree and Dipl.-Eng. professional title both in the topics of electronics engineering from the University of Antofagasta, Chile, in 2010. After finishing his studies, he joined the CDEA as a researcher collaborator. In 2014, he joined ISC Konstanz, working for the module department focusing on the development of bifacial modules and its adaptation to operate under desert climates. In 2019, he obtained a Dr.rer.nat. degree from the

University of Konstanz for his work in the field of physics, with experimental research on photovoltaics for desert applications. Since 2019, he is a full-time assistant professor at the University of Antofagasta and member of the IEEE since 2007. His current research is focused on photovoltaic applications, electromobility, and applied robotics.

Dr Fiacre Rougieux has a PhD from the Australian National University in the field of photovoltaics and semiconductor materials. Between 2012 and 2015, he was an ARENA postdoctoral fellow at the ANU where he developed high-efficiency and low-cost solar cell concepts. Between 2016 and 2018, he was an ARC DECRA fellow at the ANU where he explored the physics of defects in high-efficiency devices and successfully developed a wide range of processes to remove defects in solar cells and improve their efficiency. Since 2018, Fiacre is a senior lecturer at UNSW where he leads a team focusing on mitigating underperformance in PV cells, modules, and systems.

Prof Dr Andreas Schneider obtained his diploma in physics from the University of Freiburg in 1999. He obtained his Ph.D. at Professor Ernst Bucher's Department of Applied Solid State Physics at the University of Konstanz in 2004. From 2005, he worked for Day4Energy in Vancouver as head of the company's R&D department and established in 2008 Day4Energy's quality management department. Subsequently Dr Schneider joined the American company Jabil where he held the position of advanced engineering manager and was responsible for the scientific supervision of solar panel production. In 2011, Andreas joined ISC Konstanz where he has been responsible for the newly founded Module Development Department. Since 2016, he is a full-time professor for Electronic Parts, Circuits and Renewable Energy at the University of Applied Sciences Gelsenkirchen.

Jordi Veirman studied semiconductor physics at the National Institute for Applied Sciences (INSA) in Lyon, France, where he graduated with an engineering degree and a masters in micro-electronics in 2008. Since his PhD obtained from INSA in 2011, he has dedicated a large part of his activities to the study of the interplay between wafer quality and device performances. In parallel, he has been coordinating several national and international projects with industrial partners related to silicon-based homojunction, hetero-junction, and tandem cells.

Dr Eszter Voroshazi received her Ph.D. (cum laude) in 2012 from KULeuven in Belgium in the field of thin-film solar cells and modules and related reliability studies. From 2015, she has worked at imec as a senior researcher on module reliability and later as a R&D Manager of the PV modules and systems activities and coordinating several industrial and funded research projects. She joined Commissariat à l'Energie Atomique et aux Energies Alternatives (CEA) in 2021 as head of the PV Module Process Laboratory focusing on novel materials, module interconnection and packaging technologies for conventional and integrated PV modules from early concepts up to pilot-scale demonstration. She is a member of Scientific Committees of several international conferences and the European Technology and Innovation Platform for Photovoltaics.

Dr Guangtao Yang received his Ph.D. degree from the Photovoltaic Materials and Devices group in Delft University of Technology, the Netherlands, in 2015, for his research on n-i-p structured thin-film silicon solar cells. After his Ph.D., he worked as postdoc researcher at Delft University of Technology and Eindhoven University of Technology on the c-Si solar cells research topics. Special research focuses were given to the IBC c-Si solar cells and c-Si solar cells with passivating contacts, including passivating contact materials optimization and solar cell device fabrication. Since 2021, he is appointed as the laboratory and project manager of the TU Delft PV Technology.

Acknowledgements

The authors Delfina Muñoz and Radovan Kopecek thank enormously all the co-authors for their dedicated writing and reading during many evenings and weekend sessions to finalize this book. The original idea of this book was launched before the COVID-19 pandemic, and we went through difficult times in communication and exchanges of information. Nevertheless, the result is there. So thanks for your effort, your time and your expertise, especially to chapter leaders. Thanks for believing that this book was possible and encourage your chapter team.

Thanks to all nPV believers and supporters of the nPV workshop from the beginning. Even if we work and trust in slightly different technologies, we have the same goal, to push n-type technologies to the market. Thanks for the collaborative n-type network allowing to understanding and improving all the knowledge on n-type materials, cells, modules, systems and all technology-related topics.

In particular, we would like to thank all the dedicated companies that were and are visiting and sponsoring the nPV workshop since 2011. The nPV workshop bring the n-type community together, from fundamentals to the market. These workshops were, and hopefully will still be, a great fun and success, as the n-type community is very active and believes in the future of nPV.

Last but not least, we would like to thank our families and friends. Most of the work has been done late in the evening or during weekends. Thanks to all of you for being patient with us and understand that we are working for a better and sunny future.

Delfina dedicates this book to her parents Miguel y Carmen. They are the best combination of science and humanity. They always supported me with all my dreams and encouraged me not to give up. I want to thank my sunny children Maia, Dana and Alina who made me many designs and LEGOs while writing this book, and my husband for being always there and make my life easy and fun.

I want to thank also my PV inspiring friends form INES and from abroad: Pierre-Jean, Anis, Wilfried, Jordi, Perrine, AnSo, Charles, Christophe, Matthieu, Eszter, Elias, Pierre, Marko, Lars, and many others and especially Rado for all the time spent together as n-type friends. He is always motivating us with his energy and vision.

Radovan dedicates this book to his parents Alena and Oldrich that convinced me to study Physics instead of Art- and to be honest: it's not so bad. Thank you for all your support during my studies in Stuttgart and abroad. You were always there when I needed you. Unfortunately, my beloved mother passed away early 2020 and did not have much time to enjoy her life after her "retirement". But at least Oldrich can do it with/for her traveling with us and teaching his grandchild's math. In addition, I would like to thank my family that was very often neglected during evenings and weekends and heard many times 'This weekend I have to finalize the bifacial book'. Thank you Samuel, Noemi, and Frida for being uncomplaining during this time. Saving the world with n-type PV technology needs some time.

Last but not least, I would like to thank all my ISC Konstanz "n-type colleagues" that were/are developing this powerful technology since the beginning of ISC Konstanz like Valentin, Joris, Thomas, Razvan, Giuseppe (now at Henkel), Christoph, both Jans, Haifeng, Ning, Florian, Andreas, and many others (forgive me not mentioning you specifically). And at the very end of course, my BIG HUG goes to my female counterpart Delfi who is preaching with me "the bifacial n-type paradise" on many continents and trying to make our planet earth Earth a better place to be. Bifacial n-type technology is not only a technology- but it is also our religion.

Chapter 1

Introduction

Radovan Kopecek[1]

From 2020 on, solar PV – with bifacial passivated emitter and rear cell (PERC) technology – had been crowned to the new king of energy markets [1], reaching extremely low offers in bids in the Middle East and North African (MENA) states such as Saudi Arabia, drawing prices as low as 0.01 US$/kWh [2] in 2021. The economy of the PV market completely changed since then, as it is not dominated by the demand, but by possible supply at the moment which is mostly limited by the supply of poly-Si to about 180 GWp in 2021 [3]. This was limiting the number of installations in 2021. However, the solar cell manufacturing capacity is around 300 GW (also including old Al-BSF lines), whereas the module capacity is at 250 GW [4]. The total installed PV capacity end 2021 was about 900 GWp and, in March 2022, 1 TWp was reached [5]. About 7 years later, we will enter a yearly 1 TWp market [6], where it is not wise and also not possible to produce all poly-Si, silicon (Si), wafers, cells, and modules in China. Local production will be needed everywhere to reach our ambitious goals of a fast and sustainable energy transition.

As mentioned above, the market recently began to be dominated by limited supply rather than by demand, and it was not long before PV entered a poly-Si crisis, similar to the one in 2005, but at a completely different capacity level. In addition, the COVID-19 crisis, with some production interruptions, even shutdowns, and a ten-fold increase in transportation costs from China to Europe or US, changed the PV market completely. Already at the end of 2021, at Intersolar in Munich in October, the change in mentality of EU's EPC (engineering, procurement, and construction) companies and of the distributers became very apparent. The hike in module prices from China well over €0.20/Wp at that time, as well as the lack of availability of these modules for import, prompted PV system companies to show increased interest in locally produced modules. The huge ambitions of various EU member states, such as Germany, to install 200GW of PV systems until 2030 [7] increased the hunger for modules and the desire for security of a stable supply. Since China has a target to set up 80 GW of PV systems in 2022, the US 70 GW and the EU 70 GW as well, a severe shortage of modules and other components is expected in the coming years. The last straw that broke the camel's back was the start of the

[1]International Solar Energy Research Center Konstanz, Germany

war in Ukraine on February 22, 2022. At a historical meeting of the German Bundestag on Sunday, February 27, 2022, it was declared that renewable energies are "freedom energies" [8] and their deployment needs to be even more accelerated. Finally, as a consequence, since the EU wants to gain independence from gas imports from Russia, as well as from other energy imports, on March 31, 2022, Kadri Simson, the EU Commissioner for Energy, announced that the EU will do "whatever it takes" to bring back PV production to the EU [9]. By that is understood "vertical production", starting from ingot through to module, including the supply chain of poly-Si, glass and other necessary components. At the moment, the EU Commission and member states, such as Germany and France, are in discussion with PV networks, such as ESMC, SPE, ETIP, and EUREC, to understand what will be needed to set PV production up with the necessary "critical mass" for it to become cost effective and sustainable. PV production in EU will enter a new bifacial nPV era, as it will be described later on. The current plans are to install 30GW vertical integrated PV production until 2025 and 70 to 80GW until 2030.

1.1 PV 2022 – history, present, and future

1.1.1 How PV became the most cost effective electricity source

The history of PV is long and very dynamic. At the beginning of terrestrial applications, only ecologically aware people installed PV systems for their own use at high costs and module prices even well above 5€/Wp. Everyone was completely ignoring and later even laughing at this technology.

Figure 1.1 sketches the history, present, and a possible future of module prices depending on the demand and production capacity. PV's history can be matched to Gandhi's saying: "First they ignore you, then they laugh at you, then they fight you

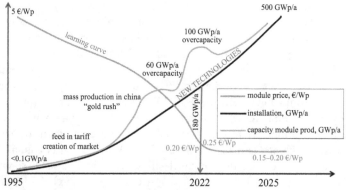

"First they ignore you, then they laugh at you, then they fight you, then you win."

Figure 1.1   History, present and future of PV module prices demand and production capacity. The schematically depicted graphs are showing the trend of each category

and then you win." Few years back, we were in the middle of "fighting" against other traditional energy sources and political hurdles influenced by fossil fuel lobbyist – such as e.g. the introduced taxes for self-consumption in Germany. These taxes were introduced to slow down the energy transition in Germany, so that the large electricity providers had more time to restructure. Similar to the 50 GW PV cap, which existed a long time in Germany. With this, we have lost about ten valuable years, which we have to catch up now. But at the end PV will win, as there is no way anymore to stop this cost effective technology.

But how did we come so fast so far? The starting point of this success story was the feed in tariff in Germany around 2000 to stimulate the market, followed by the mass production in Germany and later taken over by China, as the EU machine manufacturers transferred the technology to that flourishing and fast market. At around 2009 (after the first feedstock crisis), everyone wanted to be a part of the game and to produce cells and modules. This led quite fast to a significant overcapacity (as observed in many other industry branches, such as car or electronics industry, before) around 2010. The prices for modules dropped extremely fast at that time, modules were on stock and the big shake out began. After this consolidation phase, the capacity was again close to the demand – however the prices never went up remarkably again until 2021. Actually quite the contrary: the prices even dropped faster.

However, at the moment, we are once more in a small crisis, as we have currently once again an overcapacity of cells and modules of ca. 100 GWp, mostly because we entered in a second poly-Si feedstock crisis in 2021, limiting the module production below 200 GWp, even though the demand was and is much higher. For first time after 15 years now, modules are not always available on the PV market and the prices for modules went up from 20€ct/Wp to 25€ct/Wp between 2021 and 2022 and are still at that level. Experts believe that, due to capacity expansions that will then enter in production, end 2022 or beginning 2023 the poly-Si feedstock crisis will be over and the prices will drop again to a level of 15–20€ct/Wp.

When we look at the precise learning curve in Figure 1.2 from ITRPV 2022 [10], we see that the already steep one became even steeper after the feedstock crisis in 2005.

This extremely steep curve led to the consequence that PV became the "king of energy markets" from 2020 on. As already mentioned, this fast fall of the prices is now stopped and is believed to stabilize between 15 and 20€ct/Wp, which will be also healthy for a sustainable PV market, so that the companies can finally make good profit. However, the LCOE (levelized cost of electricity) still can be reduced on module level as well, when making the modules more efficient, with higher bifaciality, with a lower temperature coefficient and with a lower degradation rate. All this can be achieved with n-type technology.

1.1.2 *What PV technology will win?*

This question very often was raised in the past: what technology will make the race to be the most cost effective candidate for lowest cost electricity production? Many people were predicting that c-Si (crystalline silicon) cannot make it, as it is

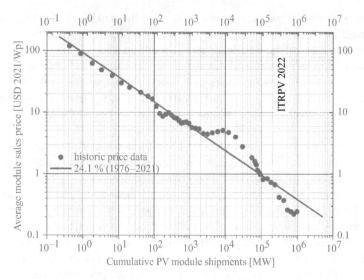

Figure 1.2   Learning curve for module price as a function of cumulative shipments since 1976 from ITRPV 2022 [10]

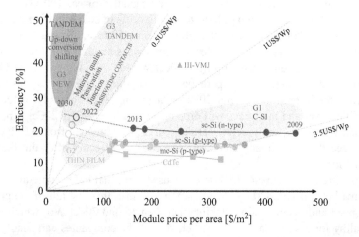

Figure 1.3   Module efficiency versus costs in dependence of different technologies. G1: c-Si. G2: thin film. G3: next generation. G3 NEW: combination of c-Si with other. Modified from [11]

an indirect semiconductor and therefore relatively thick wafers are needed (100–150 μm) to absorb the solar spectrum. Therefore, thin film solar cells (direct semiconductors), such as CdTe, GaAs, or CIGS will be needed, as thicknesses below 10 μs are sufficient. Prof. Martin Green created a comprehensive graph for this believed fact – which is depicted in Figure 1.3 [11].

He has sketched three regions in an efficiency-price graph and predicted around 2000, that c-Si (generation 1/G1) can be used to a certain extent – after that, for further lowering the costs, thin film (generation 2/G2), and then e.g. multi-junction III–V structures (generation 3 G3) will be needed. However, he, actually nobody, could not predict the unbelievable development of c-Si in respect to maturity and costs. We have updated the graph with newest efficiency and price numbers for c-Si, CdTe, and III–V multijunction PV until 2022. It is clearly visible that the c-Si technology surpassed the predicted boundary for lowest prices very quickly being well below $100\$/m^2$ and below 50\$ct/Wp. CdTe went the same direction, however at lower efficiencies. III–V MJ cells remained at high costs.

What is important for low costs in large PV systems is not only low module cost but also high power modules, as the balance of system (BOS: installation, material, inverters, etc.) represents more than 50% of the total costs. The more powerful the modules are, the lower the costs for BOS can be. More detailed considerations will be presented in Chapter 6 about LCOE (levelized cost of electricity) calculations. Therefore, the PV industry at the moment is going more and more toward the mono c-Si n-type technology and also considering more seriously bifacial technologies.

What will happen in the next 10 years in PV is quite easy to estimate. The solar cells will be more complex, mono c-Si technology at low costs will adapt evolutionary technologies such as selective diffusions, better passivations, improved metallization technologies, and carrier selective contacts. The industrial cell will slowly approach the theoretical limit of c-Si and become n-type and bifacial as will be summarized in Chapter 3. When the limit is reached it becomes unsure, which technologies could boost c-Si above 30% efficiency. However, we are very sure that it will be a combination of the most powerful and low cost c-Si technology with another technology. We call this "G3 new" (generation 3 new). Maybe we will shift the UV- and infrared parts of the solar spectrum toward c-Si band-gap with up- and down shifting or/and additional material in addition to c-Si will be used in tandem configurations such as CIGS or Perovskites. Chapter 7 will review the most promising candidates.

1.2   nPV 2022 – history, present and future

1.2.1   Short low-cost nPV history

The n-type bifacial PV history actually began with the very first solar cell processed at the Bell Labs in 1954. It was an n-type (As-doped) IBC cell with a bifacial character [12]. The development of n-type solar cells from that point was in the focus of R&D, with a fast improvement in efficiency. The technology was at that time extremely expensive and first applications of PV panels were, therefore, realized in space for powering satellites. This application was at that time actually "the dead" of n-type technology, as it was proven quite quickly that p-type Si material degrades less in space, while being bombarded by cosmic flux. From that point on, p-type technology was developed with much higher focus and when it came to terrestrial applications, p-type was the winner, as p-type devices were leading in simplicity and costs. With the development of low cost Al-paste metallization technology, instead of Al-evaporation, the p-type device became so dominant, that it was very difficult for any n-type newcomers to enter the low-cost PV market – only high efficiency device companies such as sunpower and Sanyo were able to do so.

Figure 1.4 displays the PV products for two different PV eras – (1) the monofacial mc-Si Al-BSF (back surface field) era on the top and the (2) bifacial mono c-Si era on the bottom. Between 1995 and 2017, the market was dominated by low cost mc-Si Al-BSF device, with solar cell efficiencies below 18%. The choice at that time was only between mc-Si and Cz-Si Al-BSF technology with module efficiencies at about 17%. From 2016 on, when LONGi brought low cost Cz-Si material on the market, the dominancy of bifacial Cz-Si PERC devices was initiated. But then, also the n-type devices started to gain importance, as more complex low cost processes were developed (also based e.g. on an improved surface passivation and laser technology), and module efficiency and bifaciality started to become an important factor to save balance of system (BOS), rather than the use of low cost but low efficiency devices as few years back.

The competition between p- and n-type started though in early 2000. Figure 1.5 shows the 20 years competition between p- and n-type technologies in detail.

In 2002, the development of low cost n-type devices started at different R&D groups (ECN, University of Konstanz), as it was a well-known fact that n-type material was better performing as p-type. However at that time, no low cost processes for n-type like BBr_3 tube furnace diffusion, its passivation and metallization were well developed. The only n-type solar technologies at that time in the PV market were IBC from sunpower and HIT from Sanyo with very complex processes but achieving highest premium prices for their modules. The advantage weight then slowly tipped in favor of n-type technology from 2021 on, when LID and later also high temperature stability of n-type Cz-Si was proven. In 2011, when bifaciality

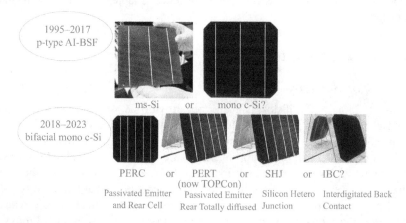

Figure 1.4   PV products in two different PV eras

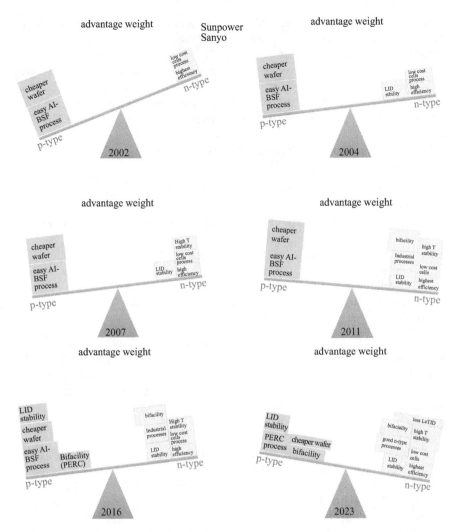

Figure 1.5 Advantage weight which compares the advantage momentum in the corresponding year

was also gaining more and more attention, n-type technology was believed to become the next big thing in the following years and companies like PVGS, Yingli, sunpreme, and LG electronics entered the nPV market. However, then p-type technology stroke back with LID stabilization of boron doped wafers with light and electricity treatment and later with the implementation of Ga-doped wafers. In addition, PERC solar cells were developed to become bifacial as well, so that bifacial PERC became in 2020 the king of energy markets. However, now, in 2022, it is clear that PERC is coming on cell level with 23.5% and on module level below

22% to its efficiency limits. In addition, light and elevated temperature-induced degradation (LeTID) became a challenge in PERC devices [13], as well as Ga-doped wafers showed some additional degradation [14] leading to underperforming PV systems, so that now there is a common agreement that n-type technologies will become more visible in few years from now. Jolywood is these days the leader of TOPCon technology, REC with alpha-Series, the leader of high efficiency SHJ technology and SPIC with ZEBRA the leader of low-cost high efficiency IBC technology.

In summary, if we classify the PV market in eras, the past, present, and the future looks the following:

1. 1995–2017 monofacial Al-BSF era (because of simplicity of Al-BSF and contact formation)
2. 2017–2025 bifacial mono c-Si PERC era (because of low-cost high efficient PERC process)
3. 2025–2035 bifacial mono c-Si nPV era (because of high-efficiency potential at low cost, higher bifaciality, and lower temperature coefficient as well as degradation)
4. 2035–2050 bifacial tandem era? (because of the possibility for overcoming the Auger limit of 29% efficiency)

If the bifacial tandem era will really come from 2035 is not yet clear, as this would mean for the first time bringing a revolution into the PV market. So far the innovation in the PV market were based on an evolution of c-Si technology improving the surface passivation and with that increasing the voltage of the device, as we will see it in the following.

1.2.2   nPV status

As PERC is now coming to its efficiency limits, the question is what next? Who will be "the emperor"? Will there be a fast switch from PERC to n-type technology, as was the case about five years ago for Al-BSF to PERC? PV Tech Research, as depicted in Figure 1.6 [15], as almost now the whole PV community, believe so.

But what will be the next mainstream technology? SHJ (silicon heterojunction) solar cell? Tunnel oxide passivated contact (TOPCon)? A combination of both in an interdigitated back contact (IBC) structure? Tandem cells? The race is on and the muscle-showing announcements are now accelerating. PERC announcements from Tier1 manufacturers on solar cell and module level advancements have the aim of demonstrating that this technology is still not at its end. n-type technology announcements, however, have the function of showing which way to go in the future.

In theory this is quite simple. Figure 1.7 depicts the linear yearly growth of cell efficiencies in industry of about 0.6% absolute, as presented first by Martin Hermle [16]. In 2016, the switch from Al-BSF to PERC was initiated, as Al-BSF technology reached its efficiency limit of below 20% in production. This is mostly because of the limited passivation of the rear side by homogeneous Al-BSF, which is depicted in the cross section on the right.

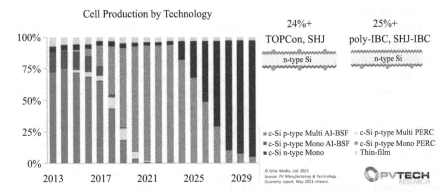

Figure 1.6 *(Left) Historical data showing the switch form Al-BSF technology (blue and red bars) to PERC (brown bars) and forecast of a switch to n-type technology (grey bars) in coming years [15]. (Right) Schematic cross sections for TOPCon, SHJ and poly-Si- and SHJ-IBC*

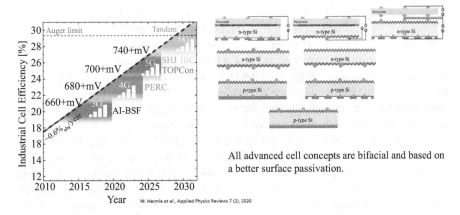

Figure 1.7 *(Left) Efficiency (voltage) improvements and limits of c-Si technologies in the past and coming years with analogy to mobile networks 3G–6G. The graph is adapted from [4]. (Right) Schematic cross sections for Al-BSF, mono and bifacial PERC, TOPCon, SHJ, IBC and 2, 3, and 4-terminal tandem configurations*

With a dielectric stack (AlO$_x$/SiN$_x$) below the Al-Paste and Al-BSF point contacts in PERC devices, a better rear side passivation can be realized, leading to average voltages of 680 mV and maximum of 690 mV. Now, to overcome 700 mV, passivating contacts with poly-Si in TOPCon or a-Si in SHJ technology have to be applied. Then, in the next step to overcome the Auger or even the Shockley-Queisser limit, tandem structures have to be used. So this linear curve is based in first order on improvements of the voltage by a better passivation with advanced cell structure.

We can compare these technologies to the mobile network and its speed. 3G (Al-BSF) is obsolete and these days the working horse is 4G (PERC), with 5G (passivating contacts) already in place. However, for most applications, 4G is still sufficient and the more expensive 5G is not needed. But in a few years, we will have 5G everywhere and be preparing the ground even for 6G (tandem).

High-efficiency announcements show nothing else than the practical limit of each technology, which is depicted in the graph in Figure 1.7. However, there is a gap of about 1% absolute between what is possible (with tricks in process flow and in measurements) and what is realistic in production. In the blog "Future industrial PV technologies: Champion cell announcements versus industrial reality" from 2019 [17], it is explained in detail, with which tricks the high efficiency solar cells are fabricated and measured. They very often do not have much to do with their industrial counterpart, even if the technology is named the same. In addition, even the measurements have very often some tricks included (0 Busbar measurements of even active area only, the use conductive and reflective chuck, etcetera). Table 1.1 summarizes this matter, depicting the record announcements versus average efficiencies in production and available module efficiencies on the market, as well as its potential.

Al-BSF (3G) is fading out of the PV market as the passivation of the rear side is strongly limiting the device voltage and as it is also monofacial. The average cell efficiencies in production are up to 20%, with voltages of around 665 mV with a record efficiency of 20.29% [18]. The module efficiencies on the market are well below 20%. The king of energy markets, bifacial PERC (4G), has average solar cell efficiencies of about 23% in production. PERC modules on the market are around 21%, but can reach up to 21.5% [19] e.g. for JA Solar's modules. The 23.03% PERC module efficiency announced by Trina recently [20] has not much to do with its standard product and the measured efficiency is even an "active area" measurement only.

While the standard module efficiency is calculated by dividing the electrical output power (Pmpp at STC) by the irradiance captured by the total area of the module (module length multiplied by the module width), the "active area module efficiency" take into account only for the total area of the cells resulting in a significantly higher efficiency value (approaching the cell efficiency) compared to the standard (total area) module efficiency. That means that for the active area module efficiency, the area of the gaps between the cells and the area occupied by bussing ribbons and junction boxes as well as the mandatory gaps between the outer solar cells and the edge of the laminate are not taken into account. Accordingly, the active area module efficiency is not relevant for the appraisal of benefits of high module efficiency in terms of savings for area related balance of system cost. The 24.06% record PERC cell efficiency reached by LONGi already in 2019 [21] used a selective poly-Si (PERC plus) on the front and also has other features, which are not yet and most likely never will be, implemented in industrial production.

PERC-based technologies like "TOPCon" and "PERC-based IBC" benefit from the PERC cost structure and are therefore low cost as well. The major challenge, as for almost all n-type technologies, is the reduction of Ag-consumption for the metal contacts. TOPCon (5G) cells reach in production an average efficiency of about 24%, whereas an efficiency of 26.4% [22] has been demonstrated by Jinko Solar recently.

Table 1.1 *High-efficiency announcements versus industrial reality of all relevant c-Si technologies on the PV market*

Technology	Equipment	Announcements of cell efficiency (%)	Average cell efficiency in production (%)	Available module efficiency (%)	Average voltage potential in production (mV)	Average efficiency potential in production (%)
Al-BSF	Standard	20.29 [18]	19.5–20.0	<20%	670	20.5
PERC	Standard PERC based	24.06 [21]	22.5–23.5	**21.5** [19]	690	23.5
TOPCon		26.4 [22]	23.5–24.5	**21.9–22.3** [19]	725	24.5
Low cost IBC		25.04 [23]	23.5–24.5	**22.1** [19]	735	25.0
Low cost SHJ	Thin film based	25.26 [24]	23.5–24.5	21.7–22.3 [19]	735	25.0
Complex SHJ	Thin film and electronic industry based	26.81 [25]	24.5–25.0	22.5 [19]	740	25.5
Complex IBC	Thin film and electronic industry based	26.1 [26] 26.6 [27]	25.0–25.5	22.3–22.8 [19]	740	26.0

Here the exact process flow and cell architecture has not been published by Jinko, however, we can assume that most likely a selective poly-Si(B) has been used, which is more complex and not yet ready for industrial mass production, in addition to other non-industrial features. The COO for a standard TOPCon cell is currently about 20–30% higher as for PERC. However, its higher efficiency, higher bifaciality, lower degradation, and lower temperature coefficient make these modules already attractive not only for roof top applications but for utility-scale solar, as well as for regions which are hot and on systems with high albedo as well.

PERC equipment-based IBC (5G) (Jolywood, SPIC, Trina, ValoeCell) has been demonstrated at 25.04% [23] from Trina and is produced at about 24% by SPIC Solar (still without passivating contact), having a potential of 25%. At the moment, such cells are mostly suited for rooftop PV applications with the reduction of Ag metallization, the bifacial version could be also used in utility-scale in the future. For low-cost SHJ (5G) (REC, Meyer Burger, Maxwell), an efficiency of 25.26% [24] was demonstrated by Maxwell, whereas complex SHJ (Panasonic, Kaneka) (5G) reached record efficiencies of 26.81% [25] as shown by LONGi. Complex IBC (5G) are produced by Sunpower and LG. 26.1% [26] was demonstrated by ISFH on a POLO structure and 26.6% [27] by Kaneka with SHJ-IBC.

Figure 1.8 summarizes the 20 modules with highest efficiency on the PV market for roof top applications (area $\leq$ 2 m^2) in December 2022 from Clean Energy Reviews [19]. The first top 15 modules are based on n-type technologies – IBC first, then SHJ, and finally TOPCon. Interesting to see that the low-cost versions of SHJ, TOPCon, and IBC from REC, Q-cells and SPIC are at exactly the same efficiency of 22.3%. As already discussed, PERC modules are limited to 22% module efficiency. n-type technologies have the potential to overcome 23% in the near future. The race is on who will have the first commercial module exceeding 23% on the market. At the moment, these modules are used for roof-top applications mainly, but slowly bifacial n-type modules, e.g., TOPCon from Jolywood, are entering the utility scale arena as well.

Figure 1.9 summarizes the capacity and distribution of the above mentioned three most important n-type technologies mid-2020.

The total capacity of n-type technologies mid 2020 was about 13 GWp. What is remarkable though is that n-type production is not yet fully dominated by Chinese producers, as it is the case for p-type. For TOPCon, it will be most likely the case again – however, for SHJ, it is only 55% and for IBC even much less. SPIC is the only Chinese producer who is already producing IBC technology in mass production. With investments in n-type technologies in EU and US, the technology dominancy of Chinese producers can be partly resolved.

1.2.3   nPV future

n-type future is beginning right now. Investments in new manufacturing lines in China and outside of China are mostly based on n-type technologies: TOPCon, SHJ, and IBC. In Chapter 3, which is about solar cell processes, we will learn more about the chances and challenges of n-type technologies. Figure 1.10 shows the capacity and production of n-type technologies from the past, present, and gives a forecast until 2025.

Manufacturer	Model	Max power (W)	Cell Type	Efficiency
SUNPOWER	Maxeon 6	440W	N-type IBC	22.8%
CanadianSolar	HiHero CS6R-H-AG	440W	N-type HJT Half-cut	22.5%
REC Solar	Alpha Pure R	430W	N-type HJT Half-cut	22.3%
SPIC	Andromeda 2.0	440W	N-type IBC Half-cut	22.3%
Q CELLS	Q.TRONG-G1+	400W	N-type TOPcon Half-cut	22.3%
Panasonic	EverVolt H	410W	N-type HJT Half-cut	22.2%
JinKO	Tiger NEO	480W	N-type TOPcon Half-cut	22.2%
belinus	M8 IBC Ultra	400W	N-type IBC Half-cut	20.0%
LONGi Solar	Hi-MO 6 Explorer	430W	N-type HPBC Half-cut	22.0%
SUMEC Phono Solar	Draco Mono-M6	430W	N-type TOPcon Half-cut	22.0%
HUASUN	Himalaya	400W	N-type HJT Half-cut	22.0%
Trina solar	Vertex S +	425W	N-Type TOPCon Third-cut	21.9%
FuturaSun	Zebra Pro	430W	N-type IBC Half-cut	21.8%
YINGLI SOLAR	Panda 3.0 Pro	425W	N-Type TOPCon Half-cut	21.8%
MEYER BURGER	White	400W	N-type HJT Half-cut	21.7%
risen	Titan	450W	P-Type PERC Half-cut	21.7%
AKCOME 爱康光电	Kookaburra Series	415W	P-type PERC Third-cut	21.6%
JA SOLAR	Deep Blue 3.0 light	420W	P-Type PERC Half-cut	21.5%
LEAPTON	LP182-M-60-MH	465W	P-Type PERC Half-cut	21.5%
Silfab	Elite SIL-380-BK	380W	P-type IBC	21.4%

Table title: **Most Efficient Solar Panels 2022*** (CLEAN ENERGY REVIEWS) — v3.5 - Dec 2022

*Residential panels - 54, 60, 66 cells (108, 120, 132HC), or 96 & 104 full cell. Does not include commercial panels >2m

Figure 1.8　List of high efficiency cell and module producers [19]

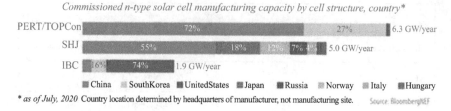

Figure 1.9   *List of bifacial cell and module producers [28]*

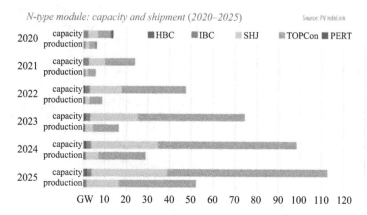

Figure 1.10   *Past, present, and future forecast of capacity and production of n-type technologies [29]*

As seen already in Figure 1.9, the capacity of n-type cell production in 2020 was about 13 GWp and will reach close to 50 GWp in 2022. This is quite a remarkable number, even though the production will still be limited to less than 10 GWp. Until 2025, the production capacity will reach 110 GWp with a yearly production of more than 50 GWp. This will be about 12.5% of the yearly market, of 400 GWp [6], by then which is rather a conservative forecast. The n-type market will be dominated by TOPCon technology, followed by SHJ and IBC. How fast then the switch from p- to n-type will be conducted depends on the challenges the n-type devices still will have by then and how fast they will be implemented into utility scale market. Figure 1.11 shows a possible, more aggressive, scenario for the switch from p-type to n-type from ITRPV2022 examining the production share of p- and n-type c-Si material.

As suggested, in 2022, already 20% of the market share is based on n-type products, whereas in 2029, almost 60% is reached. The scenario from PV Tech in Figure 1.6 [15] has a lower market share of n-type technologies in 2022, but the switch from p- to n-type going much faster – already reaching 60% market share for nPV between 2026 and 2027. Figure 1.12 depicts the current applications for n-type modules for roof-top, BIPV and VIPV and from 2025 on n-type bifacial technology is expected to enter the utility scale market with a much larger impact

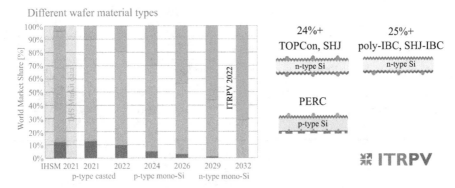

Figure 1.11   c-Si material market share forecast from ITRPV 2022 [10]

Figure 1.12   Present applications for n-type cells and modules (top) and future
applications in utility scale (bottom)

as the COO will come closer to PERC and the properties of n-type in field are much better due to its higher bifaciality and lower temperature coefficient.

As previously explained, the duration of the switch though will strongly depend on the challenges n-type devices still have as summarized in Table 1.2. It starts with (1) production of low cost n-type substrates, (2) improvement and standardization of poly-Si deposition, as well as (3) reduction of Ag metallization or even replacement by Cu and/or Al.

When n-type technology will solve all this issues, PV will enter a bifacial n-type era, also on utility scale and this will mark the big breakthrough of n-type technology due to the listed advantages in Table 1.2, leading to lowest LCOEs well below 1 ct/kWp at the end.

Figure 1.13 depicts the geographical world-wide distribution of bifacial PV systems in 2020, which was mostly based on bifacial PERC technology and cumulated total installations between 2012 and 2024 based on current forecasts.

Bifacial PV started with its first bifacial PV systems on MWp level with n-type nPERT technology with fixed tilt configurations, like the one 2013 in Japan using EarthON nPERT technology [31]. Bifacial nPERT was much more expensive at

Table 1.2   Advantages and challenges of n-type solar technology

Advantages	Challenges
High efficiency bifacial cell	Low cost n-type wafer
Low degradation	Poly-Si depositions
Low-temperature coefficient	Ag-consumption
>>>High kWh/kWp	>>> Higher COO as for PERC

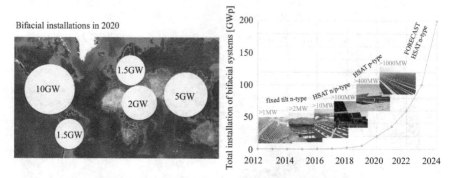

Figure 1.13   Geographical distribution of 20 GW bifacial PV systems end 2020 (left) and cumulated bifacial power (right) [30]

that time, mostly also due to the fact that n-type wafers cost has been more than 30% higher compared to p-type wafers. MegaCell in 2015 has started BiSoN (Bifacial Solar cells on N-type) cell and module production. Using BiSoN modules, MegaCell has set up in 2016 a 2.5 MWp [32] fixed tilt (La Hormiga) and ENEL a 1.75 MWp [33] HSAT system (La Silla) in Chile. 2016 was however also the time when bifacial PERC cells have entered the market and offered a more cost effective alternative for bifacial modules. While this slowed down the growth of for n-type cell technologies, it strongly pushed bifacial PV at that time. Since 2018, large HSAT bifacial PERC systems have entered the market very quickly and these days are dominating the bifacial PV market. One of the largest is the 1.3 GW HSAT bifacial PERC system in Karapinar (Turkey) which will be finalized end of 2022 [34]. However, in hot and albedo-rich regions, as in MENA countries, more and more bifacial TOPCon systems are being build, mostly using modules from Jolywood in HSAT systems, as in the Ibri II system in Oman [35], as shown in Figure 1.13 on the top. The power plant is the biggest to date to deploy the high-efficiency TOPCon technology with a power of about 500 MWp. As TOPCon technology has a higher bifacial factor as PERC it makes sense also in some cases to condition the ground to enhance the albedo. This has been done in Oman as well as in two systems by Solitek in Ukraine and Lithuania enhancing the additional bifacial power above 20%.

Figure 1.14 shows at the top the installation in Oman using the natural albedo, typical for a desert with bright sand, of about 30%. Below the smaller pictures display part of the installation in Oman (on the left) and in Lithuania from Solitek with a white foil, more than doubling the albedo to about 70%. A similar PV system from Solitek, as the first of its kind in EU, is also installed in 2019 in Ukraine [36]. As the used TOPCon modules have a bifacial factor around 0.8, the enhancement of the bifacial gain can exceed 20% as compared to a normal bifacial gain in a typical bifacial PERC system (with natural albedo of 30%) of around 10%.

In EU, we need to speed up with PV installation to become independent of fossil fuels in the coming years in order to not to depend on energy imports and, in the last consequence, to slow down global warming. Since 2021, the demand for PV modules is much higher than the actual availability, due to poly-Si crisis and

Figure 1.14   n-type bifacial utility scale systems with total power about 500 Wp using TOPCon modules in Oman [35] and a smaller installation from Solitek in Lithuania

transportation costs and challenges. Since 2022, we have entered an accelerated PV installation era, where China only cannot produce modules for the entire planet. We need local production in many countries, to satisfy the demand for PV modules in future, as we need to and we will enter a yearly 1 TW market before 2030.

In EU, we still have a cumulated capacity of PV module production of around 7–8 GWp, which has to be at least tripled in the coming 2 years and increased tenfold to at least 70 GWp until 2030. In order to get totally independent, this has to happen with the whole value chain of PV products and therefore vertical integrated production from ingots to modules has to be set up. Poly-Si and glass production has to be produced locally as well. The challenge is now to set up wafer and cell production at 30 GWp scale in the coming 3 years, as at the moment in EU, the capacity is below 1 GWp. But this is also a chance to set up next tech "5G production" in EU based on nPV, which is already happening at the moment. SHJ is produced by Meyer Burger (400 MW) and ENEL (200 MW) and IBC (ZEBRA) technology by Valoe Cell (80 MW).

Not only these manufacturers will have to increase their capacities to GW scale but there are newcomers on the market who are investing into nPV technologies as well, such as Carbon [37] and Greenland [38]. To ensure stable and sustainable PV production, institutes, companies, and several member states, under the lead of Spain, have started to put a so-called "PV-IPCEI" (Important Project of Common European Interest) in place, which will make it possible for participating member states to subsidize PV production based on innovations in their countries. Figure 1.15 depicts all so far planned PV-IPCEI projects. All cell developments are

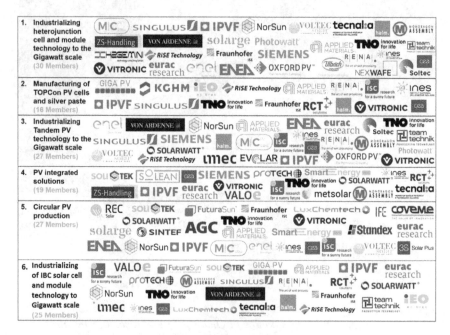

Figure 1.15   PV-IPCEI projects for EU's PV production [39]

based on n-type technologies [39]. In addition, PV integrations and circular economy are addressed.

These technologies, TOPCon, SHJ. IBC, and tandem, will be further developed in the PV-IPCEI projects and rolled out in GW scale in countries like Germany, France, Italy, Spain, Lithuania, and Poland. The future of nPV production in EU looks very bright. Politics, industry, and R&D for the first time are going hand in hand.

1.3   nPV book 2022

There were many nPV publications about n-type technologies having better properties compared to p-type in the past, which are getting more and more prominent at the moment. In addition, many announcements of nPV production in China and outside of China are being published with a higher frequency. Therefore, we have decided to write an nPV book to cover and summarize all technical and economic issues, so that it will be easier for future companies and investors to have a broad overview about upcoming next-generation technologies.

1.3.1   Latest nPV publications and presentations

The presentations are mostly done at PVSEC (EU PVSEC, Asian PVSEC, and US IEEE PVSEC) conferences, the "overview publications" in PV International, PV Tech Blogs, pv magazine as well as in Photon and other magazines. The "Taiyang News SHJ 2019 Edition" e.g. also offered a great summary of heterojunction technology.

Figure 1.16 shows some of the prominent issues dealing with n-type technology – so e.g. Photon International from 6/2015 reporting about Pandas, BiSoNs, and ZEBRAs – which are nPERT, at that time produced by Yingli and MegaCell, and IBC technologies. Taiyang News has always good summaries of new technologies and PV International is publishing many articles of devices close to the industry from prominent EU Institutes, such as FhG ISE, ISFH, CSEM, INES, IMEC, and ISC Konstanz. All these institutes realized quite quickly that n-type technology will become important in future and that is why we all started the nPV workshop

Figure 1.16   Photon International 6/2015, Taiyang News 2019, pv magazine 06/2021, and PV International 48/2022

Figure 1.17   10th nPV workshop March 30/31, 2022 in Konstanz

series in 2011 in Konstanz – in 2022 coming back as an on-site workshop again: www.2022.nPV-workshop.com (see Figure 1.17). The early presentations from the beginning on can be found on that web page (under previous workshops).

During the workshop in 2022, we celebrated the 1 TWp total installations on March 31 at a 1 TWp party at ISC Konstanz. That was also the days when the EU also announced that the EU Commission will do "whatever it takes" to bring PV production back to EU. These days even the last p-type Tier1 protagonist such as JA Solar is finally convinced that n-type the way to go [40]. We are also very sure that in the n-type arena, IBC will be finally the winner and that is why we started also to relaunch the back-contact workshops: www.backcontact-workshop.com

1.3.2   Chapters of nPV book

The following chapters of the book are structured the same way as the sessions of nPV workshops are organized. It starts with a material overview and gives insights into ingot and wafer production and properties. Then it summarizes solar cell, module, and system technologies which are important for n-type production and also discusses the difference between p- and n-type devices. These chapters are followed by COO and LCOE considerations which show that n-type devices are not only more efficient, but as PERC is coming to its limits also the costs of n-type are coming closer to p-type. In the coming years, n-type devices will also hit the efficiency limit and then the question is what will follow. The last chapter of this book deals with tandem technologies which could be the next big thing after n-type.

1.3.2.1   n-type Si material (from feedstock to wafers)

For high-efficiency devices, Si material quality is essential. Not only starting quality but also after process and in the field. This chapter therefore first describes the n-type silicon feedstock, n-type ingot pulling technology, and n-type wafer slicing technology. At the end, the Si material must result in good mechanical as well as electrical properties. The chapter continues on discussion of the impact of impurities on n-type silicon solar cells, thermal donors in n-type silicon solar cells, and gettering in n-type silicon solar cell within the solar cell process. Cost of ownership of the entire process is described as well and how the costs can be reduced in future, e.g. by faster pulling, reduction of thermal budget, and moving toward thinner wafers. The chapter is closed by the summary and outlook for n-type silicon ingots and wafers.

1.3.2.2 Solar cell technology

The n-type book is focusing on solar cell technology and therefore this chapter is by far the longest one of the book. It starts with a brief summary of industrial p-type solar cells, describing the properties of the faded out Al-BSF solar cells and the technology of the today dominant PERC solar cells.

Then the focus is on double-side contacted homojunction devices, starting with rear junction structures and then the path toward a boron emitter is discussed including the nPERT front junction fabrication process. A major focus is dedicated to polysilicon on oxide passivated contacts solar cells, starting with an introduction, historical review, and description of the working principle of the poly-Si passivated contact. Then the fabrication process of poly-Si passivated contacts is explained, describing the integration of poly-Si contacts into screen-printed or large area n-type solar cells – the so-called TOPCon solar cells. When TOPCon is implemented in the PV market, then processes toward the double-side integration of poly-Si contacts will follow.

The next solar cell technology discussed are silicon heterojunction solar cells, starting with an introduction followed by a part from birth to adulthood. Hereafter cell design, fabrication steps, and record efficiencies are reviewed. Hetero-interface properties and electronic transport for SHJ engineering and possible new process routes for SHJ mass production and next-generation PV close this section of SHJ.

The last important cell technology with the highest efficiency potential are back-contacted crystalline silicon solar cells with rear-junction structure. The paragraph describes the technology of IBC devices and processing of IBC solar cells closing with conclusion and outlook. The last important section deals with the silver usage reduction in n-type solar cells using fine line screen-printing and Cu metallization. At the end, the whole section is summarized.

1.3.2.3 Module technology

The module technology chapter starts with a description of general module requirements, discussing the potential energy generation increase compared to state of the art. Then characterization and performance monitoring is reviewed starting with electrical characterization, relevant module characterization in the field with performance monitoring. Standards and quality measure is the next important topic, first discussing the question why we actually need standards. The next topic are standards – seen from international to national level and the overview on existing standards for type approval and advanced testing and overview on existing measurement standards. Then major milestone in standardization is discussed including the question what makes current standards problematic. Hereafter standard requirements for advanced module concepts including bifacial solar modules are reviewed and steps in the right direction are described.

After the very detailed consideration of standard advanced n-type module, technologies are summarized, starting with a complete introduction, first advanced interconnection technologies are discussed followed by advanced module materials and integration of module level electronics. Finally, the end of the operational life of photovoltaic panels are reviewed.

1.3.2.4   Systems

The system chapter starts with a general description of PV systems including inverters and then gives an overview of PV systems' types and summarizes recent evolutions. Then it discusses the integration of high-efficiency modules in systems which is key for the reduction of BOS. After this, the field performances of p-type and n-type bifacial systems are compared and the chapter is closed with a summary.

1.3.2.5   COO/LCOE/LCA

Focusing on the three upcoming industrially mature cell technologies, TOPCon, ZEBRA-IBC, and SHJ as well as using n-type PERC as a benchmark, in the first part of the chapter, a scenario for industrial cell and module manufacturing is defined and the related manufacturing cost (cost of ownership, COO) is calculated. In the following, the analysis is extended to system level and – based on energy yield simulations – the levelized cost of PV generated electricity is calculated and a comparison between n-type technologies and PERC is made for three different system installation sites. The third part of the chapter is dedicated to the analysis of the environmental impact of the n-type technologies compared to the PERC mainstream technology.

1.3.2.6   Future of n-type PV

The chapter of nPV future starts with an introduction and then summarizes and reviews the carrier-selective passivated contacts technology starting with silicon heterojunction solar cells followed by TOPCon/POLO. After that bifacial solar cell development is reviewed. Next paragraph is dedicated to ultimate efficiency in c-Si PV discussing alternative carrier-selective contacts, metal oxide and alkali metal–halogen compounds and organic carrier selective contacts. The last section is reviewing state-of-the-art of silicon-based tandem sells starting with promising perovskite silicon tandems and III/V silicon tandems ending with projections and finally with a summary.

1.3.3   Summary

The book is finalized with a summary and outlook describing the state of the art and the bright future of nPV technology.

References

[1] *IEA: Solar the 'New King' of Power, Will Break Records for Decades to Come* [Online]. Available at: https://www.pv-tech.org/iea-solar-the-new-king-of-power-will-break-records-for-decades-to-come/

[2] *Saudi Arabia's Second PV Tender Draws World Record Low Bid of $0.0104/ kWh* [Online]. Available at: https://www.pv-magazine.com/2021/04/08/saudi-arabias-second-pv-tender-draws-world-record-low-bid-of-0104-kwh/

[3] *180 MW Installations in 2021*, Solar Media, PV Manufacturing and Technology Quarterly Report, May 2021.

[4] Finlay Colville, PV Cell Tech, 2021.

[5] *World has Installed 1TW of Solar Capacity* [Online]. Available at: https://www.pv-magazine.com/2022/03/15/humans-have-installed-1-terawatt-of-solar-capacity/

[6] *Global Annual Solar Deployment to Hit 1 TW by 2030* [Online]. Available at: https://www.pv-magazine.com/2022/05/17/global-annual-solar-deployment-to-hit-1-tw-by-2030/

[7] *Germany to Install 200 GW Solar by 2030. US Trade Agency Backs Extension of Import Tariffs* [Online]. Available at: https://www.reutersevents.com/renewables/solarpv/germany-install-200-gw-solar-2030-us-tradeagency-backs-extension-import-tariffs

[8] *Renewable Energy is Freedom Energy. Germany Speeds All-Green Target to 2035 to Ease Russia Grip* [Online]. Available at: https://www.rechargenews.com/energy-transition/renewable-energy-is-freedom-energy-germanyspeeds-all-green-target-to-2035-to-ease-russiagrip/2-1-1176238

[9] *EU Will Do 'Whatever It Takes' to Rebuild Solar Energy Manufacturing in Europe* [Online]. Available at: https://www.reuters.com/business/sustainable-business/eu-will-do-whatever-it-takes-bringsolar-energymanufacturing-back-europe-2022-03-31/.

[10] *ITRPV2022* [Online]. Available at: https://www.vdma.org/international-technology-roadmap-photovoltaic

[11] *Martin Green's Graph for 3 PV Generations* [Online]. Available at: https://depts.washington.edu/cmditr/modules/opv/solar_technologies.html

[12] *Patent for First Practical Solar Cell* [Online]. Available at: https://www.smithsonianmag.com/innovation/document-deep-dive-patent-first-practical-solar-cell-1-180947906/

[13] *Is LeTID Degradation in PERC Cells Another Degradation Crisis Even Worse Than PID?* [Online]. Available at: https://www.pv-tech.org/is-letid-degradation-in-perc-cells-another-degradation-crisis-even-worse-th/

[14] *Gallium Doped Silicon for High Efficiency Commercial PERC Solar Cells* [Online]. Available at: https://www.researchgate.net/publication/349446085_Gallium_Doped_Silicon_for_High_Efficiency_Commercial_PERC_Solar_Cells

[15] *Which PV Manufacturers Will Really Drive n-Type Industry Adoption?* [Online]. Available at: https://www.pv-tech.org/which-pv-manufacturers-will-really-drive-n-type-industry-adoption/

[16] M. Hermle, F. Feldmann, M. Bivour, *et al. Applied Physics Reviews* 7(2), 2020, 021305.

[17] [Online]. Available at: https://www.pv-tech.org/future-industrial-pv-technologies-champion-cell-announcements-verses-indust/

[18] K.H. Kim, C.S. Park, J.D. Lee, *et al.* Record high efficiency of screen-printed silicon aluminum back surface field solar cell: 20.29%. *Japanese Journal of Applied Physics* 56(8S2), 2017, 08MB25.

[19] *Most Efficient Solar Modules (July 2022)* [Online]. Available at: https://www.cleanenergyreviews.info/blog/most-efficient-solar-panels

[20] *Trina Solar Sets New World Record of 23.03% Aperture Efficiency for 210 Vertex P-Type PERC Module* [Online]. Available at: https://www.trinasolar.com/de/resources/newsroom/thu-07012021-1124

[21] *LONGi Solar has Bifacial Mono-PERC Solar Cell World Record Verified at 24.06%* [Online]. Available at: https://www.pv-tech.org/longi-solar-has-bifacial-mono-perc-solar-cell-world-record-verified-at-24-0/

[22] J. Solar. JinkoSolar's high-efficiency n-Type monocrystalline silicon solar cell sets our new record with maximum conversion efficiency of 26.4% [Online] 2022. https://www.jinkosolar.com/en/site/newsdetail/1827

[23] *Trina Solar Takes n-Type Mono IBC Cell to Record 25.04% Conversion Efficiency* [Online]. Available at: https://www.pv-tech.org/trina-solar-takes-n-type-mono-ibc-cell-to-record-25-04-conversion-efficienc/

[24] *Huasun Achieves 25.26% Efficiency for Heterojunction Solar Cell* [Online]. Available at: https://www.pv-magazine.com/2021/07/13/huasun-achieves-25-26-efficiency-for-heterojunction-solar-cell/

[25] *Longi's Heterojunction Solar Cell Hits 26.81%* [Online]. Available at: https://www.longi.com/en/news/propelling-the-transformation/

[26] *Separating the Two Polarities of the POLO Contacts of a 26.1%-Efficient IBC Solar Cell* [Online]. Available at: https://www.nature.com/articles/s41598-019-57310-0

[27] *Record-Breaking Silicon Solar Cell Efficiency of 26.6% Demonstrated by Japanese Researchers, Very Close to the Theoretical Limit* [Online]. Available at: https://www.zmescience.com/ecology/renewable-energy-ecology/solar-cell-close-ideal-limit/

[28] *Bloomberg's n-Type Shares* [Online]. Available at: https://www.pv-magazine.com/2021/05/08/the-weekend-read-life-after-perc/

[29] *PVinfoLink's n-Type Capacity* [Online]. Available at: https://www.pv-magazine.com/2021/06/17/n-type-solar-development/

[30] R. Kopecek and J. Libal. Entering the bifacial nPV era. *PV International*, 48, 2022, 60–67.

[31] *1.25 MWp in Japan* [Online]. Available at: http://npv-workshop.com/fileadmin/images/bifi/miyazaki/presentations/3_1_3_-_ISHIKAWA_-_World_1st_large_scale_Bifacial_PV_power_plant.pdf

[32]   *2.5 MWp in Chile* [Online]. Available at: https://www.bifipv-workshop.com/ 2017-konstanzproceedings

[33]   *1.75 MWp in Chile* [Online]. Available at: http://npv-workshop.com/fileadmin/layout/images/Konstanz-2017/9__F._Bizzarri_ENEL_Innovative_tracked_ bifacial_PV_plant_at_la_Silla_observatory_in_Chile_.pdf

[34]   *1.3 GWp in Turkey* [Online]. Available at: https://www.dailysabah.com/ business/energy/kalyon-inks-812m-deal-for-turkeys-largest-solar-energy-plant

[35]   *Ibri II System in Oman* [Online]. Available at: https://www.pv-magazine. com/2021/12/02/worlds-biggest-topcon-solar-plant-begins-generating/

[36]   *Ukraine, Lithuania Bifacial PV System* [Online]. Available at: https://smartenergydih.eu/advanced-bifacial-solitek-solar-plant/

[37]   *CARBON Plans 5 GW Production* [Online]. Available at: https://www.pv-tech.org/european-solar-manufacturing-start-up-carbon-enlists-isc-konstanz-as-technology-partner/

[38]   *Greenland Plans 5 GW Production* [Online]. Available at: https://www. photon.info/en/news/fraunhofer-ise-accompanies-construction-vertically-integrated-5-gw-photovoltaic-factory

[39]   *PV-IPCEI Projects* [Online]. Available at: https://esmc.solar/news-european-solar-manufacturing-pv/ipcei-for-pv-launched-in-brussels-eu-member-states-are-invited-to-join-the-framework/

[40]   *n-Type vs. P-Type Modules* [Online]. Available at: https://www.pv-magazine.com/2022/07/20/ja-solar-reveals-results-of-year-long-n-type-p-type-tests/

Chapter 2

n-type silicon material

Fiacre Rougieux[1], Gianluca Coletti[2], Sean Gaal,
Bowen (Harry) Guo, Nannan (Felix) Fu, Jordi Veirman,
AnYao Liu and Eivind Johannes Øvrelid

2.1   Introduction

n-type silicon feedstock and wafers are key photovoltaic (PV) enabling technologies for high-efficiency solar cells. This chapter reviews the rapidly evolving field of growth technologies, wafering technologies, and materials engineering methods. First, we review key silicon sources for n-type solar cells and present various recharging technologies. We then present the impact of impurities on silicon solar cells with a focus on metallic impurities, oxide precipitates, and thermal donors. We then present strategies to mitigate the impact of metallic impurities via gettering. Lastly, we evaluate the cost of ownership for n-type ingot growth and wafering.

2.2   n-type silicon feedstock

2.2.1   Introduction

Although the cost of high-purity silicon has steadily reduced, it is still a significant component of the overall module cost structure [1]. Many technologies have been explored to produce high-purity silicon suitable for solar applications at a lower cost than the Siemens process [2], however, the purity, quality, and form of the silicon are often not ideal. When selecting silicon feedstock, it is necessary to also understand the impact of different sources of silicon on the conversion cost to Czochralski (Cz) ingot as the dominant technology to produce n-type silicon wafers.

2.2.2   Polysilicon

The dominant method to produce polysilicon for n-type ingots has not changed for decades, starting with metallurgical grade silicon produced from lump quartz and

[1]UNSW, Kensington, Australia
[2]PV Tandem Technology and Applications at TNO, The Hague, The Netherlands

carbonaceous reductant in a submerged arc furnace operating at temperatures greater than 2,000°C [3]:

$$SiO_2 + C = Si + SiO(g) + CO(g) \tag{1}$$

The composition of the metallurgical silicon produced is primarily determined by the raw materials and the process yield, with the most significant impurities reported as iron, aluminium, calcium, and titanium, as described elsewhere [2–4]. The price of metallurgical grade silicon [5] is a significant contributor to the overall cost of poly silicon. The metallurgical grade silicon is then ground to a powder and converted to trichlorosilane either by direct chlorination or hydrochlorination.

Direct chlorination involves reacting the silicon with anhydrous hydrogen chloride at about 300°C, producing predominately trichlorosilane, silicon tetrachloride and hydrogen:

$$Si + HCl(g) = SiHCl_3(g) + SiCl_4(g) + H_2(g) \tag{2}$$

Hydrochlorination is the reaction of silicon powder with silicon tetrachloride and hydrogen at about 35 bar and 500°C, simultaneously converting silicon tetrachloride and silicon to the required trichlorosilane with a conversion of about 20% of the silicon tetrachloride to trichlorosilane:

$$Si + 3SiCl_4(g) + 2H_2(g) = 4SiHCl_3(g) \tag{3}$$

Most impurities present in the metallurgical silicon are then removed from the trichlorosilane by distillation and adsorption processes. The high-purity trichlorosilane is then thermally decomposed with hydrogen in the Siemens process to produce polysilicon or converted to silane, which is thermally decomposed to polysilicon:

$$SiHCl_3(g) + H_2(g) = Si + HCl(g) + SiCl_4(g) \tag{4}$$

$$SiH_4(g) = Si + 2H_2(g) \tag{5}$$

These are all large capital-intensive processes that require significant expertise and energy to operate but are capable of high production rates.

2.2.3   Alternative silicon sources

2.2.3.1   Granular silicon

The thermal decomposition of trichlorosilane or silane can also be accomplished in a fluidized bed reactor, producing small spherical silicon [2] at a lower capital and energy cost. Although this is not currently the dominant form of silicon used for Cz ingot production due to lower quality, it offers a significant benefit in handling; it is flowable. Issues with the quality of granular silicon result in lower ingot production rates, which invariably result in higher ingot costs regardless of the cost savings from the granular silicon, mainly due to the significant cost of conversion to ingot. If these quality issues can be resolved, it will be a very useful form of high-purity silicon.

2.2.3.2   Upgraded metallurgical silicon

There are numerous other methods to produce high purity silicon, such as Elkem Solar Silicon where the metallurgical silicon is produced in the same manner but with higher purity raw materials, then refined directly to increase the purity [2]. This can reach relatively high purity at a lower cost, but it will always contain more impurities than polysilicon. The use of upgraded metallurgical silicon to produce Cz ingot depends on the conversion cost to wafers, as well as the cell efficiency that can be produced.

2.2.3.3   Recycled silicon

After the ingot is produced, it needs to be shaped into blocks. Even a very well controlled ingot diameter and a pseudo full square wafer will yield over 35% recycled silicon just from the round section of the ingot, with additional material from the top, tail, and any other off-spec recycled ingot. Recycled ingot should have a lower level of metallic impurities than the original silicon feedstock, as long as it is processed carefully so that it does not become contaminated during handling, size reduction, screening, and addition back to the crucible. An operation capable of two or more ingots per crucible, which is desirable from a cost perspective, will create too much recycled silicon to stack in the initial crucible, requiring the recycled silicon to be converted to a shape capable of addition to a hot crucible in a furnace under vacuum.

Further complications arise when considering the impact of the disparate dopant concentration in the ingot top and tail. Even the concentration of dopant in the side cut varies along the length of the ingot. This is further complicated if p-type and n-type ingot are produced in the same location, requiring stringent separation protocols to avoid cross contamination.

2.2.4   Charging silicon

An important criterion when selecting the feedstock to produce n-type ingot and wafers is the shape and form of the silicon.

2.2.4.1   Crucible charge

The first consideration is stacking a crucible such that the maximum amount of silicon can be added, without causing operational issues such as damage to the crucible or pieces of silicon fused to the wall above the liquid level. To achieve a good initial charge, it is best to have a combination of large, medium, and small pieces of silicon. A shape and size that allows a closely packed crucible are beneficial to the initial charge, but this can be constrained by the quantity and form of silicon recycle. For example, larger pieces of polysilicon rod are helpful, but the shape requires a significant quantity of smaller silicon to fill the voids. Recycled ingot side cut stacks together well, but it needs to be an appropriate length. Cubic blocks of upgraded metallurgical silicon can be easily stacked close together.

2.2.4.2   Topping up crucibles

After the initial charge has melted, it is beneficial to top up the crucible with additional silicon, which often increases the crucible charge more than 20%. This increases the melting time, but dramatically improves the ingot productivity and reduces operating costs (crucible, gas, labor, pot scrap). The challenge to solve is how to add additional silicon from the available sources (new silicon or recycled silicon).

2.2.4.3   Recharging crucibles

After ingot production is complete, if the crucible has not been in operation beyond its lifetime and the remaining liquid in the bottom of the crucible has not accumulated excessive impurities, it is beneficial to recharge the crucible. This process involves adding additional silicon to the crucible where it melts with the small pool of liquid remaining in the crucible, spreading the cost of the crucible over multiple ingots. There is also a saving in time, as it is not necessary to cool and then reheat the furnace, further improving the productivity and reducing operating costs. The challenge is to have a method to recharge the silicon, based on the available silicon sources and methods to reduce the size of recycle streams.

2.2.5   Impurities

High minority carrier lifetime wafers require high purity silicon. Although n-type ingot may be more tolerant of some impurities such as iron, this does not imply that lower grade feedstock is acceptable. Most impurities present in the feedstock will accumulate in the remaining liquid silicon, which in the case of recharging the crucible becomes part of the next melt. The more impurities present in the silicon source, the larger the impact on subsequent ingots. A silicon feedstock that has significant levels of impurities will quickly increase the conversion costs, such that it is not feasible to use even at a substantial discount to other sources.

Another area of concern is the inclusion of dust in the silicon source, not just impurities. Small dust particles are coated in a layer of oxide which takes a long time to be dissolved into the silicon melt, so they persist in the melt longer than larger lumps of silicon. If these fine dust particles reach the solidification front, they become incorporated into the ingot and result in structure loss.

The Cz furnace itself can also be a source of impurities, with contamination from the hot zone, especially the radiation shields located close to the silicon melt. In this case, the use of low-quality silicon sources can pollute the equipment and subsequently impact future ingots. Impurities can also arise from the quartz crucible. To make crucibles, natural quartz is mined, crushed, impurities are removed by dissolution and the quartz is sent for fusion. Typical quartz purity is 2N purity which means for every 100 kg of crucible there are 100 g of impurities. There are two type of quartz impurities, either phosphorus-rich quartz or aluminium-rich quartz. For n-type ingots, phosphorus containing crucibles should be preferenced.

2.2.6   Challenges

When selecting the silicon feedstock to produce n-type ingot, there are several considerations which need to be balanced to achieve sufficient quality at the lowest cost. Although the cost of poly silicon is significant, it is similar to the conversion costs. A reduction in the cost and quality of poly silicon can easily result in a lower conversion yield and productivity, driving up the conversion cost and ultimately the final wafer cost.

Increasing the number of ingots per furnace cycle reduces the production costs but produces significant volumes of recycled silicon that must be utilized within the process. To achieve this, a simple and effective method is required to clean and shape the recycled silicon. There are several methods to charge silicon to a hot furnace, but they are generally developed internally for each enterprise to fit the available silicon sources.

Structure loss occurs when a defect enters the ingot, resulting in lost production. Silicon sources that increase the rate of structure loss are very costly, as the puller becomes a refining unit, significantly reducing the productivity. Silicon feedstock that results in structure loss can quickly increase the cost of production, such that it is not feasible to use even at a substantial discount to other sources.

2.2.7 Cost structure

It should be clear that the goal is to produce the lowest cost wafer, not to utilize the cheapest source of silicon. Although the cost of the silicon is a significant factor [1], a lower cost silicon source can have a significant impact on the conversion costs, quickly negating any benefit. Factors that always need to be considered include:

- Purity – high-purity silicon is enabling of multiple ingot production from a single cycle.
- Dust – small particles are often coated in oxide and are very slow to melt, resulting in significant structure loss. Even a few grams of silicon powder added to a crucible will quickly reduce productivity.
- Shape/form – the silicon needs to be efficiently stacked in the crucible or added to a hot crucible under vacuum.

Balancing all of these factors is critical to silicon selection.

2.3   n-type ingot pulling technology

2.3.1   Origin of Cz-growth technology

The Cz process was initially developed by Czochralski in 1918 [6]. At the beginning of the Cz process, polysilicon is loaded into a high-purity crucible and subsequently heated via graphite resistance [6]. Once the silicon melts, a seed is lowered onto the liquid level starting the process of ingot pulling. Ingot pulling can be divided into five steps: seeding, necking, crown, bodying, and tail [7]. The pulling furnace system consists of six parts, as shown in Figure 2.1, including water cooling system, vacuum argon gas system, motion control system, furnace body, temperature control and thermal system, and hydraulic lifting system.

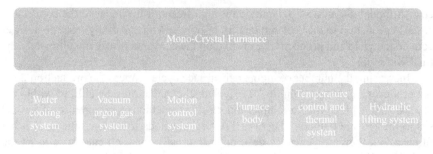

Figure 2.1   Furnace components

Figure 2.2   Process flow of Cz crystal growing process

2.3.2   Technical principle and application

Figure 2.2 shows the technological flow of the Cz growth-process. At the beginning of the Cz growth process, extending a seed into silicon melt leads to non-spontaneous nucleation. The presence of the seed provides an initial crystal nucleus and enables to reduce the energy required for mono silicon nucleation [2, 8]. The seed is monocrysatlline and it provides the crystal orientation for subsequent crystal growth [7].

For industrial production process, the ingot grows in a vacuum argon atmosphere and stable environment temperature [8]. On this basis, to guarantee the seed crystal as the only non-spontaneous nucleation in melt silicon, the overall growth environment should be kept clean [8]. Figure 2.3 shows a range of additional processes that are required on top of the process flow from Figure 2.2, to prevent spontaneous nucleation.

During an entire ingot pulling cycle, the ingot growth process only accounts for 60–70% of the total process time. With continuous improvement in growth technology, the time involved in the growth decreases even further. It is thus critical to increase the effective utilization rate of the Cz puller and improve the efficiency of the silicon growth process. To this end, two technologies have emerged: Recharge-Cz (RCz) and Continuous-Cz (CCz) ingot growth. We will describe these two technologies in the following two sections and what their impact on n-type ingot growth.

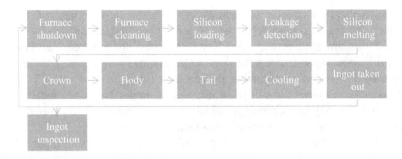

Figure 2.3   Full process flow for Cz growth

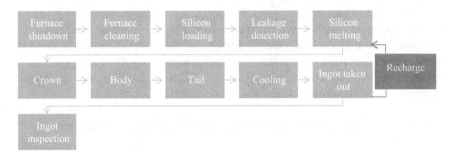

Figure 2.4.   Full ingot recharge process

2.3.3   Status of ingot pulling technology (RCz)

Figure 2.4 shows the full recharge Cz process. When ingot tailing is complete, the finished ingot is removed from the furnace. The main chamber and the pull chamber are then isolated and the ingot is removed out by rotating the pull chamber. A silicon feeder is introduced into the chamber to enable the recharging process. After recharging, the silicon feeder is dragged out from the pull chamber in the same way the ingot was removed. By avoiding furnace cooling, shutdown, disassembly, cleaning and leakage detection for every single ingot pull, the RCz process saves idle production time [8–11]. The RCz technology enables an increase of the utilization rate from 70% to more than 90% for one growth cycle.

Taking the 26-inch hot zone commonly used in the industry as an example, in the Cz scheme, the furnace feeding amount is 300 kg as comparison, the cumulative furnace feeding amount > 1,200 kg in RCz method, the production efficiency is significantly improved, and the utilization rate of crucible is also improved [12]. Thanks to the application of RCz method, the frequency of disassembly and assembly of the furnace is reduced, the frequency of the hot zone exposed in air; hot zone subjected to hot and cold alternations is also greatly reduced, and also, lifetime of hot zone failure caused by oxidation and thermal vibration are also improved with the RCz method [12].

The realization of RCz technology involves the development of auxiliary chamber and feeder, the development of long-life crucible, the development of long-life vacuum system, and other key technology nodes when introduced the recharging process. At present, the latest RCz equipment has a feeding capacity of more than 3 tons and a continuous pulling time for more than 400 h [12]. RCz technology development is still in rapid progress, larger hot zone, higher pulling speed, and low-oxygen technology which match to current furnace is the main direction of present RCz technology development and progress.

Today, similar ingot yields are achieved between gallium-doped and phosphorus-doped ingots. However, the yields are limited by two different factors. The yield of phosphorus ingots is limited by lifetime requirements while the yield of gallium-doped ingot is limited by resistivity requirements [12]. Traditionally, the yield of n-type ingots has been believed to be limited by rings at the seed-end and by metallic impurities at the tail end. Figure 2.5 shows the minority carrier lifetime in an n-type ingot after 5 regrowth. The minority lifetime degrades after subsequent regrowth toward sub ms lifetimes. Additionally, the lifetime decreases along the ingot. Figure 2.5 shows the effective defect density along the ingot, showing an effective segregation coefficient of 1×10^{-5} which is in the range of common transition metal impurities [13].

2.3.4 *Development of continuous crystal pulling technology (CCz)*

An alternative method to RCz is continuous crystal pulling technology (CCz), another instance of the Cz-growth technology [11]. CCz technology was developed in Japan and the United States before [14]. An example of hot zone composition is shown in Figure 2.6.

Although CCz remains at the pre-commercial stage, the technology has inherent advantages over Cz and RCz growth. The CCz technology makes use of double crucibles to achieve crystal pulling and feeding at the same time [15]. The key technological innovations of CCz are (1) double crucible pulling, (2) feeding device with doping function which ensure controllable feeding weight, and (3) matching pulling process and hot zone.

Ingot produced by conventional Cz technology have a wide range of resistivities as a result of impurity segregation [13]. The impurities pile up toward the tail end of the ingot due to their segregation coefficient. The selection of doping elements varies according to the different mono products. Through patent grants and other means, the industry switched from boron-doped to gallium-doped ingots around 2020 [1]. Today, gallium-doped products are the mainstream p-type products in the industry.

In the PV industry, gallium and phosphorus are the main dopant to control ingot resistivity. However, as shown in Table 2.1, both gallium and phosphorus have relatively low-segregation coefficients (compared to boron). As such, the distribution of gallium and phosphorus tends to be highly non-uniform in a Cz ingot. However, cell production requires wafers with a narrow resistivity range.

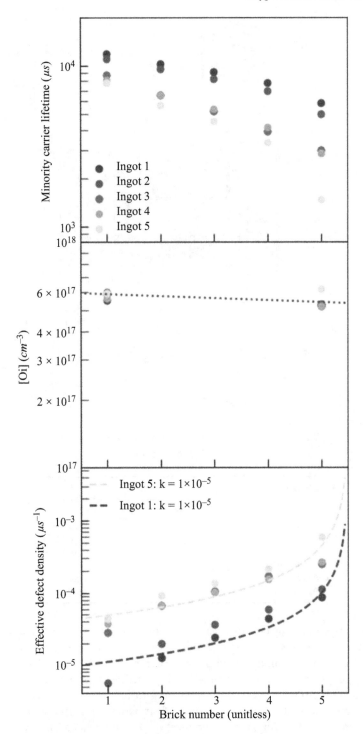

Figure 2.5   *(a) Minority carrier lifetime, (b) interstitial oxygen concentration, and (c) effective defect density in an n-type ingot after five recharges*

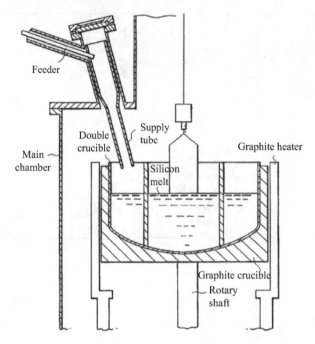

Figure 2.6   Schematic of CCz pulling technology

Table 2.1   Equilibrium segregation coefficients of the main dopants in Si

Element	k0
Boron	0.8
Gallium	0.008
Aluminium	0.002
Indium	0.0004
Phosphorus	0.35
Arsenic	0.3
Antimony	0.023

That means that using conventional Cz to grow gallium- or phosphorus-doped ingots leads to much shorter ingot length (to avoid crystallization of the impurities segregated in the remaining liquid phase). Shorter ingot length means lower body growing period and more time spent in feeding, melting, bodying, tailing processes, and other non-growth stages all compounding to increase ingot costs. CCz technology can realize the narrow resistivity distribution range through continuous feeding and crystal pulling, which can meet the requirements of narrow resistivity range by cell process. However, CCz technology is still not mature at present, and still faces issues of reliability of the feeding process, reliability of the crucible as well as serious lifetime degradation [12].

However, there are multiple issues to the commercial realization of CCz. (1) Silicon granules as CCz feed have greater surface/volume ratio, leading to more oxides increasing the risk of issues during crystallization. (2) CCz needs two crucibles which leads to more contact with crucibles and subsequently more contamination including oxygen. (3) The yield of CCz ingot is lower due to instability in the pulling process as well as metal contamination leading to lifetime degradation [12].

2.4 n-type mono-wafer slicing technology

Diamond wire slicing is the dominant slicing technology in the photovoltaic industry displacing previous slurry slicing [1, 16, 17]. As a result, wafer slicing efficiency and wafer quality has greatly improved. Its technological advantages of thin wafer and thin diamond wire on silicon materials greatly expand the space for silicon wafer cost reduction [17]. This is particularly important for n-type wafers which tend to be thinner.

The basic difference between diamond wire (Figure 2.7) and slurry slicing is that for diamond wire, the diamond is fixed on the straight-drawn steel line by bonding and electroplating to carry out high-speed back and forth slicing [17]. On the other hand, slurry slicing uses the suspension of silicon carbide through wire-net drive[16]. There are two types of diamond wire which are resin wire and electroplated wire. The difference between electroplated wire and resin wire is mainly in the method to attach the diamond particles, breaking force of finished wire and difference of finished wire diameter [16, 17]. It has become the mainstream wire in solar industry due to better slicing capability. The specific advantages of diamond wire slicing are [12, 18]:

1. *Consolidation*: Diamonds contribute more to slicing which leads to more grinding and reduces abrasives wear.

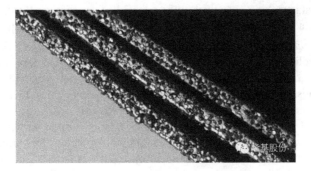

Figure 2.7   Diamond wires

2. *High hardness of diamond*: Strong diamonds are wear resistance, which will greatly prolong the lifetime of diamond wire.
3. *High slicing speed*: With the increased diamond contact area with wafers, diamond wires are able to better withstand defects brought by high wire speed which plays an advantage for high slicing speed.
4. *Silicon consuming optimization*: Less silicon material is used to produce the required wafer thickness.

The advantages of diamond wire slicing are mainly shown in three aspects: high cutting efficiency, low-cost per wafer and more controllable quality. In addition, there are other resulting advantages [12, 17, 18]:

1. *High slicing efficiency*: High slicing efficiency reduces the depreciation of the equipment, plant and other aspects, and also manual process cost per piece, and the slicing efficiency increase 3–4 times.
2. *Low-cost per piece*: Diamond wire slicing replaces the traditional slurry slicing of silicon carbide, suspension and steel wire. Diamond wire adopted by way of consolidation abrasive that more diamond can participate in slicing effectively, also, with lower curf loss, more wafer output can be get and wafer cost is greatly reduced, coupled with higher slicing efficiency advantages, slicing cost of diamond wire would be more than half compared with slurry slicing.
3. *Better quality control*:
    (a) From the perspective of slurry slicing quality control, slurry, wire, and slicing agent must be three suppliers. Compared with diamond wire, the slurry and wire are integrated, wire suppliers number is reduced, and also the production fluctuation caused by auxiliary materials can be reduced.
    (b) Wire breakage during slicing process is a big killer would affect the yield. Diamond wire would have stricter requirement and need to be electroplated, cleaned, and polished for several times.
    (c) For manufacturing process of diamond wire, it needs to pass the test of multi-channel diamond wire pull machine, and the surface electroplating, breaking force and a series of testing data.
    (d) Diamond wire performance should be established on large number of actual slicing data before providing to the customers,
    (e) Data tracking system is fully established for diamond wire production, data of each coil can be traced back to better improve the quality control.
4. *Lower silicon consumption*: More diamonds can be involved in effective slicing by consolidation, the cost of production and processing process is reduced.
5. *Environmentally friendly*: COD of slurry slicing can reach to hundreds of thousands; but the COD of diluted slicing agent can be down to 200–1,000 and the pressure of environmental treatment is reduced;
6. Diamond wire slicing does not use PEG, silicon carbide particles; also, diamond wire and silicon brick has less pressure contact, the cutting fluid and silicon powder mixture in the content of impurities reduced, potentially more conducive to the subsequent recovery and reuse of silicon powder, furnace reuse. It is still in the exploration stage to potentially benefit kerf loss recycle.

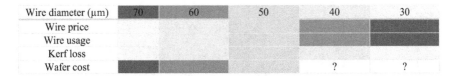

Wire diameter (μm)	70	60	50	40	30
Wire price					
Wire usage					
Kerf loss					
Wafer cost				?	?

Figure 2.8   Impact of diamond wire development on wafer cost. Lower wire diameter reduce wafer cost up to a point

Poly silicon makes up to ∼70% of the total wafer cost in wafer. The way to reduce silicon cost via changes in the wafer slicing process is by using thinner wafer, thinner diamond wires, increasing wafer output, and reducing kerf loss.

2.5   Impact of impurities on n-type solar cells

2.5.1   Metallic impurities

n-type silicon solar cells are known to be generally more immune to metallic impurities than p-type silicon cells [19]. n-type solar cells are unaffected by acceptor-related defects in particular, iron–boron pairs (FeB), chromium–boron pairs (CrB), and the boron–oxygen defect (BO) [19–21]. Though iron reduces the efficiency of n-type cells [22, 23], its impact on n-type cells is less significant than p-type cells [19]. However, this does not hold true for every impurity [19], many metallic impurities have a more significant impact in n-type silicon than p-type silicon including Co, Cr, Ni [24], and Pt and Au [25]. Metal impurities can also precipitate or decorate oxygen precipitates which can lead to a similar increase in recombination activity in n-type silicon and p-type silicon [26–29].

In order to illustrate the impact of various impurities on a 27% n-type silicon solar cell, we calculate the injection dependent lifetime and subsequently the voltage and efficiency of an idealized high-efficiency solar cell. We assume a cell thickness of 140 μm, a J_0 of 5 fA cm^{-2} per side and a short circuit current of 42.24 mA cm^{-2}. We use parameters from: Co [30], Cr [31], Ni [32], W [33], Zn [34], Mn [35, 36], Mo [37], Fe [38], V [39], Ti [36, 40], and thermal donors labeled as TD [41]. Precipitate metals were calculated using the method in [26, 42] using known precipitates stoichiometry from [43].

Figure 2.9 shows simulated efficiency of 27% n-types solar cells as a function of the concentration of various impurities.

Figure 2.9 shows that transition metals even in concentrations down to 1×10^8 cm^{-3} can affect the performance of high-efficiency solar cells. Such dilute defects are not detectable via common defect spectroscopy techniques such as deep level transient spectroscopy (DLTS) with sensitivity limited to 1×10^{10} cm^{-3} for conventional solar cell materials (about 1×10^5 cm^{-3} below the doping) but they affect the efficiency and can be observed (though not identified) using lifetime spectroscopy.

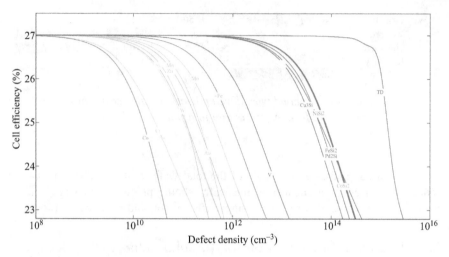

Figure 2.9   Cell efficiency as function of the impurity defect density

In the context of n-type solar cells, not only have interstitials transition metal impurities a significant impact on efficiency but so do metal silicide precipitates. The formation of precipitates at low temperature is also affected by the polarity of the bulk. Indeed, the charge state of precipitates and their nucleation rate is affected by the Fermi level. For instance, the charge state of copper precipitates is positive in p-type silicon but neutral or negative in n-type silicon as the Fermi level is close to the 0/+ level [44]. This means that a lot less copper is required to initiate nucleation in n-type silicon compared to p-type silicon [44].

2.5.2   Oxide precipitates

Interstitial oxygen formed by dissolution of the crucible during ingot growth agglomerates into precipitates and other complexes such as thermal donors during ingot cool down. Whilst complexes such as acceptor-oxygen defects are not present in n-type silicon, thermal donors and oxygen precipitates affect the minority carrier lifetime. Here we expand on oxygen precipitates as thermal donors are treated in Section 2.3 of this chapter.

Oxide precipitates leads to a new interface within the silicon wafer. The dangling bond responsible for recombination at surfaces passivated with silicon oxide is the same bond that leads to recombination at the silicon/precipitate interface [45]. Direct comparison of n-type and p-type silicon shows that for similar processing conditions, the precipitate-limited minority carrier lifetime is always higher in n-type silicon than p-type silicon [28, 46]. This is especially true at low-injection levels where oxygen precipitates have a more dramatic impact in p-type silicon [28, 46].

In the context of n-type solar cells, there are a few caveats to this finding as n-type cells have different processing requirements than p-type cells. Indeed, the recombination activity of oxygen precipitates can be enhanced by several factors:

1. Strains within precipitates: Strained precipitates are created during elevated temperature anneal (precipitate growth). Strained precipitates are more recombination than unstrained precipitates [27–29, 46]. This is particularly relevant to the operations of n-type solar cells which often require a boron diffusion leading to a higher thermal budget than typical p-type solar cells and thus a higher risk of creating strained precipitates.
2. Associated dislocations: Extended precipitate growth leads to the release of self-interstitials with subsequent formation of dislocation loops and sometimes stacking faults. These dislocations become themselves recombination centers and thus enhance the recombination activity within the sample beyond that provided by the precipitate density. As above, this is important in the context of n-type cells with higher thermal budget where the risk of dislocation formation may be higher than in p-type cells.
3. Metallic decoration: Oxygen precipitates are efficient gettering centers and have traditionally been used as such by the IC industry for decades. However, metallic decoration of precipitates enhances their recombination [46]. Metal present at the interface provide new defect levels leading to enhanced recombination activity. While n-type cells are more immune to interstitial metals, metal decoration of precipitates has a similar impact in n-type silicon and p-type silicon [28]. Thus n-type alone provides no intrinsic advantage when it comes to metallic decorated precipitates.
4. Hydrogen: Hydrogen can bind with a wide range of defects in silicon is known to accumulate at dielectric passivation layers such silicon/oxide interfaces. Results show that hydrogen effectively reduces the recombination activity of metallic precipitates [47, 48]. Hydrogenation occurring during firing is in principle independent of the polarity of the bulk as wafers are intrinsic at elevated temperatures. For low-temperature hydrogenation processes, bulk polarity influences the charge state of hydrogen and the charge states of defects to be passivated [49, 50]. In principle, this could lead to different effectiveness of hydrogen passivation between n-type and p-type silicon [50] though early measurements do not display such differences [48].

While the size of precipitates influences their recombination activity, smaller precipitates are also recombination active. Nanoprecipitates (precipitates smaller than 50 nm) have been shown to dramatically impact the lifetime of n-type silicon wafers [51].

2.5.3 Light elements and intrinsic defects

As we near the intrinsic lifetime limit, n-type cells may be limited by light elements, dopants and their complex with intrinsic defects [52]. A wide range of defects related to light-element paired with intrinsic point defects have been detected via DLTS including VO, VO_2, VH [53]. However, the impact of such complexes on the lifetime of n-type silicon wafer remains uncertain. Their dilute nature makes them hard to observe directly with techniques such as DLTS, though their effect can be seen in the lifetime. One alternative way to identify lifetime-limiting defect in high-efficiency solar cells is to measure the segregation coefficients of lifetime-limited impurities. Early results on high-lifetime n-type ingots indicate segregation coefficients of 0.6 [52], which rules out metallic impurities, pure oxygen, and carbon-related complexes and suggests that other impurities such as light elements or intrinsic defects may be involved. More recently, evidence of lifetime limiting defects related to CH complexes [54], VO-related complexes [55, 56], or VP-related complexes [56] have emerged.

2.6   Thermal Donors in n-type silicon solar cells

2.6.1   Introduction

Thermal Donors are one of the most studied defect families in silicon. A request on Web of Knowledge reveals that over 300 scientific articles do contain the keywords "Thermal Donors in Silicon" in their title. Thermal Donors bear their name from their donor character and their generation that requires heating up silicon. Historically, their existence was evidenced as early as 1954 at the Bell Telephone Laboratories [57], only one year after the first practical silicon solar cell was produced in the same lab. The pioneering studies by the Bell Telephone Laboratories, and soon after by the Signal Corps Engineering Laboratories [58, 59] already unraveled many key features of Thermal Donors, which were later progressively refined by many research groups around the world. Lately, encouraged by the increasingly stringent carrier lifetime requirements of silicon wafers for photovoltaics, Thermal Donors regained interest at the wafer but also at the device level. Hereafter we aim to provide an overview of the key knowledge related to Thermal Donors, including how they affect doping/carrier lifetime/solar cell performance, and what are the current ways available to mitigate their formation for high efficiency n-type cells.

2.6.2   Main features of Thermal Donors

At least three types of thermally activated donors have been evidenced: Thermal Donors generated in the 350–600°C range (discovered by the pioneering work of Fuller at al., often referred to as "old" Thermal Donors), "New Donors," generated at higher temperatures in the 600–900°C range [60], and "Shallow Donors" that feature particularly weak ionization energies. For the sake of brevity and due to comparatively limited information in literature about New and Shallow Donors, we shall focus on the "old" Thermal Donors in the following.

Thermal Donors were shown to form exclusively in oxygen-rich silicon (10^{17} cm^{-3}) [59] and hence most research focused on Cz-grown silicon. This discovery imparted a leading role to oxygen, the direct involvement of which was

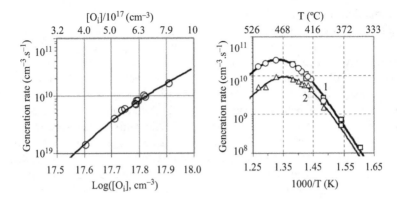

Figure 2.10   *Dependence of the Thermal Donor generation rate on the interstitial oxygen content [Oi] and the anneal temperature T (adapted with permission from [62]). [Oi] was measured by FTIR (new ASTM standard)*

demonstrated by advanced characterization techniques [electron nuclear double resonance (ENDOR) spectroscopy] [61]. Thermal Donors are often regarded as early stages of oxygen aggregation, which fostered their study in the field of microelectronics, where the control of silicon oxide precipitation is crucial.

They are most efficiently formed at around 450°C (see Figure 2.10) [62]. As such temperatures, the maximum achievable Thermal Donor concentration and the overall generation rate scale, respectively, with the third and the fourth power of the interstitial oxygen content [62, 63]. During prolonged anneal, Thermal Donors well in excess of $1,016$ cm^{-3} can be formed. High concentrations of carbon (10^{17} cm^{-3}) [64] were found to slow down the donor generation. On the contrary, high concentrations of silicon self-interstitial [65] or hydrogen were observed to enhance both kinetics and amplitude of Thermal Donor formation. This boosting effect of hydrogen has been elegantly leveraged to form deep p–n junctions at low temperatures [66]. Doping was also observed to affect the Thermal Donor generation rate. More precisely, intentional acceptor (donor) elements in excess of 10^{16} cm^{-3} were reported to speed up (slow down) the generation rate, although the effect of 10^{16} cm^{-3} donor concentrations is still controversial [67]. It was therefore inferred that electrons are involved in the sequential formation of Thermal Donors, which is supported by recent findings [68].

Wruck *et al.* revealed by Hall effect and Infrared measurements that Thermal Donors are actually double donors, introducing two energy levels approx. 50–70 meV and 130–150 meV below the bottom of the conduction band [69], getting shallower as the anneal duration increases. Wruck *et al.* additionally demonstrated that Thermal Donors should not be regarded as a single defect, but rather as a set of distinct double donor species. Wagner *et al.* [70] first reported on the formation kinetics of individual species. Their results confirmed earlier suggestions that the different Thermal Donor species form sequentially, the newly formed species with somewhat lowered ionization energies forming at the expense of previously formed species. Since then, 16 neutral thermal double donors (TDD1–TDD16) and nine positively charged species (TDD1–TDD9) were observed using Infrared spectroscopy, TDD species with small numbering being formed first in the generation sequence [71, 72]. Interestingly, the two donor species formed first in the course of annealing (TDD1 and TDD2) were found to be bistable [73]. In other words, TDD1 and TDD2 can exist in two stable different structural configurations: the historically studied configuration "H" associated with the two aforementioned energy levels, and a second one labeled "X," also introducing two distinct levels, with the particularity that – unlike TD in configuration H—the second electron is bound more strongly than the first one. Such centers are called Anderson systems with negative correlation energy (or "negative-U" systems) [74].

The composition and the exact formation sequence of Thermal Donors have now puzzled the scientific community for almost seven decades. Detailing the different models here would be excessively long and out of scope. The interested reader can refer to Ref. [75].

2.6.3   Influence on carrier lifetime and silicon solar cell efficiency

Thermal Donors can rapidly dissolve at temperatures exceeding 600°C (typically a few seconds at 700°C [76]. That way, they are readily annihilated in the course of most mainstream and historical solar cell/integrated circuit processes, which generally include process steps in excess of 700°C. As they are no longer present in the finished device, their influence on the bulk carrier lifetime has been therefore poorly investigated. Yet, the recent strong developments of n-type silicon heterojunction have triggered a renewed interest in the recombination properties of Thermal Donors. Indeed, the maximum temperature in the course of such processes is moderate (typically below 250°C), implying that Thermal Donors survive the cell (and module) processes and are therefore still present in modules under operation. This also applies to current developments on n-type transition metal oxides (TMO)-based cells, for which the cell process may be conducted at even lower temperatures [77].

In this context, intensive research on the influence of Thermal Donors on transport and recombination properties has been conducted in n-type silicon during the last decade. Using photoconductance decay techniques, Hu *et al.* reported a significant trapping activity associated with Thermal Donors concentrations in excess of 10^{14} cm^{-3} [78], building on previous observations [79]. Owing to the deep position of the observed trap energy level, the authors hypothesized that this trap could act as an effective recombination center as well. Several research groups reported defect-induced carrier recombination in Thermal Donor-rich Cz silicon. Hu *et al.* reported on an increasingly reduced carrier lifetime for Donor contents

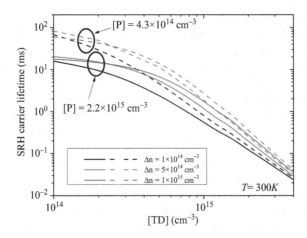

Figure 2.11 Relation between the concentration of Thermal Donors and the Shockley-Read-Hall lifetime for various injection levels (Δn). Adapted with permission from [41]

above 10^{14} cm^{-3} [80], while Thermal Donors are very well tolerated at lower concentrations (see also [81, 82]). The observed recombination activity was broadly attributed to Thermal Donors, without however specifying which species was at play. Tomassini *et al.* [41] confirmed a progressive lifetime reduction above 10^{14} cm^{-3}, with values falling below the millisecond range between 5×10^{14} cm^{-3} and 10^{15} cm^{-3} for typical resistivity and injection level values (see Figure 2.11). Using photoconductance and Hall effect measurements as a function of temperature, they tentatively attributed the observed recombination to the bistable TDD2 defect in configuration X. Markevich *et al.* challenged this assumption using junction capacitance techniques, and eventually invalidated the involvement of TDD2 [83]. Instead, they suggested that standard Thermal Donors in configuration H could be responsible for carrier recombination (in agreement with findings in [84], with a pre-requisite that they are in their singly ionized form, i.e. when the Fermi level approaches or surpasses the Donor level located 130–150 meV below the conduction band. More recently, Basnet *et al.* strengthened the earlier view that silicon oxide precipitates could form simultaneously during Thermal Donor generation, and that these precipitates could be responsible for the recombination activity observed in Thermal Donor-rich silicon [85]. The study of the influence of Thermal Donors on the carrier lifetime in p-type could bring further insight into the problem and help discriminate Basnet *et al.*'s from Markevich *et al.*'s models, as Thermal Donors cannot be in the singly ionized state in p-type silicon (Fermi level always in the lower half bandgap).

From these reports of lowered carrier lifetimes in Thermal Donors-rich wafers, it comes with no surprise that excessive concentrations of Thermal Donors are associated with reduced cell efficiencies. Regarding silicon heterojunction devices, for which Thermal Donors are retained in the bulk after cell processing, Tomassini *et al.* found a threshold Thermal Donor concentration of around 5×10^{14} cm^{-3} below which they had a negligible effect on the cell efficiency. For a concentration of 10^{15} cm^{-3}, a value that can be met in practical ingots, their findings indicate an unacceptable 0.5–0.8% abs. efficiency reduction, demonstrating how critical the control of Thermal Donors is. Similar conclusions were drawn later [84]. For solar cells fabricated with high-temperature processes (e.g. Passivated Emitter and Rear Cell (PERC), Passivated emitter rear totally diffused (PERT), Tunnel Oxide Passivated Contact (TOPCon)), although Thermal Donors are no longer present in the finished device, excessive concentrations are not desirable either, as they may give rise to additional recombination-active silicon oxide precipitates that may grow in size during the cell process [85, 86]. For instance, wafers with an as-grown Thermal donor concentration around 10^{15} cm^{-3} led to 1% abs. efficiency drop for n-PERT solar cells, with a characteristic circular region of degraded carrier lifetime ("black core") as well as striations [87]. Such features are generally attributed to recombination through silicon oxide precipitates, as was demonstrated in [88].

2.6.4   Signature of Thermal Donor formation in Cz ingots and strategies for their avoidance

Thermal Donors are preferentially formed in the part of the ingot crystallized first ("seed-end"), as it generally contains more oxygen and experiences the longest

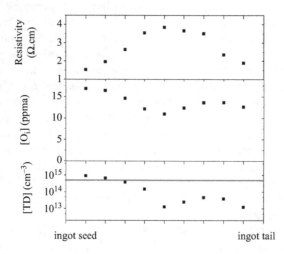

Figure 2.12   Typical resistivity, interstitial oxygen ([O$_i$]) and Thermal Donor ([TD]) concentrations along an n-type Thermal-Donor-rich Cz ingot (adapted from [J. Veirman et al., Sol. Energy Mater. Sol. Cells 158, Part 1 (2016) 55–59]). Oxygen measurements in accordance with new ASTM standard. The red line shows the upper limit for heterojunction cells determined in M. Tomassini et al., J. Appl. Phys. 119 (2016), 084508

residence time in the 350–600°C range [89, 90]. Figure 2.12 illustrates a typical example for an n-type Thermal Donor-rich Cz ingot, with a Thermal Donor concentration approaching 10^{15} cm^{-3} at the seed end. There, the authors reported effective lifetime values with state-of-the art passivation as low as 300 μs at the wafer center (microwave photoconductance decay), which is for instance well below what is required for high efficiency heterojunction cells.

Due to their double-donor character, Thermal Donors induce a downward (upward) resistivity shift in n-type (p-type) ingots compared to what is expected from segregation laws, as they add up (substract) to the intentional doping. The larger the oxygen content and the longer the dwell time in the 350–600°C range, the larger the resistivity shift for a given intentional doping. Note that the resulting shift is generally higher at the ingot center (i.e., wafer center) due to a usually larger oxygen content.

Owing mostly to the low carrier lifetimes associated with Thermal Donors, and to their intimate link with silicon oxide precipitation, the mitigation of Thermal Donor formation has been a subject of intense research in the photovoltaic community. This research was further motivated by the ongoing shift to longer and larger ingots in production, which are more prone to extensive Donor formation owing to their larger thermal mass. To keep the Thermal Donor content acceptable, several upgrades have been implemented by ingot manufacturers and are now routinely deployed (see Figure 2.13). Among them, cooling tubes can be used to reduce the time spent by silicon in the Thermal Donor formation range

Figure 2.13   Cooling tube to speed up the cooling of the upper ingot part during the pulling process. The ingot grows inside the visible cylinder formed by the cooling jacket and the cooling tube. Reproduced from Ø. Nielsen [nPV workshop 2021, held online] with courtesy of NorSun AS and TDG-MT

(350–600°C), thereby impeding their formation. Alternatively, new hot zone designs can be engineered to lower the crucible temperature and thus reduce the oxygen release into the melt [91].

At the wafer or brick level, Thermal Donor Killing (or Thermal Donor Annealing) treatments were developed in the past, involving heating the material at temperatures around 650°C for some minutes/hours to dissociate Thermal Donors. Such treatments effectively restore the resistivity to its value governed solely by the intentional doping. However, this kind of treatments also promotes the nucleation of oxide precipitate nuclei [92], which may then expand during the high tempera-ture steps of the cell process (if any), jeopardizing the cell conversion efficiency. Alternatively, shorter thermal treatments at higher temperature (typically > 1,000°C for some seconds/minutes) were also tested at the wafer level [93] before entering the solar cell process. Such treatment, referred to as "Tabula Rasa," sup-presses all Thermal Donors but also most oxide precipitate nuclei, making the wafer much more resilient to any further high-temperature processing. Yet, such additional step remains uncommon in the PV industry as per today (contamination issues, additional cost, etc.), particularly in a context where leader Cz ingot man-ufacturers have made considerable progresses in mitigating such defects at the ingot level. This is illustrated in Figure 2.14, where the latest developments enable a remarkable reduction in the interstitial oxygen and the Thermal Donor contents at

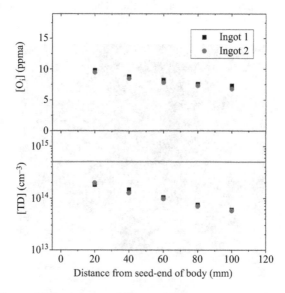

Figure 2.14   Interstitial oxygen and Thermal Donor concentrations at the ingot seed end for two state-of-the-art n-type ingots (courtesy of Norsun AS). Oxygen measurement in accordance with new ASTM standard. The red line depicts the upper limit for silicon heterojunction cells derived in [41]

the (critical) seed-end. Specifically for heterojunction cells, these developments allow to keep the Thermal Donor content safely below the threshold concentration of 5×10^{14} cm^{-3} reported in Ref. [41].

It is worth also mentioning that despite being undesirable in industry, Thermal Donors can ingeniously deliver valuable information to characterize Cz ingots and assist crystal growth optimization. For instance, in nowadays (defect-lean) typical wafers, the well-known and strong dependence of the Thermal Donor formation on the interstitial oxygen content (Figure 2.1) has been leveraged to quantify and map the interstitial oxygen concentration in thin and as-cut photovoltaic wafers [94], where traditional techniques such as FTIR spectroscopy will generally fail. Alternatively, as Thermal Donors are a relic of the ingot cooling during the growth process, their concentration has been used to develop ex situ techniques to quantify the ingot thermal history [95], which is complex to achieve by other means.

2.7 Gettering in n-type silicon solar cells

The quality of the silicon wafer bulk plays an important role in determining the efficiency potential of solar cells. Although some of the common metallic impurities are less harmful in n-type silicon compared to the p-type [19], gettering is still an essential step to significantly enhance the bulk material quality [23, 96]. Gettering is the process of removing unwanted metals from the silicon wafer bulk to a region where the impurities have less of an overall negative impact on the device efficiency. It generally involves three steps: release (from bounded forms, for instance), diffusion, and capture. In this section, the gettering approaches and effectiveness that are relevant for different n-type cell structures are outlined and discussed.

2.7.1 n-type PERT

n-type PERT solar cells feature a full-area heavy boron doping on the front and a full-area phosphorus doping on the rear, generally through thermal diffusion. Most of the gettering action likely comes from a typical phosphorus diffusion process, and typical boron diffusion conditions do not give a strong gettering effect, as will be detailed below.

Phosphorus diffusion is well known to result in an effective gettering of bulk metal impurities, as a combined result of the high "capture" efficiency (i.e. high segregation coefficient) of the heavily P-diffused regions and the large thermal budget of the phosphorus diffusion process (temperature and time) that allows for metal diffusion across the silicon wafer bulk to reach the surface gettering sinks. There are multiple mechanisms proposed in the literature to explain the gettering effect of heavy phosphorus diffusion: Fermi level effect and ion pairing [97, 98], metal-silicide formation at silicon-phosphorus precipitates [99–102], injection gettering [103–105], binding with phosphorus-vacancy (P_4V) clusters [106], and the possible involvement of oxygen from the thermal diffusion process [107, 108].

The latter two mechanisms were more recently proposed in the last decade. It is possible that several gettering mechanisms take place in parallel, and the exact contribution of each depends on the diffusion parameters and doping concentrations.

In general, the gettering effectiveness increases with increasing phosphorus concentration, and for the same phosphorus concentration, the effectiveness increases with decreasing annealing temperature (see, e.g., Fe gettering in [109]). Therefore, having a slow cool-down from high diffusion temperatures and adding a low-temperature (below 700°C) annealing step after diffusion enhances the overall gettering effect [110], as experimentally demonstrated by many (see, e.g. [111–114]).

In boron diffusion, although Fermi level and ion pairing plays a role in enhancing the solubility of most transition metals in highly boron-doped silicon [97, 115], its effect for reducing the bulk metal concentration is rather weak due to the very small thickness ratio of the diffused region to the silicon wafer bulk. Effective gettering from boron diffusion relies on special process conditions that create a high level of boron supersaturation such that boron–silicon precipitates [116–118] and/or boron-rich layers (BRL) [113, 119] are formed, which can act as strong gettering sinks. However, a high level of boron supersaturation is not ideal for the cell performance. The presence of a BRL is found to significantly reduce the surface passivation quality of the boron-doped Si and may introduce dislocations into the silicon wafer bulk [119–121]. As a result, the BRL is typically avoided in cell fabrication by an *in situ* oxidation step after boron drive-in. The BRL is oxidized into boron silicate glass, and the high-temperature oxidation of the BRL redistributes metal impurities, that were initially gettered by the BRL, back into the silicon wafer bulk [119]. It is possible to retain the strong gettering effect of the BRL through a low-temperature ex situ chemical etching of the layer, although the surface passivation quality is not as good as an in situ high-temperature oxidation, which further reduces the surface boron concentration [119].

In addition to the weak gettering effect associated with typical boron diffusion processes, the higher temperatures (above 900°C) used in boron diffusion mean that the process is more susceptible to external contamination, as well as internal contamination from the dissolution of metal precipitates in the case of cast-grown silicon materials.

In PERT solar cells, the heavily doped silicon surfaces are coated by dielectric films such as silicon nitride and aluminium oxide, for surface passivation and antireflection purposes. Such dielectric films also possess gettering effects at elevated temperatures where metals are sufficiently mobile [122–125], such as during contact firing, film deposition, and passivation activation anneals. Although the thermal budget of these processes is too small to cause a significant reduction of the bulk metals, the dielectric films may offer some additional gettering effect, especially in combination with the boron diffused regions. However, this possibility needs to be examined experimentally, as well as the impact of gettering (i.e. accumulation of metals near/in the dielectric films) on the passivation effect of the dielectric films.

2.7.2 Polysilicon/oxide passivating contact solar cells

In n-type polysilicon/oxide passivating contact (also known as TOPCon or POLO) solar cells, gettering mainly occurs through the high-temperature formation process of the full-area phosphorus-doped polysilicon/oxide structure on the rear. The formation of the phosphorus-doped polysilicon/oxide through thermal diffusion [126, 127], ion implantation and annealing [128], in situ doping by LPCVD [129] or PECVD [130, 131] and then annealing, were all reported to generate a gettering effect, although a direct comparison of their effectiveness is still lacking.

As the gettering effect of *intrinsic* polysilicon is not strong enough to cause a significant reduction of the bulk metal concentrations, the gettering effect of the structure mainly arises from a high dopant concentration [126]. In addition, it was found that the main gettering region in the polysilicon/oxide structure is the highly doped polysilicon layer [127]. Therefore, our existing knowledge of the gettering mechanisms and effectiveness associated with heavy phosphorus or boron doping of the crystalline silicon (c-Si) also applies to the heavily doped polysilicon. For instance, in boron-doped polysilicon/oxide structures, the presence of a high electrically *inactive* boron concentration (from thermal diffusion without in situ oxidation [126, 127], or from a high in-situ doping via LPCVD [131] or PECVD [129]) leads to an impressive gettering effect, whereas the absence of a high boron supersaturation results in no gettering action [128].

The ultrathin silicon oxide interlayer is found to act as a partial diffusion barrier that *slows down* the gettering rate by reducing the flux of metals from the silicon wafer bulk into the polysilicon gettering layer [126, 131]. The blocking effect of the ultrathin silicon oxide interlayers depends on the oxide properties (i.e. growth methods) [131]. The gettering strength of the heavily doped polysilicon layer, however, is not affected by the oxide interlayer [131]. In other words, if given sufficient thermal budget to allow metal diffusion across the silicon wafer bulk and then across the oxide interlayer, the maximum gettering effect of a particular heavily doped polysilicon layer can be achieved by driving the gettering reaction to steady state.

Similar to phosphorus diffusion gettering [132], an accumulation of metals within the heavily doped polysilicon layer does not compromise the surface passivation quality of the polysilicon/oxide structure [126]. This is particularly useful in heavily boron-doped polysilicon/oxide where boron-rich layers form. As mentioned in the preceding section on n-PERT, the BRL is a very effective gettering sink but is not compatible with the surface passivation schemes for c-Si, and as a result the BRL is avoided in boron-doped c-Si [119]. However, the presence of the BRL is found to *not* affect the surface passivation quality of the polysilicon/oxide [126, 127], which provides an additional benefit of the polysilicon/oxide structures.

2.7.3 Interdigitated Back Contact (IBC)

In IBC solar cells, both contacts are located on the rear and the heavily doped regions (doped c-Si or doped polysilicon/oxide) are selective. As the gettering effect comes from the heavily doped regions, previous discussions regarding the gettering actions of dopant diffusion into c-Si and heavily doped polysilicon/oxide structures are still relevant. However, because of a smaller region of the gettering sinks, the overall effect in reducing bulk impurities is not as good as a full-area doped region of the same process conditions.

Ion implantation is an additional technique to introduce dopants into the c-Si surface regions, and it is particularly attractive for IBC solar cells due to a precise control of the doping position. The gettering effect of phosphorus implanted c-Si, however, is not as good as that from phosphorus diffusion, which cannot be explained by the electrically *active* phosphorus concentrations [133–135]. Therefore, the weaker gettering effect may be caused by the lack of electrically *inactive* phosphorus [100, 106], or/and oxygen from a thermal diffusion process [107, 108], as previously outlined in the section on n-PERT.

In both phosphorus- and boron-implanted c-Si samples, adding a low-temperature annealing step enhances the gettering effect, which is mainly attributed to the formation of metal precipitates [133, 135–137]. Although this reduces the bulk impurity concentration, the formation of large metal precipitates (>1 micron) introduces shunting sites in the junction and thus degrades the cell performance [135]. A careful design is therefore required to optimize the overall device efficiency.

2.7.4   Heterojunction solar cells

In amorphous-silicon (a-Si)-based heterojunction solar cells (often referred to as HIT, HJT or SHJ), the entire cell fabrication process takes place at low temperature (200–300°C). There is therefore insufficient thermal budget for most of the metal impurities to diffuse to the wafer surfaces, perhaps except for the very fast diffusing metals such as copper, nickel, and cobalt (depending on the thermal budget of the whole process). Moreover, the gettering effect of the surface a-Si layers used in heterojunction solar cells has not been examined experimentally, although it seems plausible that the heavily doped a-Si layers may behave in a similar fashion as the heavily doped polysilicon layers discussed above.

A pre-treatment may be applied to improve the bulk material quality of the silicon wafers prior to the fabrication of heterojunction solar cells. This is demonstrated in, e.g., Refs [138,139], where a combination of gettering, hydrogenation, and *tabula rasa* processes resulted in high-performance heterojunction solar cells from initially low-quality silicon substrates.

2.7.5   Gettering in cast-grown silicon

Cast-grown silicon materials, such as multicrystalline silicon (mc-Si) and monocrystalline-like cast-grown silicon (cast-mono), generally contain a large concentration of metal precipitates [140–143]. These precipitates not only act as recombination centers themselves [26, 144–146] but also potentially dissolve into metal point defects during high-temperature cell processes [147], poisoning the silicon wafer bulk.

To effectively getter metal precipitates, a variable temperature approach was proposed to capitalize on both the high precipitate dissolution rate and high diffusion coefficient at high temperatures, and then the high "capture" efficiency (i.e., segregation coefficient) at low temperatures [110, 148]. In a simplified process, adding a high-temperature dissolution step prior to, for example, a standard phosphorus diffusion, followed by a slow cool-down and an additional low-temperature annealing step improves the overall gettering effect [111–114, 149–151].

2.7.6 Summary

In summary, this section discusses the most relevant gettering techniques that can be incorporated in the fabrication of several n-type cell structures, as well as the respective gettering effectiveness, improvement strategies, and potential impacts on cell performance. Heavy phosphorus, or in some cases boron, doping of the c-Si surfaces or polysilicon/oxide structures underpin most of the gettering approaches, and thus it will be helpful to further understand and clarify the underlying gettering mechanisms.

2.8 Cost of ownership

2.8.1 Costs in ingot production

The main variable costs that are related to production of one ingot are the following: graphite parts consumption, crucible, silicon, seed, argon, electricity, man hours for operation, and cost for maintenance of the furnace.

In addition one have to include fixed cost such as floor space, depreciation administration and management of the factory. By increasing productivity within the same factory, it has a large impact on the share of the fixed cost.

Therefore, there are two main focus areas to reduce cost:

1. Reduce OPEX cost on ingot level including increasing the yield in the process.
2. Increase the productivity/throughput for each crystal puller.

These two factors can explain the recent trends we have seen in crystal growing lately: increasing pull speed of the ingot, going to larger furnaces, pulling multiple crystals from one crucible and working towards continuous crystal growth.

2.8.2 Yield

The yield of the poly silicon is defined as kg round ingot/kg poly silicon. After crystal pulling, the silicon will be distributed between crown/shoulder, round ingot, tail, and pot scrap. The pot scrap is the remaining silicon in the crucible after each run. The theoretical yield can be optimized by making a flat crown, reduce over diameter of the cylindrical section and by reducing the weight of the tail by fast reduction of the diameter.

However, to flat crown and to fast reduction of diameter of the tail could lead to structure loss. To narrow diameter of the body increases the risk of the ingot to be under diameter if there are some problems with thermal fluctuations or with the diameter reading in the furnace. Structure loss, where the crystal loses its single crystal structure, is a major reason for yield loss. If structure loss happens one usually melts the crystal back and starts over again, depending on the position where it happens.

2.8.3   Throughput

As described in Chapter 2.3.1, the Cz process is divided into different steps: Evacuation/pressurization, melting, stabilization, neck, crown, body, tail, and cool down. Each step takes some time and defines the throughput for a crystal puller, typically measured in metric tons/crystal puller/month. In addition, there will be some non-productive time between the runs, turnover time, and scheduled stops for maintenance.

One method to increase the throughput is to increase the crystallization rate often referred to as pull-speed.

2.8.4   Recharge Cz

Recharge Cz meaning that one is pulling several ingots from the same crucible. This will give higher throughput since time is saved for evacuation and turn over time. Typically, the new material is melted during cooling of the ingot also saving time for melting and cooling.

Another important aspect with recharging is that the amount of pot scrap will be the same as for one ingot also increasing the yield in the process. Figure 2.3 illustrates the effect of recharging. The number of ingots that can be made from one crucible is restricted by the accumulation of impurities in the remaining melt and by the lifetime of the crucible mainly related to the dissolution of the crucible (the impurities from the crucible can be quantified, approximately 100 g crucible is dissolved per 100 kg ingot).

2.8.5   Continuous Cz

Continuous Cz has two advantages. The aspect ratio of the crucible allows for higher pull speed. The possibility to continuously blend with pure silicon makes it possible to have better control of the resistivity gradient in the ingot including the radial resistivity gradient. However, also here there is a limit for how long time one can operate due to the accumulation of impurities and the lifetime of the crucible.

2.8.6   Wafering

The most important parameter for cost in the wafer process is the cost of the wire and the cost of silicon. Also, for wafering the yield is important.

The cost can be reduced:

- by reducing the wear of the wire, typically done by the equipment manufacturer,
- by increasing throughput,
- increasing yield,

- increasing the number of wafers per kg ingot by
  - reducing diameter of the wire
  - reducing the thickness of the wafers

2.9   Summary and outlook for n-type silicon ingots and wafers

At the beginning of the 2010's at the second silicon material workshop held in Rome, a large community of silicon material experts met to discuss the challenges for silicon for photovoltaic applications [152]. Already at that time, it was clear that the general trend for silicon material was starting to change. On the one side, the requirements for silicon wafer were tightening up to make room for more efficiency devices and, on the other hand, the efficiency gap mono versus multicrystalline silicon was doomed to increase. This finding combined with strong potential for cost reduction for Cz ingots thanks to the introduction of recharging Cz and other improvement cause a paradigm shift from a market usually dominated by multicrystalline to a market dominated by monocrystalline wafers.

However, at those time, it was believed that the mono revolution would be dominated by n-type silicon cell architecture. Today we know that this has been instead led by p-type PERC technology that finally breakthrough in the market rapidly replacing the warhorse that dominate PV technology for decades: Al-BSF.

Recently we faced another major change in the silicon material arena. The long-standing boron doping has been replaced by gallium doping. This was driven by the absence of light-induced degradation (BO-related ref) and a mitigation of light and elevated temperature degradation and enabled by the expiration of Shin-Etsu Chemical patent in 2020 [153]. Though not directly related to n-type and its outlook, there is a similitude between Ga and P doping. The segregation coefficient of Ga (0.008) is two orders of magnitude smaller than B (0.8) resulting in a large resistivity distribution across the ingot in the range of 5-folds from about 10% to 90% of the solidified fraction. The segregation coefficient of P (0.35) is much larger than Ga but typical resistivity distribution is very similar. Despite typical resistivity variations in p- and n-type ingots are smaller thanks to the employment of recharging Cz and other technique, one of the main challenges of n-type ingots (i.e. the resistivity variation) is now shared with p-type ingot as well!

It was also about a decade ago that it become clear that n-type silicon wafers have longer diffusion length than p-type. This because typically n-type wafers were subject to higher diffusion temperatures (like for n-PERT architecture) and resulted in lifetime killing rings and disk that strongly affected the performance of this devices. This was due to the presence of oxygen and the formation of precipitates at high temperature. "Likely enough" p-type PERC cells (though not due to the same mechanism as PERC cells need lower emitter diffusion temperatures) were affected by BO light-induced degradation, therefore benefitting from silicon wafers with lower oxygen levels. This generated a driver for the silicon ingot producers to reduce the oxygen that also benefit the manufacturing of n-type ingot.

Two of the major challenges of n-type silicon are therefore being address or shared with the p-type counterpart. These combined with the advancement in n-type solar cell fabrication combined with efficiency limit for PERC cell technology that is being approached and the market requirement for high efficiency devices generates a new push for the large deployment of n-type technology in mass production.

Nevertheless, all the challenges on n-type silicon material are not entirely addressed. Hereby an overview of current and new challenges ahead of us.

1. Polysilicon feedstock. Despite the higher tolerance for common impurity for n-type silicon wafers, the quality of polysilicon feedstock needs to be further improved. This is needed in order to be able to pull multiple ingots from same crucible for cost and tack time reductions in the production of silicon ingots. This trend is in common to p-type ingot for similar reasoning therefore increasing quality (purity) requirements is must for the industry and not special development is needed for n-type ingots (in comparison to p-type one).

2. Oxygen level. The presence of oxygen in combination with high temperature processes results in the formation of oxygen precipitates. This appear on specific cell architecture because of the different thermal budget at which the wafer are subject during the manufacturing process. The main challenge is if oxygen can be further reduced and if current level are low enough for future development of cell architecture and relative advanced processing (a.o. polysilicon, P and B diffusions).

3. Intrinsic defect. Currently intrinsic defects are visible in some FZ wafers. This is because FZ wafers due to the ultra-low level of impurities and oxygen make more apparent their presence then in commercial Cz wafers. However, with increasing improvement in Cz, it is likely that intrinsic defect would become the new challenge to further boost the performance (diffusion length) of silicon wafers also in commercial wafers.

4. Dislocations. Currently the level of dislocation in commercial ingot is less than 10^3 cm^{-3}. At today, this does not represent a limit to the performance; however, its interaction with even low level of impurities can make them electrically active and therefore be to importance to the cell performance of ultra-high efficiency devices.

5. Sensitivity to hydrogenation. Despite commercial Cz wafers are increasing, "perfect" n-type is not capable to benefit from advanced hydrogenation like p-type silicon due to its Fermi level.

6. Metallic impurities. High-temperature cells process includes gettering-like process to reduce the level of metallic impurities. However, this is not the case for low-temperature heterojunction cell process. These might still benefit from a pre-gettering step that enables the last step purification of silicon wafers.

7. Manufacturing yield and kerf losses. Large impact on CO_2 content.

8. Wafer thickness reduction. Large impact on CO_2 content.

References

[1]   M. Fischer, A. Metz, and J. Trube, *International Technology Roadmap for Photovoltaic* (ITRPV), 2022.

[2]   B. Ceccaroli, E. Ovrelid, and S. Pizzini (Eds.), *Solar Silicon Processes: Technologies, Challenges, and Opportunities*, Boca Raton, FL: CRC Press, 2016, doi:10.1201/9781315369075.

[3]   A. Schei, J. K. Tuset, and H. Tveit, *Production of High Silicon Alloys*, Trondheim: Tapir, 1998.

[4]   V. Dosaj and H. Tveit, "Silicon and silicon alloys, production and uses," in *Kirk-Othmer Encyclopedia of Chemical Technology*, John Wiley & Sons, Ltd, 2016, pp. 1–21. doi:10.1002/0471238961.0308051304151901.a01. pub3.

[5]   "Commodity Statistics and Information | U.S. Geological Survey." https://www.usgs.gov/centers/national-minerals-information-center/commodity-statistics-and-information. Accessed May 23, 2022.

[6]   J. Czochralski, "A new method for the measurement of the crystallization rate of metals," *Z. Für Phys. Chem.*, vol. 92, pp. 219–221, 1918.

[7]   J. Friedrich, W. von Ammon, and G. Müller, "Czochralski growth of silicon crystals," in *Handbook of Crystal Growth*, New York, NY: Elsevier, 2015, pp. 45–104, doi:10.1016/B978-0-444-63303-3.00002-X.

[8]   G. Eranna, *Crystal Growth and Evaluation of Silicon for VLSI and ULSI*, Boca Raton, FL: CRC Press, 2014, doi:10.1201/b17812.

[9]   B. Fickett and G. Mihalik, "Multiple batch recharging for industrial CZ silicon growth," *J. Cryst. Growth*, vol. 225, no. 2, pp. 580–585, 2001, doi:10.1016/S0022-0248(01)00956-3.

[10] F. Mosel, A. V. Denisov, B. Klipp., *et al*, (2016). Cost Effective Growth of Silicon Mono Ingots by the Application of a Mobile Recharge System in CZ-Puller.

[11] G. Fiegl, "Recent advances and future directions in CZ-silicon crystal growth technology," *Solid State Technol.*, vol. 26, no. 8, pp. 121–131, 1983.

[12] "LONGI internal communications."

[13] J. R. Davis, , A. Rohatgi, R. H. Hopkins, *et al.*, "Impurities in silicon solar cells," *IEEE Trans. Electron Devices*, vol. 27, no. 4, pp. 677–687, 1980, doi:10.1109/T-ED.1980.19922.

[14] R. E. Lorenzini, A. Iwata, and K. Lorenz, "Continuous crystal growing furnace," US4036595A, July 19, 1977. Accessed May 24, 2022. Available: https://patents.google.com/patent/US4036595A/en

[15] T. J. Dewees, J. S. Fangman, and W. Lin, "Double crucible crystal growing process," US4246064A, January 20, 1981. Accessed May 24, 2022. Available: https://patents.google.com/patent/US4246064A/en

[16] H. J. Möller, "Wafering of silicon crystals," *Phys. Status Solidi A*, vol. 203, no. 4, pp. 659–669, 2006. doi:10.1002/pssa.200564508.

[17] T. Enomoto, Y. Shimazaki, Y. Tani, M. Suzuki, and Y. Kanda, "Development of a resinoid diamond wire containing metal powder for slicing a slicing ingot," *CIRP Ann.*, vol. 48, no. 1, pp. 273–276, 1999, doi:10.1016/S0007-8506(07)63182-5.

[18] H. Wu, "Wire sawing technology: a state-of-the-art review," *Precis. Eng.*, vol. 43, pp. 1–9, 2016, doi:10.1016/j.precisioneng.2015.08.008.

[19] D. Macdonald and L. J. Geerligs, "Recombination activity of interstitial iron and other transition metal point defects in p- and n-type crystalline silicon," *Appl. Phys. Lett.*, vol. 85, no. 18, pp. 4061–4063, 2004, doi:10.1063/1.1812833.

[20] K. Bothe and J. Schmidt, "Electronically activated boron-oxygen-related recombination centers in crystalline silicon," *J. Appl. Phys.*, vol. 99, no. 1, p. 013701, 2006, doi:10.1063/1.2140584.

[21] H. Conzelmann, K. Graff, and E. R. Weber, "Chromium and chromium-boron pairs in silicon," *Appl. Phys. Solids Surf.*, vol. 30, no. 3, pp. 169–175, 1983, doi:10.1007/BF00620536.

[22] A. Hajjiah, M. Soha, I. Gordon, J. Poortmans, and J. John, "The impact of interstitial Fe contamination on n-type Cz-silicon for high efficiency solar cells," *Sol. Energy Mater. Sol. Cells*, vol. 211, p. 110550, 2020, doi:10.1016/j.solmat.2020.110550.

[23] J. Schön, F. Schindler, W. Kwapil, *et al.*, "Identification of the most relevant metal impurities in mc n-type silicon for solar cells," *Sol. Energy Mater. Sol. Cells*, vol. 142, pp. 107–115, 2015, doi:10.1016/j.solmat.2015.06.028.

[24] J. Schmidt, B. Lim, D. Walter, *et al.*, "Impurity-related limitations of next-generation industrial silicon solar cells," *IEEE J. Photovolt.*, vol. 3, no. 1, pp. 114–118, 2013, doi:10.1109/JPHOTOV.2012.2210030.

[25] M.K. Juhl, F.D. Heinz, G. Coletti, *et al.*, "An open source based repository for defects in silicon," in 2018 IEEE 7th World Conference on Photovoltaic Energy Conversion (WCPEC) (A Joint Conference of 45th IEEE PVSC, 28th PVSEC 34th EU PVSEC), June 2018, pp. 0328–0332, doi:10.1109/PVSC.2018.8547621

[26] W. Kwapil, J. Schon, W. Warta, and M. C. Schubert, "Carrier recombination at metallic precipitates in p-and n-type silicon," *IEEE J. Photovolt.*, vol. 5, no. 5, pp. 1285–1292, 2015, doi:10.1109/JPHOTOV.2015.2438634.

[27] J. D. Murphy, R. E. McGuire, K. Bothe, V. V. Voronkov, and R. J. Falster, "Minority carrier lifetime in silicon photovoltaics: the effect of oxygen precipitation," *Sol. Energy Mater. Sol. Cells*, vol. 120, pp. 402–411, 2014, doi:10.1016/j.solmat.2013.06.018.

[28] J. D. Murphy, M. Al-Amin, K. Bothe, M. Olmo, V. V. Voronkov, and R. J. Falster, "The effect of oxide precipitates on minority carrier lifetime in n-type silicon," *J. Appl. Phys.*, vol. 118, no. 21, p. 215706, 2015, doi:10.1063/1.4936852.

[29] J. D. Murphy, K. Bothe, R. Krain, V. V. Voronkov, and R. J. Falster, "Parameterisation of injection-dependent lifetime measurements in semi-conductors in terms of Shockley-Read-Hall statistics: an application to oxide

precipitates in silicon," *J. Appl. Phys.*, vol. 111, no. 11, p. 113709, 2012, doi:10.1063/1.4725475.

[30] S. Diez, S. Rein, T. Roth, and S. W. Glunz, "Cobalt related defect levels in silicon analyzed by temperature- and injection-dependent lifetime spectroscopy," *J. Appl. Phys.*, vol. 101, no. 3, p. 033710, 2007, doi:10.1063/1.2433743.

[31] C. Sun, F. E. Rougieux, J. Degoulange, R. Einhaus, and D. Macdonald, "Reassessment of the recombination properties of aluminium-oxygen complexes in n- and p-type Czochralski-grown silicon: recombination properties of Al-O complexes in n- and p-type Cz-Si," *Phys. Status Solidi B*, vol. 253, no. 10, pp. 2079–2084, 2016, doi:10.1002/pssb.201600363.

[32] S. Rein (Ed.), "Defect characterization on intentionally metal-contaminated silicon samples," in *Lifetime Spectroscopy: A Method of Defect Characterization in Silicon for Photovoltaic Applications*, Berlin, Heidelberg: Springer, 2005, pp. 257–395, doi:10.1007/3-540-27922-9_5.

[33] S. Diez, S. Rein, and S. W. Glunz, (2005, June), *Analysing defects in silicon by temperature-and injection-dependent lifetime spectroscopy (T-IDLS)*. In Proceedings of the 20th European photovoltaic solar energy conference (pp. 1216–1219).

[34] C.-T. Sah, P. C. H. Chan, C.-K. Wang, R. L.-Y. Sah, K. A. Yamakawa, and R. Lutwack, "Effect of zinc impurity on silicon solar-cell efficiency," *IEEE Trans. Electron Devices*, vol. 28, no. 3, pp. 304–313, 1981, doi:10.1109/T-ED.1981.20333.

[35] P. Rosenits, T. Roth, S. Diez, D. Macdonald, and S. W. Glunz, (2007). Detailed studies of manganese in silicon using lifetime spectroscopy and deep-level transient spectroscopy. In Proceedings of the 22nd European Conference on Photo-Voltaic Solar Energy (EU-PVSEC'07) (pp. 1480-1483).

[36] K. Graff, "Properties of transition metals in silicon," in K. Graff (Ed.), *Metal Impurities in Silicon-Device Fabrication*, Berlin, Heidelberg: Springer, 1995, pp. 19–64, doi:10.1007/978-3-642-97593-6_3.

[37] B. B. Paudyal, K. R. McIntosh, D. H. Macdonald, and G. Coletti, "Temperature dependent carrier lifetime studies of Mo in crystalline silicon," *J. Appl. Phys.*, vol. 107, no. 5, p. 054511, 2010, doi:10.1063/1.3309833.

[38] G. Zoth and W. Bergholz, "A fast, preparation-free method to detect iron in silicon," *J. Appl. Phys.*, vol. 67, no. 11, pp. 6764–6771, 1990, doi:10.1063/1.345063.

[39] J. P. Kalejs, B. R. Bathey, J. T. Borenstein, and R. W. Stomont, (1993, May). Effects of transition metal impurities on solar cell performance in polycrystalline silicon. In Conference Record of the Twenty Third IEEE Photovoltaic Specialists Conference-1993 (Cat. No. 93CH3283–9) (pp. 184–189). IEEE.

[40]   B. B. Paudyal, K. R. McIntosh, and D. H. Macdonald, "Temperature dependent carrier lifetime studies on Ti-doped multicrystalline silicon," *J. Appl. Phys.*, vol. 105, no. 12, p. 124510, 2009, doi:10.1063/1.3139286.

[41]   M. Tomassini,, J. Veirman1, R. Varache, *et al.*, "Recombination activity associated with thermal donor generation in monocrystalline silicon and effect on the conversion efficiency of heterojunction solar cells," *J. Appl. Phys.*, vol. 119, no. 8, p. 084508, 2016, doi:10.1063/1.4942212.

[42]   W. Kwapil, J. Schon, W. Warta, and M. C. Schubert, "Erratum to Carrier recombination at metallic precipitates in p- and n-type silicon [Sep 15 1285–1292]," *IEEE J. Photovolt.*, vol. 6, no. 1, p. 391, 2016, doi:10.1109/JPHOTOV.2015.2496866.

[43]   K. Graff, "Requirements of modern technology," in K. Graff (Ed.), *Metal Impurities in Silicon-Device Fabrication*, Berlin, Heidelberg: Springer, 1995, pp. 154–163, doi:10.1007/978-3-642-97593-6_7.

[44]   R. Sachdeva, A. A. Istratov, and E. R. Weber, "Recombination activity of copper in silicon," *Appl. Phys. Lett.*, vol. 79, no. 18, pp. 2937–2939, 2001, doi:10.1063/1.1415350.

[45]   V. Lang, J. D. Murphy, R. J. Falster, and J. J. L. Morton, "Spin-dependent recombination in Czochralski silicon containing oxide precipitates," *J. Appl. Phys.*, vol. 111, no. 1, p. 013710, 2012, doi:10.1063/1.3675449.

[46]   J. D. Murphy, K. Bothe, M. Olmo, V. V. Voronkov, and R. J. Falster, "The effect of oxide precipitates on minority carrier lifetime in p-type silicon," *J. Appl. Phys.*, vol. 110, no. 5, p. 053713, 2011, doi:10.1063/1.3632067.

[47]   R. Basnet, F. E. Rougieux, S. Chang, *et al.*, "Methods to improve bulk lifetime in n-Type Czochralski-grown upgraded metallurgical-grade silicon wafers," *IEEE J. Photovolt.*, vol. 8, no. 4, pp. 990–996, 2018, doi:10.1109/JPHOTOV.2018.2834944.

[48]   B. Hallam, C. Chan, M. Abbott, and S. Wenham, "Hydrogen passivation of defect-rich n-type Czochralski silicon and oxygen precipitates," *Sol. Energy Mater. Sol. Cells*, vol. 141, pp. 125–131, 2015, doi:10.1016/j.solmat.2015.05.009.

[49]   C. Herring, N. M. Johnson, and C. G. Van de Walle, "Energy levels of isolated interstitial hydrogen in silicon," *Phys. Rev. B*, vol. 64, no. 12, p. 125209, 2001, doi:10.1103/PhysRevB.64.125209.

[50]   C. Sun, F. E. Rougieux, and D. Macdonald, "A unified approach to modelling the charge state of monatomic hydrogen and other defects in crystalline silicon," *J. Appl. Phys.*, vol. 117, no. 4, p. 045702, 2015, doi:10.1063/1.4906465.

[51]   A. Le Donne, S. Binetti, V. Folegatti, and G. Coletti, "On the nature of striations in n-type silicon solar cells," *Appl. Phys. Lett.*, vol. 109, no. 3, p. 033907, 2016, doi:10.1063/1.4959558.

[52]   F. E. Rougieux, N. E. Grant, D. Macdonald, and J. D. Murphy, "Can vacancies and their complexes with nonmetals prevent the lifetime reaching its intrinsic limit in silicon?," in *2015 IEEE 42nd Photovoltaic Specialist Conference (PVSC)*, June 2015, pp. 1–4, doi:10.1109/PVSC.2015.7355687.

[53]   G. D. Watkins, "Intrinsic defects in silicon," *Mater. Sci. Semicond. Process.*, vol. 3, no. 4, pp. 227–235, 2000, doi:10.1016/S1369-8001(00)00037-8.

[54]   M. Vaqueiro-Contreras, V. P. Markevich, M. P. Halsall, *et al.*, "Powerful recombination centers resulting from reactions of hydrogen with carbon–oxygen defects in n-type Czochralski-grown silicon (Phys. Status Solidi RRL 8/2017)," *Phys. Status Solidi RRL – Rapid Res. Lett.*, vol. 11, no. 8, p. 1770342, 2017, doi:10.1002/pssr.201770342.

[55]   F. E. Rougieux, N. E. Grant, and D. Macdonald, "Thermal deactivation of lifetime-limiting grown-in point defects in n-type Czochralski silicon wafers," *Phys. Status Solidi RRL – Rapid Res. Lett.*, vol. 7, no. 9, pp. 616–618, 2013, doi:10.1002/pssr.201308053.

[56]   P. Zheng, F. E. Rougieux, N. E. Grant, and D. Macdonald, "Evidence for vacancy-related recombination active defects in as-grown n-type Czochralski silicon," *IEEE J. Photovolt.*, vol. 5, no. 1, pp. 183–188, 2015, doi:10.1109/JPHOTOV.2014.2366687.

[57]   C. S. Fuller, J. A. Ditzenberger, N. B. Hannay, and E. Buehler, "Resistivity changes in silicon single crystals induced by heat treatment," *Acta Metall.*, vol. 3, no. 1, pp. 97–99, 1955.

[58]   W. Kaiser, P. H. Keck, and C. F. Lange, "Infrared absorption and oxygen content in silicon and germanium," *Phys. Rev.*, vol. 101, no. 4, p. 1264, 1956.

[59]   W. Kaiser, "Electrical and optical properties of heat-treated silicon," *Phys. Rev.*, vol. 105, no. 6, p. 1751, 1957.

[60]   V. Cazcarra and P. Zunino, "Influence of oxygen on silicon resistivity," *J. Appl. Phys.*, vol. 51, no. 8, pp. 4206–4211, 1980, doi:10.1063/1.328278.

[61]   T. Wichert, M. L. Swanson, and A. F. Quenneville, (eds.) "Defects in semiconductors, edited by HJ von Bardeleben," in *Materials Science Forum*, vol. 10, p. 12, 1986.

[62]   V. V. Voronkov, G. I. Voronkova, A. V. Batunina, *et al.*, "Properties of fast-diffusing oxygen species in silicon deduced from the generation kinetics of thermal donors," *Solid State Phenom.*, vol. 156–158, pp. 115–122, 2010, doi:10.4028/www.scientific.net/SSP.156-158.115.

[63]   W. Kaiser, H. L. Frisch, and H. Reiss, "Mechanism of the formation of donor states in heat-treated silicon," *Phys. Rev.*, vol. 112, no. 5, pp. 1546–1554, 1958, doi:10.1103/PhysRev.112.1546.

[64]   A. R. Bean and R. C. Newman, "The effect of carbon on thermal donor formation in heat treated pulled silicon crystals," *J. Phys. Chem. Solids*, vol. 33, no. 2, pp. 255–268, 1972, doi:10.1016/0022-3697(72)90004-2.

[65]   V. V. Voronkov, "Generation of thermal donors in silicon: oxygen aggregation controlled by self-interstitials," *Semicond. Sci. Technol.*, vol. 8, no. 12, pp. 2037–2047, 1993, doi:10.1088/0268-1242/8/12/001.

[66]   A. G. Ulyashin, Y. A. Bumay, R. Job, and W. R. Fahrner, "Formation of deep p-n junctions in p-type Czochralski grown silicon by hydrogen plasma treatment," *Appl. Phys. Mater. Sci. Process.*, vol. 66, no. 4, pp. 399–402, 1998, doi:10.1007/s003390050684.

[67]   K. Wada and N. Inoue, "Suppression of thermal donor formation in heavily doped n-type silicon," *J. Appl. Phys.*, vol. 57, no. 12, pp. 5145–5147, 1985, doi:10.1063/1.335248.

[68]   F. Tanay, S. Dubois, J. Veirman, N. Enjalbert, J. Stendera, and I. Perichaud, "Oxygen-related thermal donor formation in dopant-rich compensated Czochralski silicon," *IEEE Trans. Electron Devices*, vol. 61, no. 5, pp. 1241–1245, 2014, doi:10.1109/TED.2014.2311832.

[69]   D. Wruck and P. Gaworzewski, "Electrical and infrared spectroscopic investigations of oxygen-related donors in silicon," *Phys. Status Solidi A*, vol. 56, no. 2, pp. 557–564, 1979, doi:10.1002/pssa.2210560220.

[70]   P. Wagner, C. Holm, E. Sirtl, R. Oeder, and W. Zulehner, "Chalcogens as point defects in silicon," in P. Grosse (Ed.), *Advances in Solid State Physics: Plenary Lectures of the 48th Annual Meeting of the German Physical Society (DPG) and of the Divisions "Semiconductor Physics" "Metal Physics" "Low Temperature Physics" "Thermodynamics and Statistical Physics" "Thin Films" "Surface Physics" "Magnetism" "Physics of Polymers" "Molecular Physics" Münster, March 12…17, 1984*, Berlin, Heidelberg: Springer, 1984, pp. 191–228, doi:10.1007/BFb0107451.

[71]   W. Götz, G. Pensl, and W. Zulehner, "Observation of five additional thermal donor species TD12 to TD16 and of regrowth of thermal donors at initial stages of the new oxygen donor formation in Czochralski-grown silicon," *Phys. Rev. B*, vol. 46, no. 7, pp. 4312–4315, 1992, doi:10.1103/PhysRevB.46.4312.

[72]   P. Wagner and J. Hage, "Thermal double donors in silicon," *Appl. Phys. Solids Surf.*, vol. 49, no. 2, pp. 123–138, 1989, doi:10.1007/BF00616290.

[73]   V. D. Tkachev, L. F. Makarenko, V. P. Markevich, and L. I. Murin, "Modifiable thermal donors in silicon," *Sov. Phys. Semicond.*, vol. 18, pp. 324–328, 1984.

[74]   D. C. Look, "Statistics of multicharge centers in semiconductors: applications," *Phys. Rev. B*, vol. 24, no. 10, pp. 5852–5862, 1981, doi:10.1103/PhysRevB.24.5852.

[75]   R. C. Newman, "Oxygen diffusion and precipitation in Czochralski silicon," *J. Phys. Condens. Matter*, vol. 12, no. 25, pp. R335–R365, 2000, doi:10.1088/0953-8984/12/25/201.

[76]   H. J. Stein, S. K. Hahn, and S. C. Shatas, "Rapid thermal annealing and regrowth of thermal donors in silicon," *J. Appl. Phys.*, vol. 59, no. 10, pp. 3495–3502, 1986, doi:10.1063/1.336820.

[77]   J. Werner, J. Geissbühler, A. Dabirian, *et al.*, "Parasitic absorption reduction in metal oxide-based transparent electrodes: application in perovskite solar cells," *ACS Appl. Mater. Interfaces*, vol. 8, no. 27, pp. 17260–17267, 2016, doi:10.1021/acsami.6b04425.

[78]   Y. Hu, H. Schøn, Ø. Nielsen, E. Johannes Øvrelid, and L. Arnberg, "Investigating minority carrier trapping in n-type Cz silicon by transient photoconductance measurements," *J. Appl. Phys.*, vol. 111, no. 5, p. 053101, 2012, doi:10.1063/1.3689786.

[79] L. F. Makarenko and L. I. Murin, "Trapping of minority carriers in thermal U--donors in n-Si," *Phys. Status Solidi B*, vol. 145, no. 1, pp. 241–253, 1988, doi:10.1002/pssb.2221450123.

[80] Y. Hu, H. Schøn, E. J. Øvrelid, Ø. Nielsen, and L. Arnberg, "Investigating thermal donors in n-type Cz silicon with carrier density imaging," *AIP Adv.*, vol. 2, no. 3, p. 032169, 2012, doi:10.1063/1.4754276.

[81] Y. Miyamura, H. Harada, S. Nakano, S. Nishizawa, and K. Kakimoto, "Do thermal donors reduce the lifetimes of Czochralski-grown silicon crystals?," *J. Cryst. Growth*, vol. 489, pp. 1–4, 2018, doi:10.1016/j.jcrysgro.2018.02.034.

[82] F. Jay, J. Veirman, N. Najid, D. Muñoz, S. Dubois, and A. Jouini, "Exclusively Thermal Donor-doped Cz wafers for silicon heterojunction solar cell technology," *Energy Procedia*, vol. 55, pp. 533–538, 2014, doi:10.1016/j.egypro.2014.08.020.

[83] V. P. Markevich, M. Vaqueiro-Contreras, S. B. Lastovskii, L. I. Murin, M. P. Halsall, and A. R. Peaker, "Electron emission and capture by oxygen-related bistable thermal double donors in silicon studied with junction capacitance techniques," *J. Appl. Phys.*, vol. 124, no. 22, p. 225703, 2018, doi:10.1063/1.5053805.

[84] J. Li, X. Yu, S. Yuan, L. Yang, Z. Liu, and D. Yang, "Effects of oxygen related thermal donors on the performance of silicon heterojunction solar cells," *Sol. Energy Mater. Sol. Cells*, vol. 179, pp. 17–21, 2018, doi:10.1016/j.solmat.2018.02.006.

[85] R. Basnet, H. Sio, M. Siriwardhana, F. E. Rougieux, and D. Macdonald, "Ring-like defect formation in N-type Czochralski-grown silicon wafers during thermal donor formation," *Phys. Status Solidi A*, vol. 218, no. 4, p. 2000587, 2021, doi:10.1002/pssa.202000587.

[86] E. Letty, J. Veirman, W. Favre, and M. Lemiti, "Identification of lifetime-limiting defects in as-received and heat treated seed-end Czochralski wafers," *Energy Procedia*, vol. 92, pp. 845–851, 2016, doi:10.1016/j.egypro.2016.07.086.

[87] J. Veirman, B. Martela, E. Letty, *et al.*, "Thermal History Index as a bulk quality indicator for Czochralski solar wafers," *Sol. Energy Mater. Sol. Cells*, vol. 158, pp. 55–59, 2016, doi:10.1016/j.solmat.2016.05.051.

[88] R. Basnet, C. Sun, H. Wu, H. T. Nguyen, F. E. Rougieux, and D. Macdonald, "Ring defects in n-type Czochralski-grown silicon: a high spatial resolution study using Fourier-transform infrared spectroscopy, micro-photoluminescence, and micro-Raman," *J. Appl. Phys.*, vol. 124, no. 24, p. 243101, 2018, doi:10.1063/1.5057724.

[89] M. Chatelain, M. Albaric, D. Pelletier, J. Veirman, and E. Letty, "Numerical method for thermal donors formation simulation during silicon Czochralski growth," *Sol. Energy Mater. Sol. Cells*, vol. 219, p. 110785, 2021, doi:10.1016/j.solmat.2020.110785.

[90] A. Srinivasa, S. Herasimenka, A. Augusto, and S. Bowden, "Effect of ingot variability on performance of silicon heterojunction solar cells," in *2020 47th IEEE Photovoltaic Specialists Conference (PVSC)*, June 2020, pp. 2238–2241, doi:10.1109/PVSC45281.2020.9300928.

[91]   Ø. Nielsen, "Towards GW-scale sustainable ingot and wafer production in Europe," Presented at the nPV workshop, 2021.

[92]   K. F. Kelton, R. Falster, D. Gambaro, M. Olmo, M. Cornara, and P. F. Wei, "Oxygen precipitation in silicon: experimental studies and theoretical investigations within the classical theory of nucleation," *J. Appl. Phys.*, vol. 85, no. 12, pp. 8097–8111, 1999, doi:10.1063/1.370648.

[93]   V. LaSalvia, A. Youssef, M. A. Jensen, *et al.*, "Tabula Rasa for n-Cz silicon-based photovoltaics," *Prog. Photovolt. Res. Appl.*, vol. 27, no. 2, pp. 136–143, 2019, doi:10.1002/pip.3068.

[94]   J. Veirman, S. Dubois, N. Enjalbert, and M. Lemiti, "A fast and easily implemented method for interstitial oxygen concentration mapping through the activation of thermal donors in silicon," *Energy Procedia*, vol. 8, pp. 41–46, 2011, doi:10.1016/j.egypro.2011.06.099.

[95]   J. Veirman, E. Letty, W. Favre, M. Albaric, D. Pelletier, and M. Lemiti, "Novel way to assess the validity of Czochralski growth simulations," *Phys. Status Solidi A*, vol. 216, no. 17, p. 1900317, 2019, https://doi.org/10.1002/pssa.201900317.

[96]   M. M. Kivambe, J. Haschke, J. Horzel, *et al.*, "Record-efficiency n-type and high-efficiency p-type monolike silicon heterojunction solar cells with a high-temperature gettering process," *ACS Appl. Energy Mater.*, vol. 2, no. 7, pp. 4900–4906, 2019.

[97]   D. Gilles, W. Schröter, and W. Bergholz, "Impact of the electronic structure on the solubility and diffusion of 3d transition elements in silicon," *Phys. Rev. B*, vol. 41, no. 9, p. 5770, 1990.

[98]   R. N. Hall and J. H. Racette, "Diffusion and solubility of copper in extrinsic and intrinsic germanium, silicon, and gallium arsenide," *J. Appl. Phys.*, vol. 35, no. 2, pp. 379–397, 1964.

[99]   A. Ourmazd and W. Schröter, "Phosphorus gettering and intrinsic gettering of nickel in silicon," *Appl. Phys. Lett.*, vol. 45, no. 7, pp. 781–783, 1984.

[100]   A. Ourmazd and W. Schröter, "Gettering of metallic impurities in silicon," *MRS Online Proc. Libr. OPL*, vol. 36, 25–30, 1984.

[101]   A. Correia, B. Pichaud, A. Lhorte, and J. B. Quoirin, "Platinum gettering in silicon by silicon phosphide precipitates," *J. Appl. Phys.*, vol. 79, no. 4, pp. 2145–2147, 1996.

[102]   M. Seibt, A. Döller, V. Kveder, A. Sattler, and A. Zozime, "Phosphorus diffusion gettering of platinum in silicon: formation of near-surface precipitates," *Phys. Status Solidi B*, vol. 222, no. 1, pp. 327–336, 2000.

[103]   E. Spiecker, M. Seibt, and W. Schröter, "Phosphorous-diffusion gettering in the presence of a nonequilibrium concentration of silicon interstitials: a quantitative model," *Phys. Rev. B*, vol. 55, no. 15, p. 9577, 1997.

[104]   W. Schröter and R. Kühnapfel, "Model describing phosphorus diffusion gettering of transition elements in silicon," *Appl. Phys. Lett.*, vol. 56, no. 22, pp. 2207–2209, 1990.

[105]   R. Falster, "Platinum gettering in silicon by phosphorus," *Appl. Phys. Lett.*, vol. 46, no. 8, pp. 737–739, 1985.

[106] R. Chen, B. Trzynadlowski, and S. T. Dunham, "Phosphorus vacancy cluster model for phosphorus diffusion gettering of metals in Si," *J. Appl. Phys.*, vol. 115, no. 5, p. 054906, 2014.

[107] M. Syre, S. Karazhanov, B. R. Olaisen, A. Holt, and B. G. Svensson, "Evaluation of possible mechanisms behind P gettering of iron," *J. Appl. Phys.*, vol. 110, no. 2, p. 024912, 2011.

[108] J. Schön, V. Vähänissi, A. Haarahiltunen, M. C. Schubert, W. Warta, and H. Savin, "Main defect reactions behind phosphorus diffusion gettering of iron," *J. Appl. Phys.*, vol. 116, no. 24, p. 244503, 2014.

[109] A. Haarahiltunen, H. Savin, M. Yli-Koski, H. Talvitie, and J. Sinkkonen, "Modeling phosphorus diffusion gettering of iron in single crystal silicon," *J. Appl. Phys.*, vol. 105, no. 2, p. 023510, 2009, doi:10.1063/1.3068337.

[110] P. S. Plekhanov, R. Gafiteanu, U. M. Gösele, and T. Y. Tan, "Modeling of gettering of precipitated impurities from Si for carrier lifetime improvement in solar cell applications," *J. Appl. Phys.*, vol. 86, no. 5, pp. 2453–2458, 1999.

[111] J. Härkönen, V.-P. Lempinen, T. Juvonen, and J. Kylmäluoma, "Recovery of minority carrier lifetime in low-cost multicrystalline silicon," *Sol. Energy Mater. Sol. Cells*, vol. 73, no. 2, pp. 125–130, 2002.

[112] P. Manshanden and L. J. Geerligs, "Improved phosphorous gettering of multicrystalline silicon," *Sol. Energy Mater. Sol. Cells*, vol. 90, no. 7–8, pp. 998–1012, 2006.

[113] S. P. Phang and D. Macdonald, "Direct comparison of boron, phosphorus, and aluminum gettering of iron in crystalline silicon," *J. Appl. Phys.*, vol. 109, no. 7, p. 073521, 2011, doi:10.1063/1.3569890.

[114] M. Rinio, A. Yodyunyong, S. Keipert-Colberg, Y. P. B. Mouafi, D. Borchert, and A. Montesdeoca-Santana, "Improvement of multicrystalline silicon solar cells by a low temperature anneal after emitter diffusion," *Prog. Photovolt. Res. Appl.*, vol. 19, no. 2, pp. 165–169, 2011, doi:10.1002/pip.1002.

[115] S. A. McHugo, R. J. McDonald, A. R. Smith, D. L. Hurley, and E. R. Weber, "Iron solubility in highly boron-doped silicon," *Appl. Phys. Lett.*, vol. 73, no. 10, pp. 1424–1426, 1998.

[116] S. M. Myers, M. Seibt, and W. Schröter, "Mechanisms of transition-metal gettering in silicon," *J. Appl. Phys.*, vol. 88, no. 7, pp. 3795–3819, 2000.

[117] S. M. Myers, G. A. Petersen, T. J. Headley, J. R. Michael, T. L. Aselage, and C. H. Seager, "Metal gettering by boron-silicide precipitates in boron-implanted silicon," *Nucl. Instrum. Methods Phys. Res. Sect. B Beam Interact. Mater. At.*, vol. 127, pp. 291–296, 1997.

[118] T. Terakawa, D. Wang, and H. Nakashima, "Role of heavily B-doped layer on low-temperature Fe gettering in bifacial Si solar cell fabrication," *Jpn. J. Appl. Phys.*, vol. 45, no. 4R, p. 2643, 2006.

[119] S. P. Phang, W. Liang, B. Wolpensinger, M. A. Kessler, and D. Macdonald, "Tradeoffs between impurity gettering, bulk degradation, and surface

passivation of boron-rich layers on silicon solar cells," *IEEE J. Photovolt.*, vol. 3, no. 1, pp. 261–266, 2013, doi:10.1109/JPHOTOV.2012.2226332.

[120]   K. Ryu, A. Upadhyaya, H.-J. Song, C.-J. Choi, A. Rohatgi, and Y.-W. Ok, "Chemical etching of boron-rich layer and its impact on high efficiency n-type silicon solar cells," *Appl. Phys. Lett.*, vol. 101, no. 7, p. 073902, 2012, doi:10.1063/1.4746424.

[121]   M. A. Kessler, T. Ohrdes, B. Wolpensinger, and N.-P. Harder, "Charge carrier lifetime degradation in Cz silicon through the formation of a boron-rich layer during BBr$\less$sub$\greater$3$\less$/sub$\greater$diffusion processes," *Semicond. Sci. Technol.*, vol. 25, no. 5, p. 055001, 2010, doi:10.1088/0268-1242/25/5/055001.

[122]   A. Y. Liu, C. Sun, V. P. Markevich, A. R. Peaker, J. D. Murphy, and D. Macdonald, "Gettering of interstitial iron in silicon by plasma-enhanced chemical vapour deposited silicon nitride films," *J. Appl. Phys.*, vol. 120, no. 19, p. 193103, 2016, doi:10.1063/1.4967914.

[123]   A. Y. Liu and D. Macdonald, "Impurity gettering effect of atomic layer deposited aluminium oxide films on silicon wafers," *Appl. Phys. Lett.*, vol. 110, no. 19, p. 191604, 2017, doi:10.1063/1.4983380.

[124]   A. Liu and D. Macdonald, "Impurity gettering by atomic-layer-deposited aluminium oxide films on silicon at contact firing temperatures," *Phys. Status Solidi RRL – Rapid Res. Lett.*, vol. 12, no. 3, p. 1700430, 2018, doi:10.1002/pssr.201700430.

[125]   A. Liu, Z. Hameiri, Y. Wan, C. Sun, and D. Macdonald, "Gettering effects of silicon nitride films from various plasma-enhanced chemical vapor deposition conditions," *IEEE J. Photovolt.*, vol. 9, no. 1, pp. 78–81, 2019, doi:10.1109/JPHOTOV.2018.2875871.

[126]   A. Liu, D. Yan, S. P. Phang, A. Cuevas, and D. Macdonald, "Effective impurity gettering by phosphorus- and boron-diffused polysilicon passivating contacts for silicon solar cells," *Sol. Energy Mater. Sol. Cells*, vol. 179, pp. 136–141, 2018, doi:10.1016/j.solmat.2017.11.004.

[127]   A. Liu, D. Yan, Y. Wong Leung, *et al.*, "Direct observation of the impurity gettering layers in polysilicon-based passivating contacts for silicon solar cells," *ACS Appl. Energy Mater.*, vol. 1, no. 5, pp. 2275–2282, 2018, doi:10.1021/acsaem.8b00367.

[128]   J. Krügener, F. Haase, M. Rienäcker, R. Brendel, H. J. Osten, and R. Peibst, "Improvement of the SRH bulk lifetime upon formation of n-type POLO junctions for 25% efficient Si solar cells," *Sol. Energy Mater. Sol. Cells*, vol. 173, pp. 85–91, 2017, doi:10.1016/j.solmat.2017.05.055.

[129]   M. Hayes, B. Martel, G. W. Alam, *et al.*, "Impurity gettering by boron- and phosphorus-doped polysilicon passivating contacts for high-efficiency multicrystalline silicon solar cells," *Phys. Status Solidi A*, vol. 216, no. 17, p. 1900321, 2019, doi:10.1002/pssa.201900321.

[130]   Z. Wang, Z. Liu, M. Liao, *et al.*, "Effective gettering of in-situ phosphorus-doped polysilicon passivating contact prepared using plasma-enhanced

chemical-vapor deposition technique," *Sol. Energy Mater. Sol. Cells*, vol. 206, p. 110256, 2020, doi:10.1016/j.solmat.2019.110256.

[131]   A. Liu, Z. Yang, F. Feldmann, *et al.*, "Understanding the impurity gettering effect of polysilicon/oxide passivating contact structures through experiment and simulation," *Sol. Energy Mater. Sol. Cells*, vol. 230, p. 111254, 2021, doi:10.1016/j.solmat.2021.111254.

[132]   D. Macdonald, H. Mäckel, and A. Cuevas, "Effect of gettered iron on recombination in diffused regions of crystalline silicon wafers," *Appl. Phys. Lett.*, vol. 88, no. 9, p. 092105, 2006, doi:10.1063/1.2181199.

[133]   V. Vähänissi, A. Haarahiltunen, M. Yli-Koski, and H. Savin, "Gettering of iron in silicon solar cells with implanted emitters," *IEEE J. Photovolt.*, vol. 4, no. 1, pp. 142–147, 2014, doi:10.1109/JPHOTOV.2013.2285961.

[134]   E. Cho, Y.-W. Ok, L. D. Dahal, A. Das, V. Upadhyaya, and A. Rohatgi, "Comparison of POCl$_3$ diffusion and phosphorus ion-implantation induced gettering in crystalline Si solar cells," *Sol. Energy Mater. Sol. Cells*, vol. 157, pp. 245–249, 2016, doi:10.1016/j.solmat.2016.05.057.

[135]   H. S. Laine, V. Vahanissi, A. E. Morishige, *et al.*, "Impact of iron precipitation on phosphorus-implanted silicon solar cells," *IEEE J. Photovolt.*, vol. 6, no. 5, pp. 1094–1102, 2016, doi:10.1109/JPHOTOV.2016.2576680.

[136]   A. Haarahiltunen, H. Talvitie, H. Savin, *et al.*, "Gettering of iron in silicon by boron implantation," *J. Mater. Sci. Mater. Electron.*, vol. 19, no. 1, pp. 41–45, 2008, doi:10.1007/s10854-008-9640-2.

[137]   H. S. Laine, V. Vahanissi, Z. Liu, *et al.*, "Elucidation of iron gettering mechanisms in boron-implanted silicon solar cells," *IEEE J. Photovolt.*, vol. 8, no. 1, pp. 79–88, 2018, doi:10.1109/JPHOTOV.2017.2775159.

[138]   R. Basnet, W. William, J. Yu Zhengshan, *et al.*, "Impact of pre-fabrication treatments on n-type UMG wafers for 21% efficient silicon heterojunction solar cells," *Solar Energy Materials and Solar Cells*, vol. 205, p. 110287, 2020, doi:10.1016/j.solmat.2019.110287.

[139]   B. Hallam, D. Chen, J. Shi, R. Einhaus, Z. C. Holman, and S. Wenham, "Pre-fabrication gettering and hydrogenation treatments for silicon heterojunction solar cells: a possible path to >700 mV open-circuit voltages using low-lifetime commercial-grade p-type Czochralski silicon," *Sol. RRL*, vol. 2, no. 2, p. 1700221, 2018, doi:10.1002/solr.201700221.

[140]   A. A. Istratov, T. Buonassisi, R. J. McDonald, *et al.*, "Metal content of multicrystalline silicon for solar cells and its impact on minority carrier diffusion length," *J. Appl. Phys.*, vol. 94, no. 10, pp. 6552–6559, 2003, doi:10.1063/1.1618912.

[141]   D. Macdonald, A. Cuevas, A. Kinomura, Y. Nakano, and L. J. Geerligs, "Transition-metal profiles in a multicrystalline silicon ingot," *J. Appl. Phys.*, vol. 97, no. 3, p. 033523, 2005, doi:10.1063/1.1845584.

[142]   A. Liu, C. Sun, H. C. Sio, X. Zhang, H. Jin, and D. Macdonald, "Gettering of transition metals in high-performance multicrystalline silicon by silicon nitride films and phosphorus diffusion," *J. Appl. Phys.*, vol. 125, no. 4, p. 043103, 2019, doi:10.1063/1.5050566.

[143]   C. Sun, A. Liu, A. Samadi, C. Chan, A. Ciesla, and D. Macdonald, "Transition metals in a cast-monocrystalline silicon ingot studied by silicon nitride gettering," *Phys. Status Solidi RRL – Rapid Res. Lett.*, vol. 13, no. 12, p. 1900456, 2019, doi:10.1002/pssr.201900456.

[144]   M. Kittler, J. Lärz, W. Seifert, M. Seibt, and W. Schröter, "Recombination properties of structurally well defined NiSi2 precipitates in silicon," *Appl. Phys. Lett.*, vol. 58, no. 9, pp. 911–913, 1991, doi:10.1063/1.104474.

[145]   P. S. Plekhanov and T. Y. Tan, "Schottky effect model of electrical activity of metallic precipitates in silicon," *Appl. Phys. Lett.*, vol. 76, no. 25, pp. 3777–3779, 2000, doi:10.1063/1.126778.

[146]   A. E. Morishige, F. D. Heinz, H. S. Laine, *et al.*, "Moving beyond p-type mc-Si: quantified measurements of iron content and lifetime of iron-rich precipitates in n-type silicon," *IEEE J. Photovolt.*, vol. 8, no. 6, pp. 1525–1530, 2018, doi:10.1109/JPHOTOV.2018.2869544.

[147]   T. Buonassisi and A. A. Istratov, "Impact of metal silicide precipitate dissolution during rapid thermal processing of multicrystalline silicon solar cells," *Appl. Phys. Lett.*, vol. 87, no. 12, p. 121918, 2005, doi:10.1063/1.2048819.

[148]   M. Seibt, A. Sattler, C. Rudolf, O. Voß, V. Kveder, and W. Schröter, "Gettering in silicon photovoltaics: current state and future perspectives," *Phys. Status Solidi A*, vol. 203, no. 4, pp. 696–713, 2006, doi:10.1002/pssa.200664516.

[149]   J. Schön, H. Habenicht, M. C. Schubert, and W. Warta, "Understanding the distribution of iron in multicrystalline silicon after emitter formation: theoretical model and experiments," *J. Appl. Phys.*, vol. 109, no. 6, p. 063717, 2011, doi:10.1063/1.3553858.

[150]   V. Vähänissi, H. S. Laine, Z. Liu, M. Yli-Koski, A. Haarahiltunen, and H. Savin, "Full recovery of red zone in p-type high-performance multicrystalline silicon," *Sol. Energy Mater. Sol. Cells*, vol. 173, pp. 120–127, 2017, doi:10.1016/j.solmat.2017.05.016.

[151]   J. Hofstetter, D. P. Fenning, D. M. Powell, A. E. Morishige, H. Wagner, and T. Buonassisi, "Sorting metrics for customized phosphorus diffusion gettering," *IEEE J. Photovolt.*, vol. 4, no. 6, pp. 1421–1428, 2014, doi:10.1109/JPHOTOV.2014.2349736.

[152]   G. Coletti, I. Gordon, M. C. Schubert, *et al.*, "Challenges for photovoltaic silicon materials," *Sol. Energy Mater. Sol. Cells*, vol. 130, pp. 629–633, Nov. 2014, doi:10.1016/j.solmat.2014.07.045.

[153]   T. Abe, T. Hirasawa, K. Tokunaga, T. Igarashi, and M. Yamaguchi, "Silicon single crystal and wafer doped with gallium and method for producing them," US6815605B1, November 9, 2004 Accessed: May 23, 2022. Available: https://patents.google.com/patent/US6815605B1/en

Chapter 3

n-type silicon solar cells

Sébastien Dubois[1], Bertrand Paviet-Salomon[2], Jia Chen[3],
Wilfried Favre[1], Raphaël Cabal[1], Armand Bettinelli[1] and
Radovan Kopecek[4]

n-type silicon (Si) technologies played a major role in the early age of photo-voltaics (PV). Indeed, the Bell Laboratories prepared the first practical solar cells from n-type crystalline Si (c-Si) wafers (Figure 3.1) [1–3]. Therefore, the domination of p-type technologies over the last decades for the production of commercial solar cells could appear as a paradox. This is essentially explained by historical reasons. Fifty years ago, the dominant market for c-Si solar cells was space power applications. In space, solar components are affected by radiation damages (electrons, protons). Interestingly, this degradation is significantly reduced by using p-type cells instead of n-type devices [4]. Thus, the solar cell developments for space applications focused on p-type wafers. When the first commercial productions for terrestrial applications were launched, they took benefit of these early developments for space missions and were therefore naturally based on p-type devices. Then, with the rapid growth of PV, p-type solar cells were the main recipients of the industrially oriented innovations and eventually maintained their domination over the PV market.

However, at the beginning of the twenty-first century, n-type solar cells received considerable attention for two main reasons. First, several groups pointed out that n-type wafers would feature superior electric properties than their p-type counterparts, as described in the previous chapter. Second, even if p-type cells dominated the PV market, the commercial technologies with the best conversion efficiency (η) values (i.e. the Sanyo [5] and Sunpower Inc. [6] cells) were based on n-type wafers. Thus, over the last 15 years, n-type technologies experienced a fruitful revival since several high efficiency architectures were investigated, with remarkable progresses in terms of both PV performances

[1]Université Grenoble Alpes, CEA, LITEN, INES, Le Bourget du Lac, France
[2]PV-Center, Centre Suisse d'Electronique et de Microtechnique SA, Neuchatel, Switzerland
[3]Jolywood (Taizhou) Solar Technology Co., Ltd, Taizhou, Jiangsu, China
[4]International Solar Energy Research Center (ISC) Konstanz Rudolf-Diesel-Str. 15, Konstanz, Germany

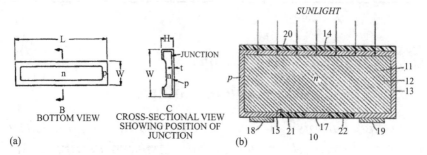

Figure 3.1   Sketches of the first modern silicon solar cells developed by the Bell Laboratories. (a) Reprinted with AIP permission from [2]. (b) Reprinted from [3]

(i.e. efficiency, bifaciality, reliability) and industrialization (i.e. mass production compliant equipment, process simplifications). Nowadays, commercial PV products are still dominated by p-type devices. Nevertheless, the market share of n-type technologies is continuously increasing and could even become dominant in the near future.

This chapter focuses on the main n-type Si cells technologies. However, it will start by presenting the current commercial p-type products. This is particularly important since the newly developed n-type cells will have to present superior performances than these p-type references in order to be largely deployed into mass production. Furthermore, the development of n-type technologies over the last two decades strongly benefited from the progresses achieved with p-type devices (e.g. Al-rear emitter, aluminum oxide films). Interestingly, the opposite is also true. The progresses related to n-type solar cells pushed p-type technologies toward constant improvements (e.g. bifacial devices, light-induced degradation mitigation). Eventually, both p-type and n-type technologies took benefit of this synergy which highly contributed to the recent outstanding progresses of the PV industry. Regarding n-type devices, the chapter will begin with double-side contacted homojunction cells, since they feature many common points with the standard p-type products (in terms of architecture and process equipment). As the efficiency of these cells is limited by metal contacts-recombination issues, the chapter will continue with the presentation of the two main families of passivated contacts cells, poly-silicon on oxide structures and amorphous/crystalline Si heterojunction cells. Eventually, back-contact technologies (either based on high- or low-temperature fabrication processes) will be presented, as they feature the highest efficiency potentials. Figure 3.2 displays the schematic views of the various solar cells architectures presented in this chapter. Following the description of the main technologies, the chapter will end with the presentation of various approaches for reducing the amount of silver in n-type solar cells, since this is one of the major remaining challenges that n-type products have to overcome to ensure their long-term competitiveness.

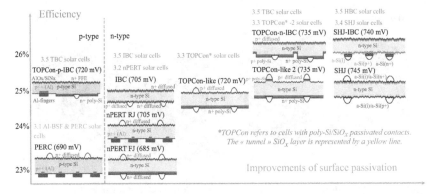

Figure 3.2   Schematic representations of the various solar cells technologies presented in this chapter. The x-axis represents the improvements in surface passivation. The y-axis corresponds to the conversion efficiency

3.1   Industrial p-type solar cells

Nowadays, the PV market is still dominated by p-type Si solar cells. This part will focus first on the aluminum (Al) back surface field (Al-BSF) cell, which reigned on the PV industry until the middle of the 2010s. Then it will address the passivated emitter and rear contact (PERC) technology, which successfully took over and largely prevails currently at the industrial level.

3.1.1   Al-BSF solar cells

The Al-BSF technology essentially relies on the so-called back surface field obtained by an Al p$^+$ doping of the whole rear surface of the device. The firing of screen-printed Al pastes creates the Al-BSF. Indeed, during the metallization firing, Al and Si form a liquid alloy. Upon cooling, an Al-doped Si layer featuring a high doping level is redeposited between the substrate and the Al contact. The Al-BSF reduces the rear surface recombination while acting as an efficient hole collector. The beneficial effects of the Al-BSF have been well known since the 1980s (e.g. [7]). However, the technology was deployed at the industrial level rather in the middle of the 1990s thanks to the emergence of cost effective Al pastes and automatic screen-printing lines [8].

Figure 3.3 presents the structure of an Al-BSF solar cell. After the KOH texturing of both surfaces, a phosphorus (P)-doped n-type emitter is usually formed by thermal diffusion in tube furnace. Following the etching of the phosphorus silicate glass (PSG), a front hydrogen (H)-rich silicon nitride (SiN$_x$:H) film acting as an anti-reflective coating (ARC) layer is deposited by plasma-enhanced chemical vapor deposition (PECVD). Then, the front (Ag grid) and rear (full Al layer with

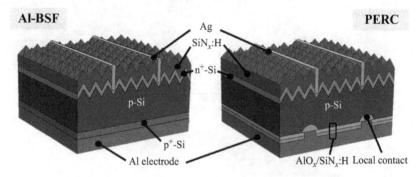

Figure 3.3   Schematics of Al-BSF and PERC solar cells with the main layers indicated

local Ag pads for the subsequent ribbon soldering) electrodes are deposited by screen-printing, before a belt furnace co-firing. Notice that because of the double-side feature of the tube furnace P diffusion, an edge isolation step has to be implemented, usually by chemical/plasma etching or by laser.

In order to limit the efficiency losses, the requirements of the P-emitter profiles are different for the metallized and non-metallized (i.e. passivated) regions of the device. For the metallized regions, high surface doping levels and deep P concentration profiles are needed for decreasing the contact resistance and *screening* the recombination active metal/silicon interface. For the passivated regions, low surface doping levels and shallow profiles reduce carrier recombination. With homogeneous emitters, a trade-off between these opposite trends has to be found, which induces efficiency losses. To overcome this issue, the concept of selective emitters (SEs) was implemented, which consists in different adapted emitter profiles for the metallized and passivated regions, respectively. SE allows efficiency gains around 0.5–0.6%$_{abs}$ [9]. Several SE technologies were developed, which can either be used in Al-BSF or in PERC devices. Currently, the main SE technology deployed at the industrial level is based on the localized laser irradiation of the PSG layer [10], approach requiring only one additional step.

Screen-printed Al-BSF cells fabricated from Czochralski (Cz) Si wafers reached PV conversion efficiency values as high as 20% [11]. The main advantages of the Al-BSF technology rely on its process simplicity, low cost (e.g. no scarce elements such as indium, relatively low silver content thanks to the rear Al metallization), and good reliability. Furthermore, it is worth mentioning that this structure is particularly well adapted to multicrystalline (mc) Si wafers, due to complementary and synergetic effects between the P-diffusion gettering and the H-passivation of surface and bulk defects (H diffusing from the front SiN$_x$:H layer) [12]. Nevertheless, beyond an efficiency limit close to 20%, the Al-BSF architecture is mono-facial and poorly compatible with thin wafers due to bowing issues related to the different thermal expansion coefficients of Al and Si.

3.1.2 PERC solar cells

3.1.2.1 Monofacial PERC solar cells

Two mechanisms inherent to the rear metallization limit the PV conversion efficiency of Al-BSF solar cells: (i) carrier recombination at the full area Al-contact and (ii) infrared light absorption by the rear electrode. Very interestingly, the PERC technology allows mitigating these efficiency losses by passivating the rear side with dielectric layers, which also increase the rear reflectivity. Locally, the dielectric layers are opened to form contacts between the Si substrate and the Al rear electrode. The PERC front architecture (Figure 3.3) is similar to that of the Al-BSF cell.

The PERC concept was proposed in 1989 by Blakers *et al.* [13] with a laboratory device featuring a silicon oxide layer on the rear surface with local Al point contacts formed by photolithography. Despite outstanding lab scale results, it took more than two decades to deploy this structure in industrial production lines. In 2021, PERC solar cells represented about 80% of the world market share for cell technologies [14]. This domination of PERC solar cells is essentially explained by the development of cost-effective processes for (i) the rear surface polishing, (ii) the rear surface passivation, and (iii) the local Al-contacts formation.

The rear surface polishing is necessary for removing the n^+ layer formed on the rear side during the P-diffusion and for reducing the surface roughness (surface passivation improvement). For this purpose, single-side polishing equipment and processes were specifically developed with successful results [15], relying either on HF/HNO_3 mixtures or alkaline solutions.

A key advance, which highly contributed to the wide adoption of the PERC technology by the industry, is related to the development of low-temperature processes for the passivation of the rear surface, as substitutes to the thermal oxidation steps used in lab-scale solar cells, which could induce degradations of the bulk carrier lifetime because of the high thermal budget. Among the different passivation approaches investigated, the most interesting, essentially used nowadays in mass production, relies on the deposition of aluminum oxide (Al_2O_3) layers. Indeed, these layers feature outstanding passivation levels when deposited on p-type Si surfaces (surface recombination velocity values below 10 cm s^{-1}) [16]. Furthermore, they contain negative charges, which do not induce efficiency losses because of the formation of an inversion layer, as it is the case with positively charged SiN_x:H films. In industrial lines, the Al_2O_3 layer is either deposited by atomic layer deposition (ALD), mainly via spatial-ALD systems, or by PECVD [14,17]. In order to maintain high passivation levels after the metallization firing and to avoid interactions with the screen-printed Al paste, the Al_2O_3 film (thickness below 20 nm) is capped by a protective SiN_x:H (deposited by PECVD) [18], which can also contribute to improved surface passivation levels by providing H to the Al_2O_3/c-Si interface.

One of the main features of the PERC technology relies on the local metal contacts formation on the rear surface. For laboratory devices, photolithography-based approaches were used, which are not compatible with mass production

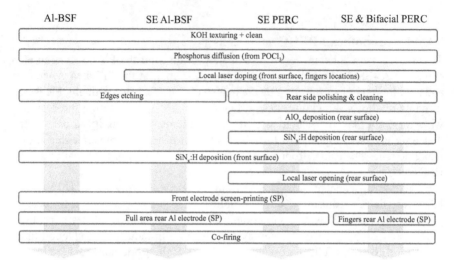

Al-BSF   SE Al-BSF   SE PERC   SE & Bifacial PERC

KOH texturing + clean

Phosphorus diffusion (from POCl₃)

Local laser doping (front surface, fingers locations)

Edges etching | Rear side polishing & cleaning

AlO$_x$ deposition (rear surface)

SiN$_x$:H deposition (rear surface)

SiN$_x$:H deposition (front surface)

Local laser opening (rear surface)

Front electrode screen-printing (SP)

Full area rear Al electrode (SP) | Fingers rear Al electrode (SP)

Co-firing

Figure 3.4   Comparison of the fabrication processes of Al-BSF cells, Al-BSF cells with selective emitters (SE), PERC cells with SE, and bifacial PERC cells with SE. Only the main steps are presented

because of cost and production throughput issues. To overcome this limitation, a cost effective solution was developed in the 2000s, which combines the laser ablation of the dielectric stack with the screen-printing of an Al paste [19,20]. During the subsequent firing, Al-BSF contacts are formed through the local openings in the dielectric layers. This approach was fostered by the development of high-speed laser equipment.

One of the reasons behind the success of the PERC at the industrial scale is related to the proximity of its fabrication process with that of the Al-BSF solar cell [21]. Therefore, most of the Al-BSF production lines could be up-graded for the processing of PERC solar cells, with relatively low expenses. Figure 3.4 presents the fabrication processes of both Al-BSF and PERC solar cells. The first steps of the fabrication sequence (texturing, P-diffusion) are common for both technologies. Then, for the PERC, the wet chemical edge isolation must be modified to include the polishing and smoothing of the rear surface (equipment upgrade needed). Then, the rear dielectric stack is deposited which imposes one or two additional equipment depending on the selected Al_2O_3 deposition technology (PECVD versus ALD). The dielectric stack is locally opened by laser ablation (additional equipment needed). The last steps, screen-printing and firing, are common to both cell technologies (Al-BSF and PERC).

3.1.2.2   Bifacial PERC solar cells

As the rear surface of conventional PERC cells is totally covered by the screen-printed Al layer, standard PERC cells are monofacial (the light illuminating the rear side of the device cannot be absorbed by the Si substrate). However, in 2015, ISFH and SolarWorld presented an evolution of the PERC technology, the so-called PERC+, for which on the rear side, the full area Al-layer is replaced by a screen-printed Al grid [22]. Thus, the resulting device is bifacial, which is particularly interesting for applications where the rear surface of the cell receives significant amounts of reflected light. Interestingly, the PERC+ fabrication process developed by ISFH and SolarWorld is close to the monofacial PERC fabrication process. The main particularities of the PERC+ process concern the alignment of the Al grid on the laser contact opening and the use of specific Al pastes, in particular compatible with the printing of fine lines [23]. This proximity between the PERC and PERC+ fabrication processes contributed to the fast deployment of the bifacial PERC into mass production, with beyond SolarWorld, pioneer companies such as Trina Solar, Longi Solar, and Neo Solar Power [23]. In terms of performances, Hanwha QCELLS showed that the conversion efficiency of bifacial PERC cells can be similar to the values obtained with monofacial PERC (gap of 0.1%$_{abs}$ only), with a bifaciality factor of 0.78 for their bifacial Q.ANTUM device [24]. Notice that simplified approaches were recently proposed to obtain bifacial PERC cells, relying on the use of firing-through Al pastes, which avoid the laser ablation step [25].

In terms of performances, the leading commercial PERC products featured in 2021 PV conversion efficiency values of about 23% and 21% with Cz-Si and mc-Si substrates, respectively [14]. Notice that LONGI holds the certified record for a large size (surface area of 244.59 cm^2) bifacial PERC solar cell with an efficiency value of 24.0% [26]. These results illustrate the strong efficiency gain offered by moving from Al-BSF to PERC.

3.1.2.3 LID and LeTID mitigation

Under carrier injection, the carrier lifetime of Cz-grown boron (B)-doped Si is subjected to severe degradations related to the formation of complexes involving B and oxygen (O) atoms [27,28]. Under illumination, this effect is referred to as light-induced degradation (LID). B-O-related LID represented a major risk for the wide deployment of the PERC technology on Cz-Si wafers. Indeed, due to the improved surface passivation and the high initial bulk carrier lifetime, the performances of Cz-Si PERC solar cells were strongly affected by LID. For instance, relative efficiency losses under illumination of about 6% were reported for PERC devices prepared from Cz-Si wafers [29]. However, efficient solutions were developed over the last 15 years to mitigate this issue. First in 2006, Herguth *et al.* showed that the B-O LID defect could be permanently deactivated by low T annealing steps under carrier injection [30,31]. This phenomenon is often referred to as "regeneration" (Figure 3.5). Recently, several companies proposed regeneration equipment, either based on intense illuminations (halogen lamps, lasers) or electric current injection [32]. Second, the B-O LID can be suppressed by substituting B by gallium (Ga) [33]. Nevertheless, Ga doping posed production yield and cost issues because of the low segregation coefficient of Ga, which induced strong resistivity variations along the height of the ingot. However, the recent emergence of new Cz pulling techniques (e.g. continuous Cz, recharged Cz) and in general the strong process improvements achieved by the main Si wafers companies, offered solutions to limit

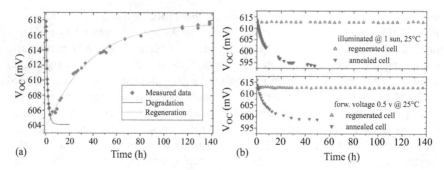

Figure 3.5   (a) Evolution under illumination of the V_{oc} of an Al-BSF Cz-Si solar cell (2.8 Ω.cm) at a temperature of about 70°C. The blue and green solid lines are single exponential fitting curves associated with the degradation and regeneration effects, respectively. (b) Evolution at 25°C of the V_{oc} of Al-BSF Cz-Si solar cells without ("annealed") and with ("regenerated") regeneration steps. During the stability test, the cells were either illuminated (top) or forward biased (bottom). Reprinted with WIP permission from [31]

these doping variations. In 2021, more than 90% of the p-type monocrystalline wafers used by the PV industry were Ga-doped. Therefore, we can consider that the B–O LID issue has been successfully solved at the industrial level.

Beyond B–O LID, another degradation mechanism can affect the performances of PERC solar cells under the combined actions of temperature and illumination, usually referred to as LeTID for light and elevated temperature-induced degradation [34]. LeTID alters the performances of both mc and Cz-Si PERC solar cells [35]. Furthermore, Ga-doped PERC devices would also suffer from LeTID [36]. The underlying mechanisms are still matter of debates, however, several studies pointed out the involvement of hydrogen atoms [37,38], essentially introduced in the Si bulk during the firing step from the hydrogen-rich dielectric layers. Several approaches were proposed for reducing LeTID, such as adaptations of the firing step and additional dark anneals [39]. In addition, similar to B–O LID (but usually with different kinetics), LeTID can be definitely suppressed by regeneration processes relying on annealing steps under carrier injection [40]. Nowadays, LeTID issues are usually suppressed in commercial PERC cells [41].

3.1.3   Conclusions

This part presented the two main p-type cell technologies and the main reasons behind their domination on the PV market over the last decades. The success of the Al-BSF was essentially explained by the simplicity of its fabrication process and its compliance with both Cz and mc Si substrates (low thermal budget, external gettering, and hydrogenation). However, the PV conversion efficiency values of Al-BSF structures were limited to about 20% essentially because of carrier recombination at the full Al rear contact. Interestingly, by providing an elegant solution to passivate the rear surface while allowing local Al contacts, the emergence of the PERC technology offered the possibility to maintain a dynamic of progress improvements for commercial cells, with record efficiencies reaching 24% for large surface area devices. The success of the PERC is also explained by its proximity with the Al-BSF technology, Al-BSF production lines could be easily upgraded for the processing of PERC cells. Eventually, it can also be pointed out that the emergence of n-type technologies pushed p-type cells to constant improvements. Indeed, to compete with the bifacial features of n-type modules, PERC cells became bifacial. Furthermore, to tackle the argument of more stable PV performances with n-type substrates, solutions were developed to neutralize LID and LeTID effects in PERC devices. However, it is also worth noting that some of the improvements behind the successes of Al-BSF and PERC cells considerably served the development of high-T processed n-type technologies, such as the cost effective formation of Al-doped collectors or the p-type surface passivation by AlO_x layers, as it will be shown in the following part.

3.2 Double-side contacted n-type homojunction devices

n-Type silicon is known to be a relevant challenger to p-type Si, especially when it comes to high efficiency cell concepts. The use of high quality n-type materials opens the door to many options in terms of cell architectures, depending on which efficiency/cost trade-off is considered. Among the several n-type cell architectures developed so far, we chose to limit the scope of this section to the high-temperature-processed homojunction devices, involving metallizations on both sides. This device family still including many members, the choice was made to sort the manifold device variations in two categories, depending on whether the emitter is located on the rear or on the front-side of the cell.

Keeping in mind the objective of deploying PV into mass production, we focus in this section rather on the cell technologies that were designed with the aim of facilitating a transfer into production lines. In this context of industrial feasibility, we will see that the rear-emitter architectures are mostly monofacial-oriented devices (i.e. with a bifaciality factor η_{rear}/η_{front} below 80%, where η_{rear} and η_{front} stand for the rear and front PV conversion efficiency values, respectively) while a front-emitter configuration is naturally better suited to highly bifacial concepts (bifaciality factor superior to 90%).

3.2.1 Rear-junction structures

Considering n-type Si, the processing of a p–n junction implies to incorporate a p^+-doping on one surface of the substrate. Aluminium (Al) and boron (B) are usually the most used p-type dopants. Whereas the techniques available to incorporate B atoms into the Si crystalline lattice are numerous [42–44], it makes no doubt that the most cost-effective option in the case of Al is the use of Al screen-printed pastes, inherited from the classical p-type Al-BSF structure.

The requirements in terms of Al surface coverage for granting sufficiently low series resistances affect the optical losses due to shadowing. Consequently, the Al-emitter is better suited on the rear side of the device (the bifaciality factor will be therefore limited to values usually below 80%).

For a long time, the passivation of a B-doped p^+-surface posed more issues than its n^+- counterpart, especially in the case of textured surfaces, partly due to the formation of a highly defective boron-rich layer (BRL) during the B diffusion process [45]. Locating the B emitter on the rear side (i.e. far from the front and bulk regions where most of the carriers are photogenerated) is a way to promote the open-circuit voltage (V_{oc}) of the cell. Analogically to p-type PERC, the use of flat surfaces at the rear side of the cell was preferred (at the expense of bifaciality), such defect-poor surfaces facilitating the early boron emitter integrations, from a surface passivation point of view.

3.2.1.1   Al-rear emitter as a start

The first n-type cell concept considered here (see Figure 3.7) integrates a full sheet Al-doped p^+ emitter on the rear side of the cell, which is a transposition of the p-type Al-BSF reference structure. This *aluminum alloyed full area rear emitter* concept was particularly investigated in the early 2000s [46] as a low-cost and industrially compliant n-type cell option. In this n^+np^+ cell structure, the front surface field (FSF) is formed by phosphorus (P) thermal diffusion and the p^+-emitter is obtained from screen-printed Al alloying on the rear. The devices quickly reached efficiencies over 18% on Czochralski (Cz) substrates [47] and close to 15% on multicrystalline (mc) Si 150 cm^2 wafers [48]. Thanks to the continuous improvements brought to the structure (notably the optimizations of the FSF, of the anti-reflective stack and of the Al-emitter depth) Fraunhofer ISE presented a 19.8% efficient lab-scale solar cell on a 4 cm^2 float-zone (FZ) substrate, involving Ag-plated front contacts [49]. Soon after, groups from Fraunhofer ISE and Konstanz University reported solar cells with surface areas higher than 156×156 mm^2 and η of 19.3% and 19.4% using respectively plated [50] and screen-printed front contacts [51].

With the *PLUTO* concept, *Suntech* pushed further the structure: a conversion efficiency above 19.9% was obtained on 238.95 cm^2 Cz-Si wafers by combining the Al rear emitter with a phosphorus-implanted selective FSF and light-induced-plated (LIP) Ni/Cu/Ag front contacts [52]. Despite all the front-side improvements brought to the full area Al rear emitter structure, the conversion efficiency remained below an upper limit of 20%. At this stage, the efficiency was limited by carrier recombination in both the emitter and at the Al-p^+ surface [53]. As illustrated in Figure 3.6, when not passivated, the lowest saturation current densities reported for Al-doped emitters (J_{0Al}) remain higher than 300 fA/cm^2. The path toward higher efficiencies requires shifting to more sophisticated cell architectures, introducing a passivation of the rear side and the use of local Al-emitter in combination with boron doping.

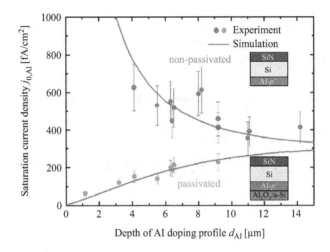

Figure 3.6   Saturation current densities as a function of the depth of the Al doping profile (d_{Al}) for non-passivated and passivated Al-p^+ regions, either experimentally determined from carrier lifetime measurements (symbols) or computed via Sentaurus simulations (lines). From [53]

3.2.1.2   The path toward a boron rear emitter

In the process of improving the performance of the full area Al-emitter, significant modifications of the rear side structure were proposed. The first one consisted in removing chemically the Al overlying layer after the Al-p^+ emitter formation so that a passivating stack could mitigate its surface recombination activity. For example, the passivation of the Al-p^+ surface by an AlO$_x$/SiN stack allowed to curb $J_{0,Al}$ below 100 fA/cm^2 when the thickness of the emitter was lower than 3 μm. Such rear emitter could be contacted afterward by passivation dielectric local opening and additional screen-printing of Al-based paste. The so called *ALU+* solar cells resulting from this approach reached 20% conversion efficiency on 143.5 cm^2 high resistivity (10 Ω cm) mono-crystalline wafers [53,54].

An evolution of the *ALU+* concept consisted in localizing the Al-emitter by local contact opening (LCO) made by laser through a stack of dielectric layers, while maintaining the passivating stack between the contacted regions. With such features, the cell rear side was made quite similar to the rear side of actual p-type PERC structures. This approach showed no significant advantages in terms of performance when compared to full area passivated Al-emitter [52]. Indeed, as for p-type PERC, the firing of screen-printed Al pastes onto locally opened passivating layers raises the issue of contacts with unwanted shallow Al-p^+ doping, increasing the recombination at contact borders and enhancing the risks of emitter local shunting [55–57]. The introduction of Si additives into Al metallization pastes was proposed to reduce the point contact Al–Si layer thickness and increase the local Al-p^+ region depth [58,59]. Additionally, adapted firing conditions were proposed

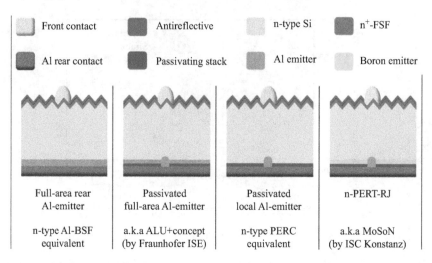

Figure 3.7   Comparison of n-type rear junction solar cell schemes: focus on the rear side evolution

with the aim of preventing the unwanted formation of voids in the local Al-emitter [60]. All these features led to an improvement of p-type PERC cells performances. However, no significant efficiency progresses were reported after 2015 on the equivalent n-type structure.

The next evolution consisted in combining features of both p-PERT (i.e. rear-side blanket boron doping) and p-PERC (localized Al-p^+ regions under the contacts) into one cell concept referred to as n-PERT-RJ (for n-type-passivated emitter rear totally diffused rear junction) whose scheme is given on the right-hand side of Figure 3.7. Among the many groups working on this approach, *Hanwha Q-cells* demonstrated conversion efficiencies of 21.3% on 6 in. mono-Si wafers as soon as 2013, using screen-printed front contacts and the physical vapor deposition (PVD) of Al on the rear [61]. Two years after, *IMEC* pushed the concept up to a record efficiency of 22.5% (on a three busbars 223 cm^2 device) by integrating additional features such as a laser-doped selective FSF and Ni\Cu\Ag plated contacts at the cell front-side [62]. In 2016, IMEC was able to reach 21.7% efficient large area devices with a V_{oc} over 690 mV combining front-side plated contacts with rear side screen-printing [63]. In the meantime, *Hanwha Q-cells* had proposed its own industrial version of n-PERT-RJ featuring up to 21.8% conversion efficiency, on large area (243.4 cm^2) cells processed through a variation of the *Q.ANTUM* production process-flow [64,65]. Similar to what had been achieved on p-type with the bifacial *PERC+* concept [66], *ISFH* teams developed n-PERT-RJ cells exhibiting improved bifacial properties (η_{rear}/η_{front} = 78%) [67].

In 2019, the *ISC Konstanz* eventually demonstrated that a highly efficient n-PERT-RJ was industrially feasible with its 5 busbars M2-sized (area 244.3 cm^2) fully screen-printed *MoSoN* devices approaching 22% efficiency with a bifaciality factor of 60% [68].

Thanks to the steady improvements implemented since then, V_{oc} and efficiencies up to 695 mV and 22.8%, respectively, have been reported with the *MoSoN* approach in a 0-busbar configuration [69], getting closer to the 23.0% efficiency record obtained as soon as 2018 by *IMEC* on a monofacial version of n-PERT-RJ which was also fully screen-printed [70].

3.2.2 Front-junction structures

3.2.2.1 The n-PERT FJ structure

Initially developed on p-type wafers for space applications, the PERT-FJ (for front junction) was transposed to n-type substrates for terrestrial applications in the early 2000s. In this version of PERT, the emitter (boron-doped in the case of n-type solar cells) is brought back to the front-side while a phosphorus-doped region is formed on the rear side as a BSF (Figure 3.8). Efficiency values close to 22% on FZ-Si (21.1% on Cz-Si) were obtained by the *UNSW* on lab-scale solar cells (few cm^2) and Ti/Pd/Ag front contacts [71,72].

Some groups proposed to transpose the historical world record 24.7%-efficient p-type PERL (passivated emitter rear locally diffused [73]) to n-type substrates with the goal of simplifying the fabrication process. Some approaches, such as the *PassDop* (developed at *Fraunhofer ISE*, combining doping and passivating layers with laser local doping) showed interesting results in this sense [74]. However, it is commonly admitted that the full area rear doping of PERT makes it a much simpler and far more relevant structure for a future mass production deployment.

Since 2004, several groups (from *ECN, Konstanz University, CEA,* etc.) worked on n-PERT-FJ with an industrially oriented approach, ruling out the microelectronics inherited tricks (e.g. photolithography, evaporated contacts, inverted pyramids, etc.) to focus on fully screen-printed devices fabricated from large area ($\geq$ 100 cm^2) wafers: from a 16.6% starting efficiency obtained on Cz-Si in 2005 [75], the n-PERT-FJ cell exceeded 18% in 2008 [76]. The conversion efficiencies kept increasing as well as the size of the device: the following results were all obtained on 239 cm^2 solar cells. In 2011, *ECN* presented its *n-Pasha*

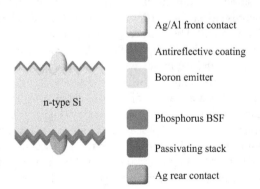

Figure 3.8   n-PERT-FJ solar cell scheme

concept reaching 20% efficiency [77]. In 2017, *ISC Konstanz* improved its version of n-PERT-FJ (the so-called *BiSoN* cell) up to 20.9% using Ag-based screen-printed pastes on both sides of the cell [78]. The year after, *SERIS* presented its own fully screen-printed n-PERT-FJ solar cells with a best efficiency of 21.4% [79].

Since 2017, *IMEC* holds the record for a large area bifacial n-PERT-FJ with its 22.8%-efficient device. Such a high efficiency was made possible by using direct plating (Ni/Ag contacts) instead of screen-printing, with a 0-busbar metallization scheme [80].

In the world of n-type PV, the n-PERT-FJ structure may be the closest thing to what we may call a mainstream cell architecture, if any. Jolywood has implemented the n-PERT-FJ structure and process-flow with ion implantation since 2016 with 2GW capacity and achieved mass-production efficiency around 22% [81].

In what follows, the processing sequence used for the fabrication of the n-PERT-FJ solar cells will be presented, with a focus on the key feature of this type of cell, the boron emitter.

3.2.2.2   The n-PERT FJ main fabrication process

The core of n-PERT-FJ solar cells is the p^+nn^+ structure. The considered approaches to obtain it have a strong influence over the whole cell process. The integration of the p^+ (boron-doped) and n^+ (phosphorus-doped) regions is actually highly depending on the doping techniques involved. Although the n-PERT-FJ processing options are numerous, all of them in the end can be set in two categories, depending on whether the doping technique involved is single or double side (see Figure 3.9).

We will recall the main features of the n-PERT-FJ process as designed in its early stages. Analogically to the $POCl_3$ diffusion, which is widely used for phosphorus emitters, a gaseous diffusion technique involving a BBr_3 precursor was first used for the boron emitter formation. As for $POCl_3$, this double-sided doping technique implies the preliminary deposition of a diffusion barrier on the rear side. Both the barrier and boron silicate glass (BSG, grown during the diffusion step) had to be etched-off afterwards in hydrofluoric acid (HF). The same sequence, involving barrier deposition, gaseous diffusion, and subsequent etch of the undesired barrier and phosphorus silicate glass (PSG) layers in HF, had to be reiterated for the phosphorus BSF formation. The surface passivation usually involved a thermal oxide (SiO_2) layer on the emitter side, and PECVD silicon nitride (SiN:H) layers on both sides. Metal Ag/Al and Ag pastes were screen-printed onto the front and rear sides to contact the emitter and the BSF, respectively, via a co-firing step. The resulting PERT devices had already a bifacial-oriented structure, both sides being texturized and featuring grid metallization schemes (Figure 3.8). It is worth noticing that the thermal budget of this process was quite high, involving three steps at elevated temperatures for the BBr_3 diffusion (in the 850–1,000°C range), the $POCl_3$ diffusion (in the 800–880°C range), and the thermal oxidation (in the 700–1,000°C range). The exposure to a high thermal budget could be detrimental to the bulk quality of wafers, as it may generate silicon oxide precipitates and/or oxygen induced stacking faults [82]. In this frame, wafers featuring high concentrations of

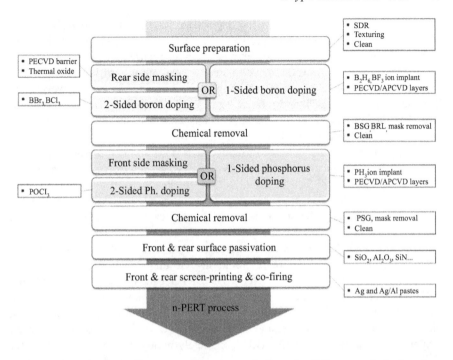

Figure 3.9   n-PERT cell process sequence including the different technical options for each step

interstitial oxygen and/or elevated THI (for Thermal History Index, as defined in [83]) should be avoided for PERT processing.

Manifold processing tricks have been developed through the history of n-PERT in order to make the fabrication of the solar cell simpler, cost-effective and to reduce its process thermal budget. After a short focus on the boron emitter (the core of the device) specificities, some examples of process simplifications will be presented in the following sections.

3.2.2.3   Focus on the boron emitter

The introduction of a boron-doped region into a silicon solar cell comes along with specific issues to be addressed properly, even more in the case of a n-PERT-FJ device, in which the boron-doped region acts as the cell emitter. This part focuses on three non-trivial topics: (i) the diffusion induced BRL, (ii) the p^+-surface passivation, and (iii) the B-emitter contacting by screen-printing.

3.2.2.3.1   Gaseous diffusion

The liquid source diffusion process is still today one of the most widely used techniques for boron emitter processing. It involves a BBr_3 (or BCl_3) liquid precursor and is mainly based on two high temperature steps, the first called "deposition" and the second "drive-in." The BSG layer is a mixed-phase of

B_2O_3–SiO_2 that is formed during the deposition. The boron oxide B_2O_3 results from the reaction of BBr_3 vapor (conveyed by a flow of N_2 gas carrier) with O_2, while SiO_2 comes from the subsequent reaction of condensed B_2O_3 with the silicon wafer surface [84]. During the drive-in step, the "piling up" of boron dopants at the wafer surface leads to the formation of a Si–B compound that is likely to be SiB_6 according to [85,86], and referred to as BRL (for B-rich layer) in the literature. The thicknesses of the BSG and BRL are related to the flow of BBr_3 introduced during the process, as shown in Figure 3.10.

The BRL has been highlighted to alter the surface and boron emitter passivation [87]. The chemical removal of the BRL was shown to induce a drastic reduction of the J_{0e} [88]. In addition, a BRL thicker than 10 nm was shown to degrade the bulk lifetime by generating dislocations in the substrate during the diffusion cooling step (the stress inducing the dislocation generation is due to a mismatch of BRL and Si thermal expansion coefficients) [89]. The first n-PERT-FJ process flows included a chemical sequence devoted to (1) deglaze the sample (the BSG was etched in HF) before (2) removing the underneath BRL in a bath of boiling HNO_3 (alternatively HF: HNO_3). Then, an elegant approach to get rid of the BRL was proposed, which consisted in performing an in situ oxidation step in the BBr_3 tube furnace, right after the boron diffusion. In this configuration, the oxygen diffuses through the boron glass and oxidizes the BRL. With this process sequence, a hydrophobic wafer surface (i.e. BRL completely removed) can be obtained after

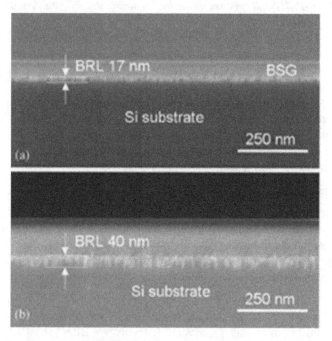

Figure 3.10   Cross-section SEM images of the BSG-BRL/Si interface for a N_2 carrier gas flow of (a) 15 sccm and (b) 60 sccm [88]

an HF dip [90]. Since boron has a higher solubility in SiO_2 than in Si, the resulting boron emitter concentration profile exhibited a depleted region at the wafer surface. This feature (though it must be controlled to maintain low emitter contact resistance values) demonstrated drastic gains in terms of emitter passivation: J_{0e} close to 140 fA/cm^2 dropped down to <30 fA/cm^2 when performing the temperature ramp-down ending the diffusion process under pure O_2 [88].

3.2.2.3.2 B-doped surface passivation

For long, there was no industrial solution for the passivation of (p^+) boron-doped surfaces used as emitters in n-type solar cells. For the passivation of phosphorus (n^+) emitters used in p-type solar cells, hydrogen-rich SiN_x layers are commonly used. However, when applied on p^+ doped surfaces, the SiN_x layers result in low or even no passivation at all due to their large positive fixed charge density ($Q_f > 2 \times 10^{12}$ cm^{-2}). This increases the minority carrier (i.e. electron) density near the surface in a p^+-emitter enhancing carrier recombination.

The first effective passivating stacks developed for boron emitters involved a thin thermal oxide (SiO_2) layer, which features a low interface states density (D_{it}) and a low fixed charge density (Q_f between 10^{10} and 10^{11} cm^{-2}) and is capped by an H-rich antireflective SiN_x layer that advantageously acts as an efficient hydrogen source [91].

In 2008, *ECN* proposed to replace the high-temperature thermally grown SiO_2 layer by an ultra-thin (1.5 nm) chemical oxide obtained from a nitric acid oxidation wet bench process [92], competing the thermal oxide passivation properties (J_{0e} values down to 23 fA/cm^2 were reported on polished p^+np^+ samples involving 60 Ω/sq boron emitters) once capped with SiN_x. This approach greatly contributed to the reduction of the n-PERT-FJ thermal budget and process costs [93].

Other effective passivation schemes were reported for p^+-emitters. *Fraunhofer ISE* considered the use of a passivating floating junction in-between the front contact fingers. With this approach, the boron emitter surface doping is over-compensated by n^+-doping (by $POCl_3$ diffusion), opening the door to the use of existing layers already known to have excellent passivation properties on n^+-surfaces (e.g. SiN_x) [94]. The introduction of such a floating junction on polished p^+np^+ samples (140 Ω/sq boron emitter) resulted in an outstanding reduction of J_{0e} from 390 fA/cm^2 down to 49 fA/cm^2 and promising results were obtained with lab-scale solar cells (21.7% reached on 4 cm^2 FZ-Si n-PERL devices). *ISC Konstanz* investigated the passivating properties of the BSG [95]. Properly tuned and capped with a 65 nm-thick layer of SiN_x, the BSG demonstrated passivating properties competing the referent thermal SiO_2/SiN_x passivation scheme, with J_{0e} values as low as 29 fA/cm^2 on polished p^+np^+ structures (with a 70 Ω/sq boron emitter).

An alternative passivation method involving aluminum oxide (Al_2O_3) was proposed in the 2010s. Al_2O_3 can be prepared by atomic layer deposition (ALD, either thermally or plasma-assisted) [96,97], more recently by PECVD [98]. The excellent passivation properties of Al_2O_3 are attributed to a very low D_{it} and a high density of negative fixed charges ($Q_f \sim 5 \times 10^{12}$ cm^{-2}). The activation of the fixed charges usually requires a thermal anneal at $T \geq 400°C$, possibly provided by the

SiN_x deposition step. Very low J_{0e} values could be reached (around 10 fA/cm^2 and 30 fA/cm^2 for 100 Ω/sq and 54 Ω/sq emitters, respectively) when implementing this passivation layer onto p$^+$-surfaces. Since then, Al_2O_3 passivation has become more and more prominent in PV manufacturing as it is perfectly suited to passivate both p$^+$-emitters for n-PERT-FJ devices and the rear side of p-type PERC cells.

3.2.2.3.3   Specific contact-induced degradation

The metallization of n-PERT-FJ consists in printing both sides of the devices with silver-based fingers and busbars following a so-called H-pattern. A subsequent firing process, performed in an IR belt furnace, allows the glass-frit contained in the metal paste to etch through the ARC (on the front) and the passivating layers (on the rear) and ensures the contact formation between the silver electrodes and the silicon substrate. The addition of aluminum into the silver-based paste lowers the contact resistance on boron emitters but leads to additional recombination losses due to the spiking of Al [99]. The recombination beneath the contact is associated with a specific recombination current density (J_{0met}) and contributes to the increase of the global diode saturation current density of the solar cell (J_{01}) limiting the final cell V_{oc}. Losses of 15–45 mV in V_{oc} were reported between un-metallized and metallized n-PERT-FJ solar cells [100]. So far, the options reported to effectively reduce the J_{0met} component are: (i) the reduction of the metallization fraction via new printing techniques (e.g. stencil based) or via the improvements of the existing ones; (ii) the use of deeper emitter profiles in order to preserve the space charge region from metal induced recombination: J_{0met} above 13,000 fA/cm^2 on a boron shallow emitter (0.3 μm deep) could be reduced below 4,300 fA/cm^2 using a deeper (>0.5 μm) emitter [99,101]; (iii) some Al-free Ag pastes were reported to be capable of contacting properly boron emitters (i.e. introducing specific contact resistances below 10 mΩ cm) [102].

3.2.2.3.4   Alternative doping and simplified n-PERT-FJ fabrication processes

The easiest way to reduce the n-PERT-FJ cell production cost is to simplify the cell process. In this purpose, single-side doping techniques were considered as interesting alternatives to BBr$_3$ or BCl$_3$ gaseous diffusions. The use of single-side doping techniques offers the opportunity to apply one single high-temperature step (referred to as *co-diffusion* or *co-annealing* in the literature) for the simultaneous formation of the boron emitter and the phosphorus BSF. Among the technical options available, two single-side doping techniques demonstrated the potential to compete with the historical gaseous diffusion process: ion implantation and the use of doping layers.

The implantation of dopants into silicon can be done by the microelectronics inherited beam line ion implantation (BLII) technique: the tuning of both implantation dose (cm^{-2}) and energy (keV) enables to precisely control the dopants profile. Furthermore, highly reproducible emitters can be obtained. A very high boron emitter quality was reported using BLII of B$^+$ ions [103] as well as highly efficient screen-printed n-PERT-FJ devices (with efficiency values up to 21.8%) [104]. Once implanted, dopants still need a thermal activation to occupy substitutional sites in the Si lattice and become electrically active. In the case of boron, a temperature above 1,050°C is required for this activation annealing, in order to remove the BLII-induced crystal defects and boron clusters, and thus reach the targeted emitter quality [105]. As mentioned before, such a high temperature may alter the bulk quality by promoting the formation of oxygen-related defects but also by increasing the risk of potential metal contaminations. More recently, an alternative to historical BLII was developed. This multi-energetic implant technique referred to as plasma immersion ion implantation (PIII) features shallow dopant profiles before activation, dopants remaining located at the surface of the wafer. Advantageously, plasma immersion tools are associated with lower maintenance costs and reduced footprint compared to beam line. In addition, this technique induces less implantation defects (located rather at the surface thus easily removed by an HF etching) and facilitates the formation of high-quality B-emitters at lower temperatures than BLII. Using B_2H_6 as a boron precursor during PIII process, J_{0e} values below 30 fA/cm^2 were reported on resulting p^+np^+ textured and passivated samples annealed at 950°C (against 400 fA/cm^2 with *BLII* emitters) [106]. The most advanced process simplification made possible by implantation techniques would consist in implanting phosphorus and boron ions on the rear and front-sides respectively, before co-annealing the implanted regions, and eventually proceeding to the passivation and screen-printing/firing steps. *CEA-INES* evidenced the relevance of this approach, using PH_3 and B_2H_6 precursors for the BSF and emitter PIII implantation, respectively. n-PERT-FJ solar cells made using this process and from 239 cm^2 Cz-Si wafers reached an efficiency of 20.7%, which competes the efficiencies of referent BBr$_3$ devices [106].

Doping layers deposition can be obtained by many techniques (e.g. spin-on coating [107], screen-printing [108]). One technique sparking strong interests relies on the use of atmospheric pressure CVD and PECVD processes. In the case of boron doping, silicon oxide-based layers (such as SiO_xB and BSG) are preferred [109,110]. Many research groups worked on the combination of boron doping layers with classical gaseous POCl$_3$ diffusion. This approach gave birth to several co-diffused n-PERT (either FJ or RJ) cell concepts (*CoBiN, SolENNA$_1$, BiCore*, etc.) [111–113]. *CEA-INES* proposed further simplifications of its *SolENNA* concept, developing boron-containing SiO_xN_y and phosphorus-containing SiN_x layers, with optical and passivating properties allowing to keep these layers thorough the whole cell process, therefore avoiding any barrier deposition and/or HF etching steps [114]. All the aforementioned simplified n-PERT-FJ concepts based on co-diffusion lie in the 20.5–21% efficiency range.

3.2.3 Conclusion

The n-type solar devices were seen as a relevant alternative to the historical p-type technologies, based on the higher electric performances that n-type silicon offered, and the specific advantage of n-type cell architectures for bifacial PV deployment. It has permitted to new players (as big as Yingli or Jolywood, etc.) to enter the PV market.

Nonetheless, the PERC cell was the structure largely chosen for a deployment to mass production, as this technology could be implemented in the existing lines with few investments. Rapidly, PERC has become the new PV industrial standard. This position provided a clear advantage to the PERC structure, which benefited from scale-related costs reductions as well as a wider range of innovations, notably in the field of metallization pastes. In this propitious context, both R&D institutes and industrials did a great job to push further and further the PERC concept. They developed bifacial PERC variations and demonstrated record efficiencies reaching 24% [26]. In this context, the interest of moving to n-type cell production was logically questioned, as it required investments in new tools and processes, with no guarantee of maintaining a significant difference in terms of cell efficiency and/or cost of ownership (CoO) on the long term.

Now that PERC has presumably reached its upper limit and that next-generation n-type PV devices (described in the following parts) aim at reaching over 25% efficiencies in single junction (>30% in tandem configuration), the relevancy of n-type PV makes no doubt.

3.3   Polysilicon on oxide-passivated contacts solar cells

3.3.1   Introduction

Despite the aforementioned outstanding progresses, the PV performances of homojunction n-PERT solar cells suffer from carrier recombination losses due to the presence of both highly doped n^+ and p^+ regions within the absorber material and direct interfaces between the metal electrodes and the silicon substrate.

Indeed, Lim *et al.* [109] and Pen *et al.* [115] investigated the various sources of efficiency losses of screen-printed back-junction n-PERT solar cells. They concluded that the reduction of carrier recombination losses was the main lever to improve the performances of such devices [109]. Both studies showed that the main source of recombination losses was the front n^+ region. These losses can be limited by adapting the n^+ doping profile: lower surface concentration, shallower collector. However, the downside of such advanced doping profiles is that they increase the metallization-induced recombination losses [116]. This last issue is particularly important with screen-printed electrodes, since beyond silver the metal pastes contain highly reactive metal oxides, which can enhance space–charge–region recombination losses [117]. Furthermore, these losses are increased by the glass frit-induced etching of both the passivation layers [118] and the highly doped regions [119]. Selective n^+ doping (i.e. heavier doping under metal electrodes and shallower doping elsewhere) could partly solve this issue. However, this interesting approach features several drawbacks, among which more complex fabrication processes and alignment challenges that impose the highly doped area to be three to four times wider than the actual metal finger, enhancing recombination losses [115].

An alternative approach has received increasing interest over the last few years, as it solves the surface recombination issue, while maintaining a fabrication process compliant with the use of firing through metallization pastes. This approach consists in implementing carrier selective passivated contacts combining an ultra-thin ($\sim$15 Å) silicon oxide (SiO_x) and a phosphorus-doped polycrystalline silicon (poly-Si) layer, on the rear side of front junction n-PERT solar cells. This solution offers the opportunity to eliminate the recombination losses due to both the metal/semiconductor interface and the highly doped region by physically displacing them outside the Si substrate.

3.3.2   Historical review

The emergence of poly-Si contacts for solar cells is strongly related to the important role played by poly-Si films in integrated circuit (IC) devices. In 1972, Kamioka *et al.* proposed a new approach for reducing the dimensions of the emitter of microwave transistors, based on the use of n-type poly-Si films as emitter diffusion sources [120]. Since this pioneering work, poly-Si emitters became key components of high-speed bipolar and CMOS devices, essentially due to their ability to reduce the junction depth, while maintaining (sometimes even improving) the current gain [121]. Interestingly, in relation with the future applications of poly-Si in solar cells, in 1979, De Graaff *et al.* highlighted that the properties of n-p-n transistors with poly-Si layers as emitters could be improved when a thin (20–60 Å) insulating layer is present between the poly-Si film and the mono-crystalline Si substrate [122].

Beyond poly-Si, an alternative material sparked strong interests at the end of the 1970s, called SIPOS for semi-insulating polysilicon. This material, which will also play an important role in the development of high-temperature passivated contacts for solar cells, is obtained by adding oxygen to the poly-Si layer and was first used as a passivating coating for high voltage devices [123]. Later, the interest of SIPOS as the emitter material of n–p–n heterojunction transistors was highlighted, with a current gain 50 times higher than the value found for its homojunction counterpart [124].

These appealing properties of poly-Si films raised interests in the PV community (Figure 3.11). In 1981, Van Overstraeten claimed that it could be interesting to implement a poly-Si film in between the metal grid and the silicon substrate to decrease the overall surface recombination velocity [125]. A couple of years later,

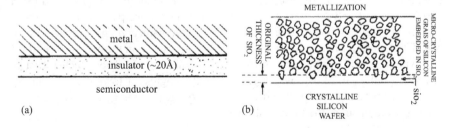

Figure 3.11   Pioneer works on poly-Si contacts for solar cells. (a) Schematic diagram of the structure of a MIS contact. Reprinted with Elsevier permission from [126]. (b) Model of the SIPOS-Si heterojunction. Reprinted with AIP permission from [128]

Green *et al.* reported on the fabrication of metal–insulator–semiconductor (MIS)-like cells with degenerately doped poly-Si as the metal layer, featuring open-circuit voltage (V_{oc}) values in the range 660–665 mV [126]. Green *et al.* suggested that one advantage of the poly-Si approach for this advanced cell concept concerns its compatibility with high temperature ($\geq 800°C$) treatments, particularly for the subsequent screen-printing of metal contacts. Another interesting integration of poly-Si films in c-Si solar cells was reported by Tarr *et al.* in 1985 [127]. They implemented n-type poly-Si films (thickness in the range 50–80 nm) on the front surface of p-type c-Si solar cells. The films were deposited by low-pressure chemical vapor deposition (LPCVD), and the best device featured a V_{oc} of 652 mV (authors deliberately avoided the growth of an interfacial oxide). The same year, remarkable results were presented by Yablonovitch *et al.* [128], who fabricated a double heterostructure solar cell by integrating a stack made of SIPOS and SiO_2 between the metallization and a p-type c-Si wafer. They measured a V_{oc} value of 720 mV and the n-type SIPOS/SiO_2 contact provided a saturation current density (J_0) of only 10^{-14} A cm^{-2}. Interestingly, they proposed hypotheses for explaining the carrier transport throughout the SiO_2 layer (thickness in the range 25–40 Å). More precisely, they argued that transport by carrier tunneling is unlikely and that the electrical contact in between the SIPOS film and the substrate is probably created by thin spots and pinholes in the SiO_2. Following these promising results achieved with SIPOS, Kwark *et al.* showed that poly-Si emitters could achieve similar performances (i.e. similar J_0) than their SIPOS counterpart, if an intentionally grown chemical oxide is present under the poly-Si layer [129]. The importance of the interfacial oxide layer was further confirmed by Gan *et al.*, who obtained poly-Si/SiO_x emitters featuring $J_0 \leq 5 \times 10^{-14}$ cm^{-2} [130]. They also demonstrated that the surface passivation properties could be improved by conducting forming gas anneal (FGA) steps.

All the pioneering works conducted during the 1980s and 1990s highlighted the promising surface passivation and carrier transport properties of the poly-Si contacts. Nevertheless, the first successes regarding significant improvements in conversion efficiency only appeared at the beginning of the twenty-first century, with the demonstration of a 24.2% efficient Interdigitated back contact (IBC) n-type solar cell structure implementing both p-type and n-type poly-Si passivated contacts [131,132]. Shortly after, researchers from Fraunhofer-ISE [133] and ISFH [134] developed poly-Si contacts coupling excellent surface passivation properties and low poly-Si/c-Si interface resistivity values. Feldmann *et al.* integrated their so-called TOPCon (for tunnel oxide passivated contact) stack on the rear surface of FZ Si n-type solar cells with a boron-diffused front emitter [133]. They reached first an efficiency value of 23.0%, which subsequently progressed toward 25.7% [135], opening the road to the extensive investigations of the technology and eventually its current deployment at the industrial level.

3.3.3   Working principle of the poly-Si-passivated contact

Glunz *et al.* [136] presented the three prerequisites for an ideal contact: (i) surface passivation (i.e. reduction of carrier recombination at the metal contact), (ii) carrier selectivity (i.e. allow one carrier type to pass through the contact while blocking the other one), and (iii) low resistive losses. Interestingly, these three features can be successfully gathered with « passivating and carrier selective contacts », for which the metal electrode is spatially separated from the silicon substrate. According to Glunz *et al.*, in the most successful implementations, the passivating and carrier selective contact is composed of (i) an interface passivating layer (IPL) and (ii) a carrier separation layer (CSL) between the IPL and the metal contact.

Regarding the poly-Si passivated contact, the IPL is usually made of an ultra-thin (e.g. ~2.0 nm) SiO_x layer (Figure 3.12). Glunz reviewed the passivation properties of SiO_x films [137]. As SiO_x layers feature a rather low density of fixed charges, the surface passivation essentially relies on the reduction of the interface states density. This passivation mechanism is referred to as chemical passivation. Usually, the chemical passivation provided only by the SiO_x film is not good enough to reach the passivation levels required in high-efficiency solar cells. Therefore, the fabrication processes of poly-Si contacts frequently integrate inter-face hydrogenation treatments, essentially by capping the poly-Si film with a hydrogen-containing dielectric layer such as AlO_x or SiN_x, followed by a thermal

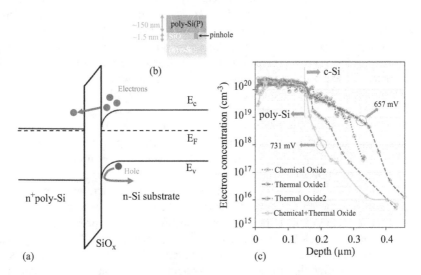

Figure 3.12   *(a) Schematic band diagram of a poly-Si(n^+)/SiO_x passivated contact. (b) Schematic representation of a poly-Si/SiO_x passivated contact featuring a pinhole through the SiO_x layer. The typical dimensions of the different layers are indicated. (b) Electron concentration profiles through a poly-Si(n^+)/SiO_x/c-Si stack for various kinds of interfacial oxides (LPCVD deposition with in situ phosphorus doping). The implied open-circuit voltages related to two doping profiles are specified. These results highlight both the influence of the SiO_x layer on the dopant diffusion into the c-Si substrate and its effect on the surface passivation level*

anneal [138]. It was recently demonstrated that the SiO_x film would retain hydrogen atoms at the SiO_x/substrate interface, improving the overall passivation quality [139].

The CSL is associated with a highly doped poly-Si layer (Figure 3.12). The high doping level generates a strong band bending in the silicon absorber creating an induced junction, which leads to carrier-selectivity. A high doping of the poly-Si layer is beneficial for reducing the surface recombination by field effect passivation [140,141]. Usually, an additional diffusion of dopants from the poly-Si films into the Si substrate is observed (Figure 3.12). The extent of this in-diffusion needs to be carefully controlled. Indeed, for an n-type poly-Si contact formed on an n-type wafer, on the one hand, it can provides both surface passivation [140] and fill factor improvements [142] by reducing the density of minority charge carriers at the SiO_x/substrate interface and enhancing the extraction of majority charge carriers, respectively. On the other hand, highly doped and deep in-diffused regions can enhance Auger recombination (Figure 3.12).

Interestingly, despite the presence of ultra-thin dielectric layers, poly-Si contacts feature low junction (poly-Si/substrate interface) resistivity (ρ_j). As an example, Rienäcker *et al.* [143] fabricated n-type poly-Si/SiO_x structures combining a low saturation current density of 6.2 fA cm^{-2} with a ρ_j of only 0.6 mΩ cm^2. The carrier transport mechanisms through the SiO_x film are still matter of debate. In some cases, particularly when the SiO_x thickness is below 20 Å, quantum tunneling would be the dominant mechanism [144]. For instance, Steinkemper *et al.* [140] reproduced the excellent performances of solar cells with poly-Si contacts by considering only transport by tunneling through the SiO_x films in their numerical simulations. In other cases, particularly for SiO_x thicknesses higher than 20 Å and structures exposed to high T (>900°C) anneals, alternative mechanisms would be involved. As an example, Yablonobitch *et al.* [128] suggested that the carrier transport throughout the SiO_2 layer in their SIPOS solar cells is probably enhanced by the presence of thin spots and pinholes in the dielectric film (Figure 3.11). Later, Peibst *et al.* [145] presented strong arguments regarding the fact that homogeneous tunneling is probably not the only transport mechanism in poly-Si contacts. More precisely, they showed that the junction (depending on the process conditions) could be considered as a "classical, locally contacted emitter," assuming local current flows through the SiO_x film at locations where the oxide thickness is reduced or where pinholes are formed. In particular, they elegantly brought strong experimental arguments toward the presence of such pinholes, via the observation of direct local contacts between the poly-Si and the c-Si substrate by high-resolution transmission electron microscopy (HR-TEM). The presence of local conduction paths through the SiO_x film was highlighted by various studies via conductive atomic force microscopy (c-AFM), see for instance the study published by Zhang *et al.* [146] which points out the importance of the transport by pinholes for SiO_x thicknesses higher than 1.2 nm. For such thicknesses, pinholes would be essential for having devices with good fill factor. However, their density/diameter needs to be carefully controlled since a too high surface area covered by pinholes would affect the overall surface passivation. In contrast, the presence of local conduction paths is not systematically in contradiction with a transport by pure tunneling. As an example, Kale *et al.* [147] studied poly-Si contacts with two SiO_x thicknesses: 1.5 nm and 2.2 nm. They observed by electron beam-induced current (EBIC) analyses an homogeneous transport for the 1.5 nm SiO_x sample attributed to uniform tunneling, and an heterogeneous transport for the 2.2 nm SiO_x sample, attributed to tunneling through locally thin SiO_x regions. As a conclusion, the transport mechanisms of charge carriers in the poly-Si contact strongly depend on its fabrication process (especially the SiO_x growth method and the poly-Si annealing temperature).

Beyond their excellent properties in terms of surface recombination, charge separation and carrier transport, poly-Si contacts feature an interesting added value in comparison for instance with low-temperature processed silicon heterojunctions, which is their ability to develop strong external gettering effects, as shown by several studies [148,149]. As an example, Hayes *et al.* [149] demonstrated that the fabrication process of in situ doped n-type poly-Si/SiO_x structures would extract more than 99% of the interstitial iron atoms initially present within the Si bulk. Liu showed that this gettering effect does not affect the quality of the passivated contact [148]. They also investigated the mechanisms behind this strong gettering effect. They demonstrated that for phosphorus-doped poly-Si passivating contacts, metal impurities are relocated from the silicon wafer bulk to the heavily phosphorus-doped poly-Si layer [150]. The gettering effect would be dominated by segregation gettering. The heavily doped poly-Si layer acts as the main gettering sink and the features of the SiO_x interlayer such as its growth method or stoichiometry deter-mine the gettering kinetics [151]. This makes poly-Si contacts interesting for using cost-efficient but usually non-perfect Si materials, such as advanced cast-mono wafers [152].

3.3.4 Fabrication processes of poly-Si-passivated contacts

3.3.4.1 SiO_x formation

The formation of the SiO_x layer plays a critical role on the properties of the poly-Si contact. The thickness and stoichiometry of the SiO_x layer have a major influence on the performances of poly-Si passivated contacts (Figure 3.12). Indeed, the lower the thickness, the stronger the diffusion of dopants from the poly-Si layer into the substrate (which also usually leads to lower dopant concentrations within the poly-Si film) [153]. Regarding the stoichiometry, Moldovan *et al.* [154] showed that less stoichiometric SiO_x films are more prone to disruptions during high temperature annealing (which improve the carrier transport, but affect the chemical passivation and enhance dopant diffusion into the substrate). The oxide stoichiometry is dependent on the oxidation mechanisms [154] and therefore on the technique used to form this ultrathin layer.

The formation techniques of the SiO_x can be divided in two categories, wet-chemical processing or dry methods (Table 3.1). The two main wet-chemical pro-cesses are nitric acid oxidation of silicon (NAOS) [155] and ozonized DI-H_2O rinsing [156]. The NAOS process is usually conducted at temperatures close to the

Table 3.1   *Main positive (+) and negative (−) features of various techniques*
investigated for forming the ultra-thin SiO$_x$ layer

Technology	Wet/dry	Lab scale demonstration	Compliant with existing PV equipment and production	In situ oxidation
NAOS	Wet	+	−	−
CNS	Wet	+	+	−
DI-O$_3$	Wet	+	+	−
FIA	Wet	+	−	−
UV-O$_3$	Dry	+	−	−
ALD	Dry	+	+	+
Thermal oxidation	Dry	+	+	+
Plasma-assisted oxidation	Dry	+	+	+

boiling point of the chemical solution (i.e. 121°C). At such temperatures, HNO$_3$ volatilizes and the composition of the solution changes during the process, degrading the reproducibility, uniformity and quality of the SiO$_x$ films, making this approach not industrially suitable [157]. To solve this issue, Tong *et al.* [157] investigated a mixture of HNO$_3$ and H$_2$SO$_4$ referred to as CNS for concentrated nitric and sulfuric acids, which offers high-quality SiO$_x$ layers with a chemical process conducted at lower temperatures, close to 60°C. Beyond these both approaches, Tong *et al.* [158] proposed an alternative technique based on field-induced anodization (FIA). This cost-effective technique would offer a good controllability of the SiO$_x$ properties, but such a process is not yet deployed at the industrial level for the fabrication of solar cells.

Even if high-quality oxides were obtained with wet-chemical processes, "dry" approaches offer strong interests in terms of reduced water and chemical products consumptions. An interesting option was proposed by Moldovan *et al.* [154], for which the oxide is grown from O$_3$ by photo-oxidation with an UV excimer source. The ability of this cost-effective method to grow thermally stable stoichiometric oxides was demonstrated [154]. Nevertheless, it would require an additional step after the wafer surface preparation and specific equipment adapted to the PV industry requirements. Thus, approaches compatible with equipment already used in existing PV production lines were investigated.

First, high-quality n-type poly-Si contacts can be obtained by using thermal dry oxidation (wafers are placed in a quartz tube furnace under pure oxygen) to form the interfacial oxide [159]. Second, several studies demonstrated the potential of plasma-assisted oxidation, either from N$_2$O [160,161] or from CO$_2$ [162] precursors. Last, Lozac'h *et al.* formed poly-Si contacts by depositing the ultrathin SiO$_x$ with plasma-assisted ALD [163]. This technique is particularly interesting to form passivated contacts on textured surfaces, thanks to the excellent conformity and homogeneity of the deposited films. Interestingly, the SiO$_x$ formation by thermal oxidation, plasma-assisted oxidation and ALD deposition, can be conducted in the equipment used to deposit the amorphous Si (a-Si) or poly-Si layers, during the same technological step. Such an approach is referred to as in situ oxidation. The potential of in situ oxidation was demonstrated either for thermal oxidation treatments followed by the LPCV-deposition of Si films [164] or for plasma-assisted oxidations followed by plasma-enhanced chemical vapor deposition (PECVD) of a-Si films [165]. According to the ITRPV road map, 80% of the poly-Si contact solar cells produced at the industrial level would rely on in situ oxidation [14]. Therefore, the future dominant technologies for the oxide formation will strongly depend on the choice of the processes used to deposit the a-Si/poly-Si layers [166].

3.3.4.2 a-Si/poly-Si deposition

Following the formation of the SiO_x layer, a-Si or poly-Si films are deposited before a high temperature crystallization step essentially used to electrically activate the dopants within the film. The lab scale-processed poly-Si contacts at FhG-ISE and ISFH, which contributed to the current revival of the technology, were based on PECVD or LPCVD depositions of Si layers (Figure 3.13).

LPCVD is one of the dominant process in the semiconductor industry for depositing a-Si/poly-Si films. Therefore, the first industrialization attempts of the poly-Si contact for PV applications were essentially based on this technology. Nowadays, LPCVD is the main process for the a-Si/poly-Si deposition in mass production (about 84% of the world market share in 2021) [14]. LPCVD proceeds by pyrolyzing silane between 575°C and 650°C in a low-pressure reactor. Depending on the process conditions (e.g. temperature, doping), the film properties can vary from amorphous layers with no detectable structure to fully polycrystalline materials featuring a columnar structure [167]. LPCVD allows the deposition

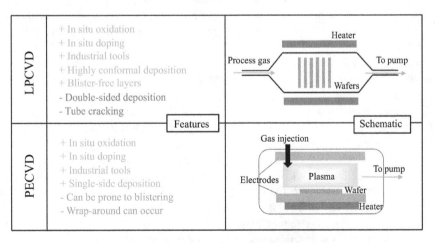

Figure 3.13   Features and schematic representations of the two main techniques used at the industrial level for the deposition of poly-Si layers. Adapted from [169]

of blister-free and highly conformal films. The process is also compatible with in situ doping, for instance via the addition of phosphine during the deposition run. It is worth noting that initially the addition of phosphine could limit the throughput due to thickness and doping uniformity issues. Nevertheless, solutions were developed to overcome this issue, for instance by increasing the deposition temperature [168]. Several equipment suppliers propose LPCVD furnaces providing high quality in situ doped poly-Si layers and meeting the PV industry throughput requirements [141,168]. A weakness of LPCVD is inherent to the regular replacement of the quartz tube due to the mechanical stress induced by thick (e.g. ~100 μm) poly-Si films enhancing the risk of tube breakage [169]. Another disadvantage of LPCVD when the poly-Si contact has to be formed on only one side of the wafer is related to the double-side deposition of the technique. This leads to the parasitic deposition of a layer on the edges and the periphery of the undesired side (wrap-around effects), even if the wafers are placed in a back-to-back configuration [170]. In order to avoid the presence of this undesired poly-Si layer, additional steps are required (e.g. barrier layers deposition, etching steps), which make the cell fabrication process more complex and usually reduce the production yield.

Interestingly, such wrap-around issues can be reduced by using PECVD (especially in-line type PECVD), which is a well-proven technique in the PV industry for the single-side deposition of dielectric layers. PECVD offers other advantages, such as higher deposition rates than LPCVD. Deposition rates are highly dependent on the process parameters and can significantly vary from one study to another. Nevertheless, as an example for intrinsic Si films, the LPCVD deposition rate reported in [171] is about 6 nm/min, whereas the PECVD deposition rate reported in [172] is about 22 nm/min. One drawback of the PECVD technique is due to the high hydrogen content within the a-Si film, which can induce blistering effects. However, this issue can be ruled out by adapting the process parameters, for instance by increasing the deposition temperature [156]. Recently, by using industrial PECVD equipment, several groups demonstrated the possibility to form highly passivating 170 nm-thick blister-free poly-Si layers [141,173]. ITRPV predicts a fast adoption of PECVD for poly-Si deposition in mass-production, since in 2026, this technique could represent about 40% of the world market share [14].

Beyond these dominating technologies, alternative approaches were considered these past few years for the single-side deposition of a-Si/poly-Si layers. Several studies reported the processing of poly-Si contacts by hot-wire CVD (HWCVD) and atmospheric-pressure CVD (APCVD). The first allows the formation of high quality poly-Si contacts with a high deposition rate of 42 nm/min for intrinsic films [174]. HWCVD is compatible with in situ doping, but industrial tools are not available yet. On the other hand, industrial APCVD equipment are already available and offer cost-reduction opportunities since the process does not require any vacuum systems or plasma sources [175]. Furthermore, APCVD has already been used for several applications in PV, including the deposition of anti-reflective layers. One drawback of both HWCVD and APCVD is that they would not permit in situ oxidation. Physical vapor deposition (PVD) techniques were also recently proposed for depositing a-Si layers to form poly-Si/SiO$_x$ passivated contacts with high surface passivation properties. PVD can occur through evaporation or sputtering (magnetron or ion beam) and could provide cost reduction opportunities since no silane infrastructure is needed. Furthermore, due to their low hydrogen content, the PVD-deposited films are less prone to blistering issues. Lossen *et al.* investigated the use of electron-beam PVD (EB-PVD) for the formation of n-type poly-Si contacts [176]. The deposition rate of intrinsic Si layers could be as high as 500 nm/min while obtaining high surface passivation properties (iV_{oc}>715 mV) with an ex situ doping procedure. They also demonstrated that this deposition process could be compatible with in situ doping. However, PV-dedicated industrial tools are not available yet. This last issue could be overcome with the deposition of Si layers by magnetron sputtering since promising demonstrations of the formation of poly-Si contacts were conducted with industrial sputtering platforms with deposition rates as high as 141 nm/min for intrinsic layers [177]. Interestingly, sputtering is also compatible with in situ doping [178], even if these past few years, the doping levels reached with PVD approaches could be slightly below the threshold values required for screen-printed metallization. However, the solar module producer JOLYWOOD recently used PVD with an improved in situ doping process for the successful production of poly-Si contacts solar cells in an industrial environment. The shift from LPCVD to PVD for the poly-Si deposition allowed JOLYWOOD to solve wrap-around issues and therefore to significantly simplify the cell fabrication process while improving the production yield. Thus, PVD could play an important role for the massive industrial deployment of poly-Si contacts cells. However, in general, PVD techniques require an ex situ oxidation step. A last and *a priori* interesting approach for the deposition of a-Si films to form poly-Si contacts is the plasma-enhanced atomic layer deposition (PEALD) technique. Particularly, this technique would be compliant with the in situ deposition of SiO_x film and an equipment supplier would already propose mass production equipment [179]. Nevertheless, few details are currently accessible regarding the features of the process and the quality of the deposited layers.

3.3.4.3 Poly-Si doping and crystallization

As previously mentioned, dopants can be introduced during the deposition step of the a-Si/poly-Si layers (in situ doping). This approach is usually targeted at the industrial level (process simplification). However, for different reasons related to throughput or to specific device features (e.g. selective doping), ex situ doping can sometimes be preferred. The main ex situ doping technique relies on the thermal diffusion of dopants from gaseous precursors. n-type poly-Si contacts with iV_{oc} values of 735 mV were obtained with this technique [141]. However, this double-side doping technique requires the use of barrier layers and subsequent chemical etching steps. As alternatives, single-side doping techniques were proposed, such as thermal diffusion from dopant-containing dielectric layers [139] or spin-coated dopant-rich solutions [180], or even ion-implantation [181]. As an example, Feldmann *et al.* obtained n-type poly-Si contacts featuring iV_{oc} of 725 mV by using beam line ion implantation [181]. To overcome the main drawbacks related to beam line implantation (i.e. equipment cost, high footprint) while keeping its advantage regarding the fine-tuning of the doping profile, several studies assessed the potential of PIII for doping poly-Si contacts [182,183] with very positive results. As an example, Young *et al.* elaborated n-type PIII-doped poly-Si contacts with iV_{oc} of 730 mV.

In situ doped or ion implanted a-Si/poly-Si films have to experience high temperature annealing steps in order to electrically activate the dopants. Usually, furnace tube anneals are conducted under nitrogen or argon atmospheres, in the 700–900°C temperature range. Depending on the temperature, the anneal duration typically lies between 10 min and 60 min. In order to simplify the fabrication process and limit the thermal budgets, it was shown that rapid thermal anneals (RTA) performed in lamp furnaces could also be efficient for activating dopants in poly-Si layers [184,185]. Particularly, Ingenito *et al.* formed a high-temperature carbon-rich silicon passivated contact integrated in a proof-of-concept device featuring an efficiency of 21.9%, by using a RTA with a temperature versus time profile corresponding to the industrial metallization firing of silicon solar cells. This opens the way for significant simplifications of the fabrication process of solar cells with poly-Si contacts.

3.3.5 Integration of poly-Si contacts into screen-printed or large area solar cells

The first integrations of poly-Si contacts in solar cells essentially used PVD steps for contacting the poly-Si layers, such as the thermal evaporation of Ag. However, to deploy the technology into mass production, it was particularly important to develop metallization processes based on screen-printing, preferably via firing-through pastes since H-rich silicon nitride layers usually cover the poly-Si film to enhance the chemical passivation by providing H atoms to cure interface defects.

However, the screen-printing metallization of such thin layers posed techno-logical challenges, expressed by instance by Jiang *et al.* [186], who pointed out that developments were needed to find a good trade-off between low contact resistances while restricting the paste penetration within the thin nano-scale poly-Si layer. This was confirmed by Çiftpınar *et al.* [187], who revealed that the firing-through metallization with industrial pastes locally removed large fractions of the poly-Si layers and even damaged the underlying substrate. These studies highlighted that relatively thick (i.e. >100 nm) poly-Si layers are required to screen the damages due to firing-through pastes.

Nevertheless, thick poly-Si layers induce strong parasitic absorption losses. For instance, Feldmann *et al.* [188] showed that poly-Si films located at the front-side of a solar cell induce short-circuit current density losses of about 0.5 mA cm^{-2}/10 nm of poly-Si. Therefore, in order to get high efficiency values, thick poly-Si layers (required for firing-through metallization) have to be located on the rear side of the device only, even if thick poly-Si films placed on the rear side will affect the red response of the solar cell because of free carrier absorption (FCA) losses [189].

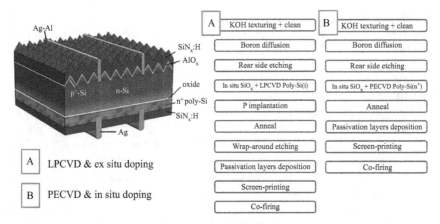

Figure 3.14   Schematic structure of an industrial screen-printed n-type solar cell with a front boron emitter and a rear poly-Si/SiO$_x$ passivated contact. Examples of two fabrication process flows (only the main steps are displayed) based on LPCVD (with ex situ doping) and PECVD (with in situ doping and assuming no wraparound) depositions of the poly-Si layer, respectively

Furthermore, it was shown by several groups (e.g. [190]) that the surface passivation properties of poly-Si contacts on polished surfaces are better than the passivation levels obtained with textured surfaces.

Therefore, the first integrations of poly-Si contacts in large area cells with firing-through screen-printed metallization addressed device architectures featuring a conventional homojunction doping on the front and a thick poly-Si passivated contact implemented on the chemically polished rear surface (Figure 3.14). Stodonly *et al.* published in 2016 the first results on such screen-printed devices [190]. They started from 6 in. Cz-Si wafers. The cells featured a boron-diffused emitter on the front-side and a poly-Si/SiO$_x$ contact on the rear side. The 200 nm-thick poly-Si layer was deposited by LPCVD and ex situ doped by a POCl$_3$-thermal diffusion process. It was covered by a SiNx:H layer before the screen-printing step. They achieved a maximum efficiency value of 20.7% with a V_{oc} of about 675 mV. The V_{oc} was significantly lower than the implied-V_{oc} monitored on non-metallized precursors, showing that the metallization procedure strongly affected the surface passivation properties (front and rear surfaces). By using a similar fabrication process flow, Duttagupta *et al.* [191] published a couple of years later an efficiency value of 21.4% for a 244.3 cm^2 large area screen-printed (Ag–Al and Ag pastes for the front and rear electrodes, respectively) solar cell. In parallel of these promising pioneering works on front-junction cells with n-type poly-Si films on the rear side, Ok *et al.* [192] investigated screen-printed rear-junction n-type cells with p-type poly-Si films on the rear side. Their devices featured a selective n$^+$ front-surface-field and on the rear a 200 nm-thick boron-doped poly-Si deposited by LPCVD with an in situ doping, coated by a SiN:H layer. Ag and Ag–Al pastes were used for the formation of the front and rear electrodes, respectively. A maximum efficiency of 20.9% (V_{oc} of 672 mV) was achieved. From 2016 to 2018, as mentioned, the developments concerned both front- and rear-junction structures. However, p-type poly-Si/SiO_x stacks usually feature weaker surface passivation properties than their n-type counterparts. Ok *et al.* presented several mechanisms explaining this difference such as the high solubility of boron atoms in SiO_x, higher interface defects densities and an enhanced carrier recombination in p-type poly-Si due to the increased tunneling probability of electrons into the p-type poly-Si layer [192]. Furthermore, Ling *et al.* [193] highlighted that the screen-printing metallization of p-type poly-Si would be more challenging than the metallization of n-type poly-Si. Indeed, by using screen-printed fired-through Ag–Al pastes for contacting p-type poly-Si layers, they observed several local "punch-through" contact regions where the paste completely consumed the underlying poly-Si, issue probably related to local Al-alloying mechanisms. Therefore, in the following years, the developments essentially focused on front-junction structures (i.e. front B-emitter and rear n-type poly-Si), as shown in Figure 3.14.

From 2019, the efficiency values achieved with screen-printed front B-emitter and rear poly-Si(n^+) n-type solar cells made remarkable progresses. This is explained by two main factors. First, the paste industrial suppliers developed specific products for poly-Si films that allowed strong reductions of both the metal contact recombination current (J_{0met}) and specific contact resistivity (ρ_c). As an example for Ag-screen-printed n-type poly-Si contacts, in 2017, Çiftpinar *et al.* [187] presented J_{0met} and ρ_c values of 400 fA cm^{-2} and 3–5 mΩ cm^2, respectively. Only 3 years later, Padhammat *et al.* [194] achieved J_{0met} and ρ_c values as low as 30 fA cm^{-2} and 0.3 mΩ cm^2, respectively. The second reason is that major industrial players started to run pilot lines fully dedicated to this new technology, pushing the efficiency values. Among these pioneer industrial actors, JOLYWOOD and TRINA SOLAR presented in 2019 efficiency values of 23.15% and 23.57% (V_{oc} of 716.7 mV), respectively [195,196]. Interestingly, the TRINA SOLAR cells were processed by using an old multicrystalline-Si solar cells production line that was converted for the fabrication of poly-Si-passivated contacts solar cells [196], demonstrating that the current PERC production lines could benefit from this advanced technology at minimum expenses. Both JOLYWOOD and TRINA SOLAR fabrication processes used a front B emitter formed by thermal diffusion before the LPCVD deposition of intrinsic Si films with an in situ thermal oxidation procedure. Both processes differ however in the way the poly-Si is ex situ doped, by ion implantation at JOLYWOOD and by $POCl_3$-thermal diffusion at TRINA SOLAR. Efficiency values reported by industrial players continued to increase. As an example, JINKO SOLAR reached an efficiency value of 24.58% in 2021, in particular via the implementation of an additional hydrogenation treatment (Table 3.2) [197]. In 2021 and 2022, several players announced externally certified efficiencies higher than 25.0% (e.g. LONGI [198] and TRINA SOLAR [199]). The record value is currently held by JINKO SOLAR, with an efficiency conversion of 26.4% for a 182 mm-size Cz wafer [200]. As previously mentioned, thick poly-Si

Table 3.2 *Examples of externally certified I–V parameters of double-side contacted n-type solar cells featuring a front boron emitter and a rear poly-Si/SiO$_x$ contact. The results from Richter [135] were obtained with lab-scale processes. The results from Chen [197] concern a commercial-like cell with screen-printed contacts*

Area (cm^2)	Metallization	J_{sc} (mA cm^{-2})	V_{oc} (mV)	FF (%)	(%)	Reference
4.0	Evaporation	42.5	724.9	83.3	25.7	[135]
251.99	Screen-printing	41.5	718.5	82.5	24.6	[197]

layers limit the damages due to fired-through pastes but, on the other hand, thinner poly-Si films are preferred to reduce FCA effects (even on the rear side, for long-wavelength) and to reach higher throughput and lower CAPEX related to the poly-Si deposition. In industrial environments, the poly-Si thickness decreased from 200 nm in 2018 to 100 nm in 2021, and further reductions below 80 nm are expected in 2022.

The cell fabrication processes previously presented were essentially based on LPCVD and ex situ doping procedures for the formation of the poly-Si film. However, in 2019, Nandakumar *et al.* [201] and Feldmann *et al.* [202] presented first results regarding screen-printed M2 solar cells (front B-emitter, rear n-type poly-Si contact) using PECVD with in situ doping to form the poly-Si film. They achieved efficiency values of 23.05% and 22.5%, respectively. The PECVD-based processing of poly-Si cells is particularly appealing since the wrap-around issue can be minimized, which offers the opportunity of significant process simplifications (avoidance of wet processes to remove the poly-Si on the front), as shown in Figure 3.14. However, it has to be noticed that even with inline PECVD, there can be a slight wrap-around located only along the edges, which may not significantly affect the efficiency, but could alter the appearance of the cell and possibly increase the reverse current (risk of non-acceptance at the moduling stage). Still in order to avoid the wrap-around issues related to the LPCVD deposition of poly-Si layers, the module producer JOLYWOOD adopted recently a simplified fabrication process based on the PVD deposition with in situ doping of poly-Si films. Beyond the process simplification, this approach led to strong improvements of the production yield (with yield values exceeding 97%). Outstanding efficiency results were achieved, with a record value at 25.4% obtained with M10 (182 mm) wafers [203].

3.3.6 Toward the double-side integration of poly-Si contacts

Remarkable progresses were achieved with solar cells integrating poly-Si passivated contacts on the rear surface with a conventional homojunction on the front-side of the device. However, despite outstanding performances, these architectures still suffer from metal contact recombination due to the direct interface between the front Ag–Al electrode and the boron emitter. Therefore, in order to maintain efficiency improvements, the front-side integration of poly-Si layers currently sparks a lot of interest. Nevertheless, the implementation of poly-Si films on the front-side could induce significant parasitic absorption losses. In order to solve this issue, two approaches are investigated: (i) selective poly-Si contacts with higher thicknesses underneath the metal grid (and in some cases without poly-Si between the fingers) and (ii) formation of ultra-thin (thickness below 20 nm) homogeneous poly-Si layers.

Ingenito *et al.* [204] presented one of the first studies about cells featuring a double-side integration of poly-Si contacts with selective poly-Si layers on the front-side (i.e. poly-Si only below the metal grid). They highlighted the high efficiency potential of such structures, since efficiency values beyond 26% could be achievable. The poly-Si patterning was obtained with a SiN_x mask created by photolithography. The poly-Si between the front grid pattern was etched by alkaline texturing. First lab-scale solar cells (2.8×2.8 cm^2) were prepared with an encouraging maximum efficiency of 20.1%.

Following this pioneering work, several research groups focused on approaches for the poly-Si patterning that would be more suitable for mass-production: (i) local laser oxidation of poly-Si [205]; (ii) laser ablation of the inter-finger region [205]; (iii) PECVD deposition through shadow masks [206]; (iv) reactive ion etching between metal contacts as shown in Figure 3.15(a) [207]; (v) inkjet masking [208]. Notice that these studies essentially focused on the development of the specific building blocks and results concerning complete solar cells were rarely reported. However, recently Yu *et al.* [209] fabricated complete devices by using an alternative approach for which the poly-Si is patterned via the screen-printing deposition of an anti-acid mask followed by a single-side HF/HFNO$_3$ etching step, as shown in Figure 3.15(b). Interestingly, their precursors were metallized with firing-through screen-printed electrodes and they achieved a promising efficiency value of 23.21%.

Regarding the double-side integration of poly-Si contacts with an ultra-thin homogeneous poly-Si layer on the front-side, the following results only focus on screen-printed n-type solar cells. As firing-through pastes are usually not suitable for poly-Si films with thicknesses below 50 nm, the metallization of such solar cells essentially relies on the combination of transparent conductive oxides (TCO) and low-temperature silver pastes. With such an approach, by using front and rear indium tin oxide (ITO) layers, Larionova *et al.* [210] and Desrues *et al.* [211] presented devices with PV conversion efficiency values of 22.3% and 22.7%, respectively. It is worth noting that by replacing the front ITO by an In-free aluminum-doped zinc oxide (AZO) layer, Desrues *et al.* achieved an efficiency of 22.4% [211]. Both studies developed lean fabrication processes, by taking advantage of the inherent double-side LPCVD deposition of thin (≤ 15 nm) intrinsic poly-Si layers, ex situ doped by ion implantation (beam line or PIII) before a co-activation annealing step (Figure 3.16). The n$^+$ poly-Si layer is located on the front-textured surface and the p$^+$ poly-Si layer on the rear polished surface. However, one of the drawbacks of this TCO-based approach concerns the lack of SiN:H layer capping the poly-Si film to provide hydrogen. To overcome this issue, Larionova *et al.* [210] deposited sacrificial H-rich AlO_x layers, subsequently chemically etched. Another interesting hydrogenation process was proposed by Meyer *et al.* [212], which

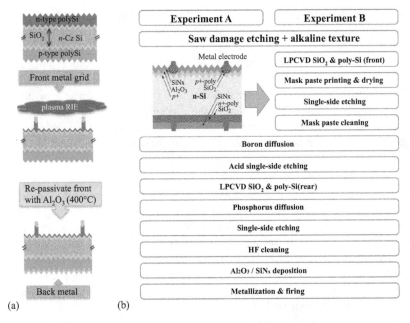

(a)   (b)

Figure 3.15   Examples of two approaches for implementing selective poly-Si/SiO$_x$ contacts on the front surface: (a) via reactive ion etching between metal contacts [207]; (b) via the screen-printing deposition of an anti-acid mask followed by a single-side HF/HFNO$_3$ etching step [209]. In (b) are presented the process steps to be added (Experiment B) to the reference fabrication process (Experiment A with a poly-Si/ SiO$_x$ contact only at the rear surface) in order to obtain the cell structure represented in the figure with a selective poly-Si(p$^+$)/SiO$_x$ on the front surface. Adapted and reprinted with Elsevier permissions from [207] and [209]

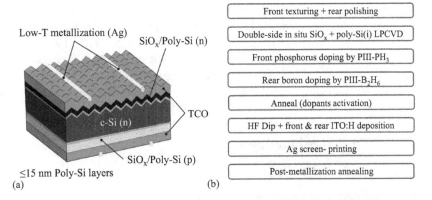

(a)   (b)

Figure 3.16   (a) Schematic representation of a double-side poly-Si contacts n-type solar cell featuring homogeneous and thin (≤ 15 nm) poly-Si layers. (b) Example of a related fabrication process flow

combines an ultra-thin (15 nm) ITO film and a non-sacrificial SiN:H layer (the last, improving the optical properties, is directly deposited on the Ag-metallized cell). Beyond these promising TCO-based structures, Yan *et al.* [213] recently brought the proof-of-concept of a double-side poly-Si contacts cell featuring a thin (20 nm) homogeneous front n^+ poly-Si layer, metallized with a co-firing step by using a firing-through Ag paste for the front-side (the rear side was contacted with an Al paste, after a local laser ablation step of the underlying SiN:H layer). The obtained efficiency is still modest (i.e. 19.6%) but opens the road for the double-side integration of poly-Si contacts with cost effective processes.

3.3.7   Conclusion

The poly-Si/SiO$_x$ contact is a good example of interactions between the developments of IC devices and solar cells. Indeed these structures were initially implemented in transistors in the 1970s with successful results, before pioneer investigations regarding their interest for solar cells in the 1980s and 1990s. Poly-Si contacts can be seen as ideal structures for solar cells since they provide all the important features of an ideal contact (high surface passivation levels, high carrier selectivity, efficient carrier transport) combined with important added values for their industrial deployment (absence of toxic/scarce materials, gettering effects, compliancy with screen-printed firing through pastes). Despite all these advantages, their massive development only started in the 2010s after the remarkable efficiency results presented by Sunpower and Fraunhofer. From this moment, the progresses regarding their integration in double-side contacted n-type solar cells became incredibly fast, since in less than a decade, their use moved from lab-scale devices to commercial products (Table 3.2), with outstanding results recently communicated by several major industrial actors and an efficiency record at 26.4% for a large area device [200]. These rapid developments took essentially benefit from the progresses previously achieved with conventional n-PERT devices (the front-side of industrial poly-Si contact cells features a conventional homojunction structure) and of the strong involvement of both equipment suppliers for depositing the poly-Si films and major industrial cell producers. Interestingly, the room for improvement is still significant regarding either advanced approaches for simplifying the fabrication processes or the development of industrial technologies for the double-side implementation of poly-Si/SiO$_x$ passivated contacts. Last, poly-Si contacts are particularly suitable for the passivation of the front surface of c-Si bottom cells used in 2-terminal monolithic multi-junction devices (e.g. perovskite on silicon tandem cells) [214]. Therefore, they will definitely continue to play a major role in the future developments of c-Si-based solar cells.

3.4   Silicon heterojunction solar cells

3.4.1   Introduction

Both double sides contacted (DSC) and IBC amorphous/crystalline silicon-based heterojunction (HJT) solar cells technologies are at the highest level in the laboratory best cells efficiency tables with reported values of 26.8% and 26.7%, respectively [215]. DSC HJT cells accounted for less than 2% market share in 2021 and could reach up to 20% in the next decade according to the International Technology Roadmap of PV [14]. The main advantages/particularities of this passivated and carriers selective contact technology (i.e. non-direct contact between metal and silicon) are:

- Very high efficiency, and open-circuit voltages > 750 mV,
- Compatible with thin wafers (<100 μm),
- Favorable T-coefficients (<-0.27%/°C) and bifaciality (>90%),
- Low-temperature fabrication processes (<230°C)
- Cell and module efficiencies improve upon light-soaking (up to +0.6%abs.)

It is worth noting that even though HJT cells current mainstream design does involve the use of larger amounts of indium and silver than PERC and TOPCon (i.e. poly-Si contacts devices) cells, which could prevent mass production deployment, this technology is compliant with indium-free and silver-free processes [216,217].

3.4.2 From birth to adulthood

First studies of devices combining amorphous silicon (a-Si) obtained from the evaporation technique and crystalline silicon (c-Si) have been reported at the end of the 1960s by Grigorovici *et al.* [218]. In 1974, Fuhs *et al.* deposited hydrogenated amorphous silicon (a-Si:H) layers by the plasma-enhanced chemical vapor deposition (PECVD) technique on top of c-Si wafer and studied the rectifying properties of this intrinsic a-Si:H/(p) c-Si device [219]. The so-called silicon heterojunction solar cell (also referred to as SHJ, or more recently as HJT) combining a thin doped a-Si:H layer with a c-Si absorber was mainly developed in the 1990s by Sanyo pushing its efficiency from 12.3% in 1990 to 24.7% in 2014. The company notably improved the solar cell performance by introducing a thin intrinsic (i) a-Si:H layer at the (p) a-Si:H/(n) c-Si interface, structure also referred to as the heterojunction with intrinsic thin-layer (HIT). The final device integrates a (n) c-Si/ (i) a-Si:H/(n) a-Si:H design at the rear surface, acting as an electron selective layer. On top of the amorphous layers, TCO layers are deposited by the sputtering physical vapor deposition (PVD) or reactive plasma deposition (RPD) technique and metallization grids are prepared by screen-printing of low temperature silver pastes. More details on the Sanyo (now Panasonic) heterojunction development history can be found in the document recently published by Taguchi [220]. In 2010, the original patent of Sanyo expired [221] and several groups joined the race of high efficiency HJT fabrication at both R&D and production levels. As an example, several thin films factories found into HJT an appealing pathway to leverage part of their knowledge and existing tools [222–224]. In 2022, HJT and TOPCon (i.e. cells with poly-Si contacts) technologies compete as the most promising post-PERC mainstream silicon-based solar products. HJT represented about 2% of the world market share in 2021 and could grow up to 20% in 2032 [14].

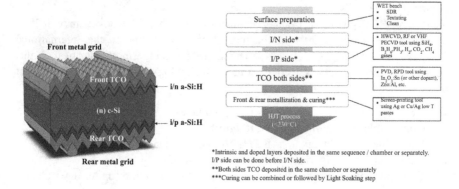

Figure 3.17   (Left) Sketch of a double sides contacted and bifacial heterojunction solar cell. (Right) Main fabrication process steps and materials

3.4.3   Cell design, fabrication steps, and record efficiencies

Different HJT architectures have been developed and studied in the literature and very high efficiencies could be reached for several of them [225–227]. A rear junction HJT cell is depicted in Figure 3.17 (left). One can rapidly denote that this cell structure is quite symmetric: both sides of the c-Si wafer are textured, then covered by thin undoped and doped a-Si:H layers stack (<20 nm), TCO (<100 nm), and metal. Front-side layers properties and metal coverage design have to be compatible with a high transparency to ensure that most of the incoming light will reach the c-Si absorber. Rear side layers properties can be tuned depending on the module design, module bifaciality being higher than 90% for utility-scale use in the glass/glass configuration. In Figure 3.17 (right), we present the corresponding fabrication process flow and materials. Excluding incoming inspection system from the sequence, it consists of various steps performed by a set of four main types of tools: wet bench, CVD, PVD, and screen-printer (combined with a curing furnace) tools. All of them can be installed in-line and process temperatures are generally lower than 230°C.

WET bench main steps are similar to those used for other advanced silicon-based PV cells preparation: saw damage removal (SDR), texturing, and cleaning steps. The a-Si:H thin layers (<20 nm on each side) can be deposited using CVD techniques (i.e. Hotwire or PECVD in RF or VHF modes). The deposition sequence presented in Figure 3.17 is the simplest one and new generation tools for mass production may propose more complex configurations with "I" layer deposited on P side, then "IN" then "P" for possible further efficiency increase and takt time optimization, with each layer actually consisting of multiple sub-layers [228]. The a-Si:H layers transparency can be tuned incorporating oxygen (O) or carbon (C) which can lead to a bandgap opening from around 1.7 eV to more than 2.2 eV [229,230]. Partially crystallized thin layers such as nanocrystalline (nc-Si:H, grain size < 10 nm) or microcrystalline (μc-Si:H, grain size > 10 nm) prepared with C or O addition can also improve transparency and carrier transport [231–234]. The PVD technique is generally used for TCO layers deposition. Rotary targets placed at top and bottom of the process chamber together with open trays allow the simultaneous deposition on both cell sides without the need of flipping the cells, while providing the necessary edge exclusion to electrically insulate the front and back electrodes. A tool composed of several successive chambers enables multilayer approaches or layers properties fine-tuning. Indium-based TCO or zinc oxides are generally used as transparent electrodes in the HJT structure and the use of H_2 gas during deposition can enhance carriers mobility and transparency of some TCO materials [235–240]. RPD process is however generally used for record cells devices as it prevents any passivation losses due to sputter damages and allows to use highly transparent TCO with high carrier mobility. The CVD and PVD steps currently involve the use of single horizontal trays on which the wafers are loaded. The maximum production tool throughput then depends on the tray dimension for a given process recipe. Finally, screen-printing line composed of several printing heads, drying ovens, and a curing furnace ($T < 230°C$) enables metal contact deposition, with lines only (so-called busbarless approach) or lines and busbars at both the front and back sides, using low temperature silver pastes. More details are provided in Section 3.6 of this chapter. Other technologies for metallization such as plating, inkjet, or laser printing are also proposed as alternatives to the screen-printing technique [241–244]. If not perfectly controlled, fabrication processes and automation may generate local defects which will affect the passivation and thus the charge transport properties. The so-called "defectivity," attributed to locally depassivated regions at front or rear sides and well-identified using photo-luminescence imaging, lead to bias-dependent minority carrier flow toward these regions and can seriously affect the solar cell fill factor [245–247]. The cell edge properties can also impact the final device performance and a particular attention must be paid to ensure a good passivating layer coverage in this region and that TCO edge exclusion is correctly tuned [248,249]. Core tools footprint and capital expenditure (CAPEX) have often been considered as main drawbacks impeding the HJT mass production deployment, but clear technology roadmap for high efficiency at production level, compatibility with thinner wafers, In and Ag-free concepts, low temperature coefficient in operation and entrance of new tools manufacturers are strong balancing arguments.

Table 3.3 lists the best confirmed large area IBC and DSC n-type HJT laboratory champion devices sorted by power conversion efficiency. Several new records have been obtained in 2021 and 2022 with the best DSC efficiency at 26.8% held by Longi, on a total area device fabricated from a M6 industrial size wafer and featuring a bifacial 9 busbars design [215,250,251]. This laboratory device has an open-circuit voltage higher than 750 mV together with a very high fill factor (FF) value of 86.6%, both approaching the theoretical intrinsic limits of 760 mV and 89%, respectively, for a 110-μm thick (n) type silicon wafer [252,253]. The device from Maxwell and Sundrive is not so far behind, reaching 26.41% on the same M6 wafer size with a bifacial 12 copper plated busbars design [254]. These record cells integrate highly transparent hydrogenated microcrystalline silicon layers with possibly some oxygen and/or carbon addition (μc-SiO:H or μc-SiC:H) at the front-side while the rear junction may be provided by an hydrogenated nanocrystalline layer (nc-Si:H) [255]. This type of layers also lowers the series

Table 3.3   IBC and double side contacted large area n-type heterojunction laboratory champion devices sorted by power conversion efficiency. (t): total area, (ap): aperture area, (da): designated illumination area. Only the most recent result for a given type of area definition is reported for each company. *IBC devices

Company	Year	PCE (%)	V_{oc} (mV)	J_{sc} (mA/cm^2)	FF (%)	Area (cm^2)
Longi	2022	26.81	751.4	41.45	86.1	274.4 (t)
Kaneka*	2017	26.7	738.	42.7	84.9	79.0 (da)
Maxwell / Sundrive	2022	26.41	750.2	40.8	86.3	274.5 (t)
Meyer Burger / CSEM*	2021	25.35	745	41.2	82.6	200.5 (?)
Huasun / Maxwell	2021	25.26	746.2	40.0	84.6	274.5 (t)
Hanergy	2019	25.11	747.0	39.5	85.0	244.4 (t)
Kaneka	2015	25.1	738.0	40.8	83.5	151.9 (ap)
CEA / EGP	2020	25.03	740.1	40.0	84.6	213.7 (da)
CEA / EGP	2021	24.86	742.6	39.7	84.2	244.8 (t)
CSEM	2019	24.73	736.0	40.4	83.3	222.8 (da)
Panasonic	2014	24.7	750.0	39.5	83.2	101.8 (t)
Kaneka	2015	24.5	741.0	40.1	82.5	239.0 (t)
Zhongwei	2020	24.05	744.6	38.6	83.7	244.4 (t)
CSEM	2019	24.03	742.0	39.0	83.1	244.3 (t)

resistance losses thanks to reduced contact resistivity with respect to a-Si:H, thereby unlocking higher FF values [256]. One should note that the majority of these record cells have a double antireflective coating (DARC) for improved light transmission at the air/TCO interface. Some changes in efficiency due to short-circuit current (I_{sc}) variations may be observed for industrial devices once in the module environment (i.e. air/glass/encapsulant/TCO interfaces) in addition to possible FF losses after interconnection. One should also note the IBC Meyer Burger/CSEM device based on tunnel and In-free approaches in the fourth position [257].

3.4.4   Hetero-interface properties and electronic transport for HJT engineering

Deposition of thin a-Si:H layers on top of c-Si wet chemically prepared surface enables to reach outstanding surface passivation properties (i.e. ultra low minority carrier surface recombination velocity). This is possible thanks to two main phe-nomena. The first one, generally referred to as "chemical passivation," is related to the mobile hydrogen reservoir of the a-Si:H layers which can diffuse to the c-Si surface and passivate recombination-active dangling bonds [225]. The second one, referred to as "field effect passivation," is related to local electrical field close to the a-Si:H/c-Si interface (often simplified as a surface-fixed charge concentration) which prevents minority carriers accumulation in a region where they could otherwise recombine on interface defects. These surface-limited recombination

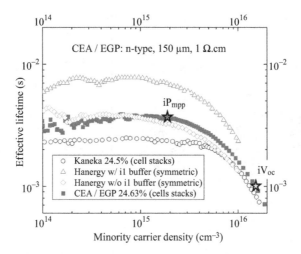

Figure 3.18   Minority carriers' effective lifetime obtained for state-of-the-art HJT precursors (symmetric or typical IN/IP cells stacks post PECVD). iV$_{oc}$ and iFF value for the CEA/EGP device are 747 mV and 86.7%, respectively. The other lifetime data are taken from [260,266]

phenomena are dominant and well evidenced using effective lifetime measurement technique (QSSPC, transient-PCD, μ-PCD, photoluminescence, etc.) for HJT precursors (i.e. non-metallized HJT cells) prepared with high bulk lifetime crystalline silicon [258–261]. Several minority carriers lifetime models have been proposed, considering different strategies to account for the influence of interface defects densities (D_{it}) and corresponding fixed charge density (Q_{it}) at the hetero-interface [262–264]. It has been shown that D_{it} values are affected by the a-Si:H deposition conditions but found to be directly linked to the a-Si:H bulk defects densities after a-Si:H/c-Si interface equilibration [265].

On n-type c-Si-based HJT, the "ip" stack generally limits the effective lifetime compared to pure intrinsic or "in" stacks. Passivation properties can be affected by various post treatments such as thermal annealing, light exposure or by the TCO deposition step (i.e. ion bombardment and/or UV emission from the plasma onto the selective contact thin layers) with the PVD technique but can recover with low-temperature annealing [267–269]. Examples of minority carriers' effective lifetime values versus minority carriers' density obtained for state-of-the-art HJT precursors are presented in Figure 3.18.

Band alignment in the HJT structure and properties has been widely investigated using advanced electrical characterizations [270–272]. A sketch of the HJT band lineup at equilibrium (dark, no bias) is presented in Figure 3.19. It is graphically constructed based on the bandgap and Fermi level position in the individual materials both affected by doping and disorder in amorphous material and interface defects (D_{it}). ITO is considered here as a highly degenerated semiconductor. If well chosen, the thin doped a-Si:H layers properties (doping and

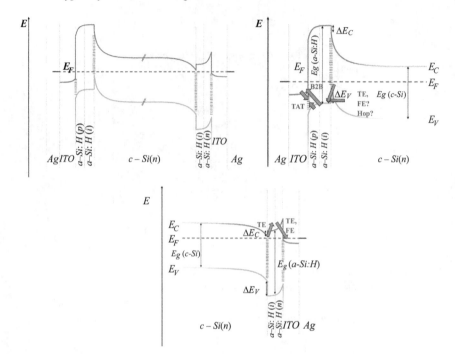

Figure 3.19   Courtesy of Léo Basset. (Top) Band diagram of a typical HJT cell at equilibrium (dark, no bias). Depending on the electronic affinities and bandgaps of each layer, the graphs above may change. Top of the ITO valence band is not represented for clarity purposes. (Bottom left) Transport mechanisms at the hole contact, (bottom right) at the electron contact

thicknesses especially) screens the TCO-induced field effect from the (i) a-Si:H/(n) c-Si/(i) a-Si:H hetero-interfaces and ensures a strong charge separation across the junction together with reduced contact resistance [273,274]. On the figure, one can easily observe the presence of several band offsets at the hetero-interfaces, between TCO and doped a-Si:H, and undoped a-Si:H and c-Si, respectively. A strong inversion layer in the c-Si wafer is also clearly visible close to the surface at the pn junction side. This inversion layer, induced by the field effect coming from the amorphous layers and dependent on interface defect properties, was evidenced directly or indirectly using various electrical characterization techniques in both p-type and n-type wafers [270,272,275–278]. There are various possible transport mechanisms at play in this structure. Band offsets are potential barriers for electrons and holes which can be overcome by thermionic emission (TE) or intra-band tunneling (FE). Band-to-band (B2B) tunneling can occur in regions were valence and conduction bands are close enough in space to generate a strong electron-hole recombination (mainly at the ITO/(p) a-Si:H interface). The presence of possible states (associated to defects and/or disorder) in the materials bandgap or at the interfaces can enhance charges tunneling through trap-assisted tunneling (TAT).

The high selectivity of the electron contact is provided by the built-in potential in c-Si and by both large valence band offsets at the (n) c-Si/(i) a-Si:H and (n) a-Si: H/ITO interfaces estimated at $\Delta E_V \sim 0.36$eV and $\Delta E_V > 2$eV, respectively, together with low potential barriers for electrons. The conduction band offset at the (n) a-Si:H/(n) c-Si interface was estimated at $\Delta E_C \sim 0.15$eV [278] and should allow for TE at room temperature, possibly in combination with tunneling. Strong band bending within the a-Si:H layers may create spikes in the conduction band and enhance tunneling at the (i) a-Si:H/(n) c-Si and at the (n) a-Si:H/ITO interfaces [274]. Some studies found that the electron contact is mainly dominated by TE mechanism [279]. At the hole contact, which is actually a n/p/n junction, two large valence band offsets can dramatically affect the hole collection. TE, intra-band tunneling and hopping through defects are possible transport mechanisms at the (n) c-Si/(i) a-Si:H interface [279,280]. An optimized ITO/(p) a-Si:H interface acts as a recombination junction, dominated by B2B possibly enhanced by interface states [269,281,282]. Finally, contacts between ITO and screen-printed metallization are considered as ohmic.

Charge transport mechanisms described before do not account for lateral transport paths. While exposing the crystalline silicon based cell to illumination, minority charges carriers photogenerated distribute across the structure both vertically and laterally (the PV cell is a 3D object!) and must diffuse to the localized metallization (fingers and busbars/wires) for collection. This lateral diffusion in the HJT structure is possible in the most conductive layers/materials (i.e. c-Si bulk wafer, TCO, and metal). The strong inversion layer within the c-Si at the p/n junction side is much too resistive (~ 10–250 kΩ/square, i.e. ~ 100–$2,500$ times the ITO sheet resistance) to provide an effective path for hole lateral conduction [283,284]. The doped amorphous layers are even less conductive with estimated sheet resistances higher than ~ 50 MΩ/square. The lateral transport below the metallization is then not only provided by the highly conductive TCO layers but also by the illuminated well passivated and high lifetime c-Si wafer with a contribution share depending on the device architecture and contact resistances [285]. The values extracted for contact resistivity (in dark conditions) are $\rho_C(e^-) = 50 - 140$ mΩcm^2 and $\rho_C(h^+) = 200 - 500$ mΩcm^2, respectively for electron and hole contacts [284,286]. Haschke *et al.* proposed a model for the calculation of effective series resistance (R_S) of two sheet resistances (i.e. the TCO and the wafer) coupled via a contact resistance (the a-Si:H layers and related interfaces), successfully applied to (n) c-Si based rear and front emitter HJT configurations [256]. The authors suggest that for n-type silicon HJT solar cells, the observed advantage of the rear junction over the front junction architecture is due to the lower contact resistance at the interface with the electron selective layer and higher mobility of electrons versus holes. The impact of electron and hole contacts on R_S value for n-type rear junction HJT cells and corresponding current densities are presented in Figure 3.20. Figure 3.20 (left) shows that an increase of $\rho_C(e^-)$ leads to higher losses in the ITO as lateral transport in c-Si is progressively

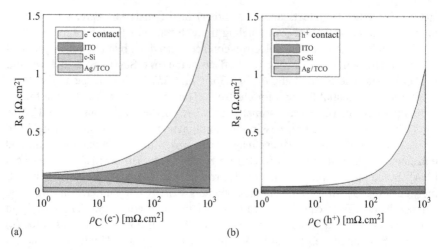

Figure 3.20   Courtesy of Léo Basset. (Left) Contribution to series resistance from the front lateral transport as a function of electron contact resistivity. (Right) Contribution to series resistance from the rear lateral transport as a function of hole contact resistivity. Data obtained from simulation using Haschke's model with the following parameters for the front lateral transport $R_{Sh}(ITO) = 200\ \Omega/sq$, $\rho_{c-Si} = 1\ \Omega$ cm, $t_{c-Si} = 160\ \mu m$, $\Delta p_{MPP} = 10^{15} cm^{-3}$, $width_{finger} = 50\ \mu m$, $pitch_{front} = 1.8$ mm, $pitch_{rear} = 0.6$ mm, and $\rho_C(Ag/ITO) = 1\ m\Omega\ cm^2$

disabled. Excluding the electron contact contribution (yellow area), R_S increases with $\rho_C(e^-)$ over the studied range. For the same ITO and contact properties, we find that lateral losses are lower at the rear side than at the front-side mainly due to the use of a smaller metallization pitch in this bifacial HJT structure example (Figure 3.20, right). The ITO and c-Si coupling is reduced at this side but not completely disabled and reaching very low values for $\rho_C(h^+)$ could lead to similar performance potential for both rear and front junction architectures.

These findings notably allow to relax the constraints on the conduction properties of the front-side TCO and the metal grid for the rear junction HJT structure and enhance photons collection in the absorber thanks to less reflective layers/grid. Leveraging the lateral majority carriers charge transport in the well-passivated c-Si wafer, some groups also introduced new HJT architectures with reduced front-side TCO thickness or with TCO localized only underneath the metal pads [257,287]. For device electrical optimization, one should then consider the varying contributions of bulk layers and interfaces properties under illumination, which can be somehow challenging to characterize [288,289].

The optical properties of HJT devices have been widely discussed in the literature and several authors proposed improvement strategies [229,237–240,290]. Parasitic absorptions in the TCO and selective contact layers are the main sources of optical losses: photogenerated electrons and holes in these layers do not

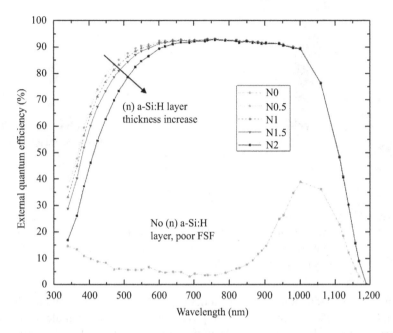

Figure 3.21   *Experimental external quantum efficiency data for HJT cells with increasing front side (n) a-Si:H layer thicknesses (from N0 to N2) showing parasitic absorption in the low wavelength range. The condition N0 without the electron selective layer suffers from poor front-side passivation properties*

contribute to the short-circuit current, except a small portion in the less defective interface layer [290]. In Figure 3.21, we can observe the influence of the (n) a-Si:H front layer thickness on external quantum efficiency (EQE) of rear junction HJT devices, the other fabrication/device parameters being kept constant. In the absence of (n) a-Si:H layer ("N0"), the front-side passivation properties are too low to enable good carrier transport and collection outside a small region close to the junction. For bifacial applications, one should consider both front and back layers' optimizations [291].

The HJT device stability under carrier injection has been studied by several groups [292–298]. Unlike mainstream technologies based on p-type silicon wafer, HJT cell & module efficiencies improve upon carrier injection (via light or current soaking), leading to absolute efficiency gains up to 0.7% for some devices after short intense illumination steps [298], but this gain can considerably vary with the cell design/process. The main contributors to this efficiency increase are systematically FF and V_{oc}. A cell efficiency improvement pareto after light soaking step is presented in Figure 3.22: one can find that a passivation boost and a simultaneous series resistance reduction are governing the efficiency change in the studied HJT device. The origin of these phenomena is not fully understood but could be

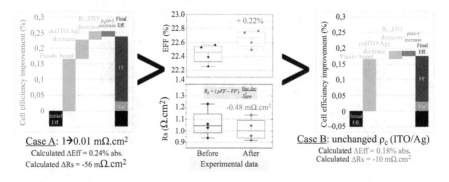

Figure 3.22   Courtesy of J. Veirman. Pareto of cell improvement after light soaking treatment on a HJT device: simulation results considering two extreme cases (A and B) and comparing with experimental data (center) taken from [296]

attributed to the reduction of interfacial defects with the crystalline silicon absorber [292]. Long exposure of non-optimized p-type layers to UV light could however be detrimental to this layer carriers' selectivity [295], and ultimately to the efficiency. Light soaking tools may become a standard step in the HJT manufacturing lines.

3.4.5   Possible new process routes for HJT mass production and next-generation PV

As PV is considered as a serious candidate for the required massive and short-term reduction of greenhouse gases emissions, new PV technologies should thus demonstrate higher efficiency than current technologies and be scalable to mass production. There is no doubt that HJT devices can reach very high efficiencies and are promising candidates to bring silicon-based PV toward its maximum efficiency limit. Regarding mass production, the use of n-type silicon and raw materials such as In and Ag is often considered as growth-limiting factors [299]. Recent studies and announcements show very high potential on high quality p-doped (Gallium doped) silicon wafers, demonstrating the versatility and robustness of the HJT technology, with a record laboratory efficiency at 26.6%, only 0.2% lower than the one of the n-type based device [215,300–302]. N-doped wafers may remain the standard for some time, with a slightly higher cost compared to p-doped wafers, but wafer thickness reduction could be a way to reduce production costs of the HJT cell as it is particularly compatible with thinner wafers [303,304]. Regarding In and Ag usage, it was already demonstrated that the HJT technology is compatible with In-free and Ag-free approaches at very high-efficiency levels but there is a need to assess the viability of these options in terms of fabrication costs and module reliability [217,237–242,250,305]. The strong market growth will also participate to equipment price reductions. Finally, some authors consider that next generation PV structures could combine p-type contact from HJT and n-type contact from TOPCon in a hybrid structure that could reach 28.9% according to advanced simulation and/or be In-free IBC structures [257], while the best efficiency for two terminals perovskite/silicon tandem devices have been obtained with HJT bottom cells [217,306,307].

3.5   Back-contacted crystalline silicon solar cells

3.5.1   Rear-junction structures

The very concept of back-contacting silicon solar cells dates back 1975, with seminal works by Schwartz and Lammert [308,309], originally intended for applications under concentrated sunlight. These pioneering contributions already disclosed the most commonly used back-contacted design nowadays, namely the so-called interdigitated back-contact (IBC) design, where the electron and the hole collectors have the shape of two interdigitated combs (see Figure 3.23). In 1985, the company SunPower was officially incorporated by Swanson and Crane [310], and has been championing the industrial production of IBC devices since then (see Figure 3.24). For almost 25 years, IBC devices were processed using the same technologies as standard double-side-contacted devices, convincingly evidenced by the ZEBRA concept from ISC, Germany, reaching up to 24% conversion efficiency [311]. In 2007, however, the first IBC with silicon heterojunction (SHJ)-based carrier-selective passivated contacts (CSPC) was demonstrated by Lu *et al.* [312]. The same year, several patents by SunPower confirmed the growing interest around the use of CSPC in back-contacted devices, be it with SHJ-based CPSC [313] or

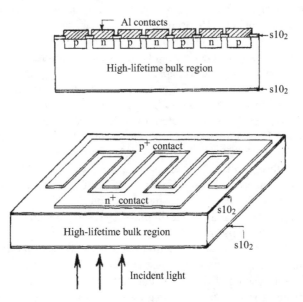

Figure 3.23   The original design of Schwartz's and Lammert's IBC solar cell of 1977. Reprinted with IEEE permission from [309]

Figure 3.24   Front (left) and rear (right) side of a SunPower's Maxeon II IBC
solar cell from 2000 featuring 22.7% efficiency. Pictures ©Museum
of Solar Energy [327]

tunneling oxide/poly-Si-based ones [314]. The year 2014 was the *annus mirabilis* for IBC devices with CSPC. That year indeed, not least than three companies released IBC devices with ≥ 25% independently certified conversion efficiency: in Japan, Panasonic and Sharp demonstrated 25.6% and 25.1% efficient IBC device with SHJ-based CSPC, respectively [315,316], whereas SunPower, USA, released a 25.0% efficient IBC device using an undisclosed CSPC technology [317]. Remarkably, these devices were the first ones to equal or outperform the long-lasting UNSW's PERL solar cell conversion efficiency record dating back 1999 [318]. In 2017, Kaneka, Japan, impressively improved the conversion efficiency mark of IBC devices using SHJ-based CSPC with a 26.7% efficient device on an n-type wafer [319,320]. Equally impressive, the company SunPower claimed in 2017 to mass-produce back-contacted c-Si solar cells (still with an undisclosed CSPC technology) with an *average* conversion efficiency over 25% [321]. In 2018, ISFH, Germany, applied the POLO CSPC technology to an IBC device on a p-type wafer and demonstrated 26.1% conversion efficiency [322]. Finally, in 2021, using the "tunnel-IBC" concept [323] pioneered by EPFL and CSEM, Switzerland, the company Meyer Burger released a 200 cm^2-large IBC device with a certified conversion efficiency of 25.35% [324], as well as a single-cell laminate based on such devices with a certified conversion efficiency of 24.7% [325]. Doubtlessly therefore, back-contacted c-Si solar cells with CSPC are the best candidates to reach the practical efficiency limit of single junction c-Si-based solar cells, even if they are recently challenged by the impressive results obtained with two side-contacted CSPC solar cells [215,326].

The typical cross-sectional and bottom-view schematics of the IBC design are given in Figure 3.25 (here assuming classical diffused contacts). As the front-side of IBC devices is exempt from metal contacts, it can be entirely tuned toward efficient light absorption and highest passivation. In Figure 3.25, the passivation layer and the anti-reflective coating (ARC) are represented as two separate layers, but they can also be combined. The rear side of IBC devices features alternating

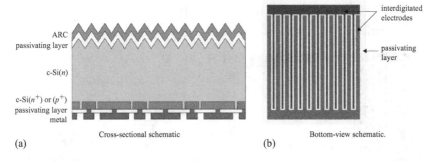

Figure 3.25 Cross-section and bottom-view schematics of a generic IBC solar cell. ©Andrea Tomasi, PhD thesis, EPFL, 2016 [334]

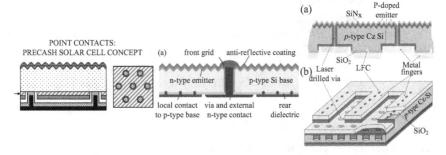

Figure 3.26 Cross-sectional and/or bottom-view schematics of: left: the "PRECASH" concept for point-contacted solar cells. Reprinted with IEEE permission from [328]. Middle: the "HIP-MWT" concept. Reprinted with WIP permission from [335]. Right: the "RISE-EWT" concept. Reprinted with Elsevier permission from [332]

stripes of electron- and hole-collecting regions, which have to be insulated from each other (either physically or electrically) to avoid lateral shunt. The rear side is usually completed with a passivating, IR-transparent layer and the metal contacts.

Several alternatives to the standard IBC design have been proposed. Most of them are very close to the original one, and often intend to balance the ratio between passivated and metallized areas by using a point-contact approach (see e.g. the "PRECASH" concept [328] in Figure 3.26 left). Note that the metal wrap-through (MWT) [329] and the emitter wrap-through (EWT) [330] approaches, as well as all their variations (HIP-MWT [331], RISE-EWT [332], etc.), belong *stricto sensu* to back-contacted devices, but are out of the scope of this review, mainly owing to their moderate efficiency gain compared to current best-in-class two side-contacted devices, lengthy processing, and limited market applications. An insightful review of such devices can be found in [333], to which the interested reader is referred for further details.

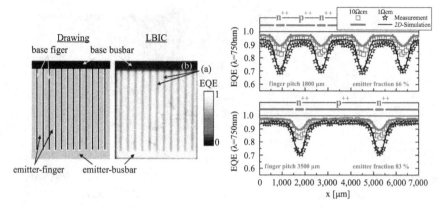

Figure 3.27   Illustration of the electrical shading effect in IBC devices. Left: light beam induced current (LBIC) image of the rear side of an IBC device, evidencing the reduced collection probability below the base fingers (lower LBIC signal). Right: simulated and measured external quantum efficiency (EQE) scans perpendicular to the IBC fingers. The reduced collection probability below the base fingers is also apparent here. Both pictures reprinted with IEEE permission from [336]

3.5.2   Technology of IBC devices

The main peculiarity of the IBC architecture lies in the fact that the rear area of the device is shared between the electron and the hole collectors. This directly constraints the fabrication of such devices, as well as their intrinsic functioning. First, it means that patterning technologies are required to locally define the electron and the hole collecting *regions* (in case e.g. of diffused regions) or *materials* (in case e.g. of thin film based CSPC). Second, IBC devices undergo a 2D carrier flow within the absorber during their operation, and consequently the wafer resistivity, the electron-to-hole collector ratio, the pitch and the specific contact resistivity of these regions need to be properly balanced to reach the best trade-off between J_{sc}, V_{oc}, and FF, and hence obtain the highest conversion efficiency. State-of-the-art methods to address these challenges are detailed in this section.

3.5.2.1   Design of IBC solar cells

Numerous finite-element simulations and analytical modeling have been conducted to identify the optimal geometry for the rear side of IBC devices. A well-known pitfall is to mitigate the lateral transport losses and the electrical shading losses [336,337]. The former are inducing series resistance losses, hence decreasing the IBC device FF, and the latter are leading to a reduced collection probability below the majority carrier collector, hence decreasing the J_{sc}. They both stem from the lateral path the photogenerated carriers have to travel to reach their respective contact, and their magnitude is hence linked to the pitch and the dimensions of the minority and majority carrier collectors. Regardless of the chosen IBC technology

Table 3.4 Comparison of the main design features of SHJ- versus poly-Si-based IBC solar cells. Data taken from [322,338,341,342]

CSPC technology	SHJ	Poly-Si
Wafer type	n-type	p-type
Minority carriers	Holes	Electrons
Majority carriers	Electrons	Holes
Optimum minority carriers collector fraction	60%	90%
Pitch	Up to several mm	Usually 400–900 μm
Minority carriers specific contact resistivity	Down to 40 mΩ·cm^2	Down to 0.1 mΩ·cm^2
Majority carriers specific contact resistivity	Down to 20 mΩ·cm^2	Down to 2 mΩ·cm^2
Gap	Unnecessary	Required (+ trench passivation)

(CSPC or non-CSPC-based), the smaller the pitch and the minority and majority carrier collectors width, the smaller the lateral transport losses and the electrical shadowing losses, and hence the higher the IBC device FF and J_{sc}. More precisely, the width of the majority carrier collector has to be twice smaller than the minority carriers diffusion length to keep the electrical shading to a neglectable level. Consequently, IBC devices based on technologies enabling high passivation quality (e.g. TOPCon, POLO, SHJ), and hence long carrier diffusion length, can tolerate larger lateral dimensions and pitches without suffering from too severe lateral transport losses and electrical shadowing losses.

In the specific case of IBC devices with SHJ-based CPSC, recent works identified the optimal minority carriers (here holes) collector fraction to be of about 60%, and pitches below 1 mm were also found to be a pre-requisite to reach high FF [338,339]. Under these assumptions, the specific contact resistivity must be lower than 40 mΩ·cm^2 (resp. 50 mΩ·cm^2) on the electron (resp. hole) collector to keep the contact resistance losses at a neglectable level [340]. Note that these values markedly differ from the ones relevant to IBC devices using diffused regions or poly-Si-based CSPC. Indeed, for such devices, minority carrier collector fractions over 90% are usually sought for, and pitches are usually much smaller than 1 mm. These differences mainly owe to the longer carrier diffusion length offered by the SHJ technology, allowing the minority carriers to travel over mm-wide distances without being recombined, and therefore relaxing the constraints on the pitch dimension. In contrast, the specific contact resistivities of diffused and poly-Si-based IBC devices are usually ten to hundred times lower than those of SHJ-based devices (Table 3.4).

3.5.2.2 Patterning technologies for IBC devices with CSPC

The proper functioning of IBC devices requires the electron and the hole electrodes to be patterned with the required design (most frequently, two interdigitated combs) and to be electrically insulated from each other (using a gap or a trench). As described above, the relevant CSPC technologies for c-Si-based IBC devices consist in stacks of different materials. For instance, for SHJ-based IBC, starting from the c-Si surface, these materials are: a thin silicon film (a-Si:H or similar), a TCO layer (e.g. ITO), and a metal layer (most frequently Ag, sometimes Cu). Practically, this means that these three different materials must be patterned to obtain the required rear electrode design. These layers can be patterned altogether or separately. Interestingly, it is not necessary to physically separate all these materials to ensure the electrical insulation of the electron and the hole electrodes: indeed, several materials, such as a-Si:H, feature a very low lateral conductivity, and therefore the electron and the hole collectors can overlap at this stage without detrimental effects to the device performance. Leveraging on this peculiarity opens interesting opportunities to simplify IBC devices fabrication [323].

Regardless of their nature, the layers to be patterned – hereafter referred to as the target layers – can be patterned using either subtractive or additive approaches. In subtractive patterning techniques, the target layer is full area deposited, and subsequently locally removed. This can be done either by etching it from the top, or by lift-offing it. In contrast, in additive patterning techniques, the target layer is directly deposited solely at the required places. Note that subtractive and additive techniques are often combined to perform the complete patterning of full CSPC stacks.

3.5.2.2.1   Subtractive patterning techniques

Photolithography combined with wet chemical or dry etching has been for decades the workhorse of thin film patterning. In this approach, the target layer is first deposited over the full area of interest and then covered with a light-sensitive resist. Selected regions of the resist are then illuminated through a photomask using a dedicated optical setup. The illuminated (resp. shadowed) regions are then etched away in the case of a positive (resp. negative) resist, leaving the target layer exposed through these apertures. The target layer is subsequently selectively etched away at these regions using dedicated wet chemicals or dry etching approaches. The patterned resist can also be used as a blocking mask during the ion implantation steps of poly-Si-based CSPC. Eventually, the remaining resist is stripped off. Alternatively, the photoresist can be patterned first, and then covered by the target layer to perform the lift-off of this latter. The use of photolithography has been demonstrated for the patterning of the thin silicon film, the TCO, and the metal layer of SHJ-based CSPC, separately or altogether, as well as for the patterning of the n- and p-type ion-implanted regions and metal grid for IBC devices based on the POLO CSPC technology. The major asset of photolithography patterning is doubtlessly its accuracy: features with a size of 1 μm or less can be routinely patterned. In contrast, the lengthy and numerous process steps required by this technique are probably its main weakness regarding its potential upscaling. Recent examples of IBC-SHJ devices fabricated using photolithography are [343,344], and POLO-like IBC devices are [322,345]. An illustration of the working principle of photolithographic patterning is proposed in Figure 3.28 (left).

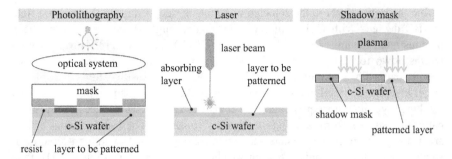

Figure 3.28   Illustration of different patterning techniques of relevance to IBC solar cells

Laser-based patterning is also a subtractive lithography technique and is in substance very close to photolithography itself. Here again, the target layer is first full area deposited and capped with a second layer. However, in contrast to photolithography, this second layer is not necessarily a light-sensitive resist, but rather a material featuring suitable light absorption at the wavelength of the chosen laser beam. This absorbing layer can be made of a single layer of e.g. SiN_x [346] or in the form of a multilayer stack forming a distributed Bragg reflector [347]. In any case, the absorbing layer is then locally opened using a laser beam. The following steps are alike those of classical photolithography, namely wet or dry etching of the target layer through the opened areas (or alternatively lift-off) and stripping of the remaining absorbing layer. Importantly, this method has not been demonstrated yet for the patterning of the CSPC metal layers and is most frequently used to pattern the thin silicon film parts of SHJ-based CSPC or the doped poly-Si regions in POLO-like CSPC. Though laser ablation of TCO is well-known from thin-film silicon solar cells [348], its applications to TCO patterning in SHJ-IBC solar cells is not widespread. Laser-based patterning techniques are intended to feature the same assets as photolithography without its weaknesses. Features with 10-µm width have been demonstrated, and the throughput is compatible with mass-production. However, the major challenge of laser-based techniques for the patterning of CPSC is to fully open the absorbing layer without damaging the sensitive passivation at the c-Si/CSPC interface [346]. Recent examples of IBC-SHJ devices fabricated using laser-based patterning are [347,349] and POLO-like IBC devices are [350]. An illustration of the working principle of laser-based patterning is proposed in Figure 3.28(middle).

Alternatively, metallic hard masks can be used in combination with dry etching to locally remove parts of the target layers. This approach has been demonstrated to pattern the n-type a-Si:H layers of SHJ-based IBC [351].

3.5.2.2.2   Additive patterning techniques

The direct additive patterning of target layers can be realized *in situ* using shadow masks placed on top of the c-Si wafer surface. Apertures in the shadow mask define the regions where the selected materials are deposited. In the specific field of IBC devices with CSPC, direct patterning through shadow masks has been demonstrated for thin silicon films [352,353], TCO and metals [354], with feature widths down to 300 μm. Markedly, shadow mask patterning is doubtlessly the leanest patterning technique available for IBC devices with CSPC: indeed, the target layer is patterned in a single step, and neither the deposition of an additional masking layer nor subsequent cleaning steps are required. However, shadow mask patterning suffers from tapering and under-deposition. These detrimental effects are usually more critical when shadow masking is used with non-directional vacuum deposition techniques, such as chemical vapor deposition (CVD), than with directional deposition methods, for instance physical vapor deposition (PVD) or thermal evaporation. Tapering effects stem from the adsorption of gaseous radicals on the shadow mask sidewalls, and usually result in patterned features being thinner at their edges – e.g. close to the shadow mask sidewalls – than at their center [355]. This effect has been demonstrated to be more and more stringent for narrower mask apertures [356,357]. Under-deposition results in the patterned features to be larger than expected, owing to the tailing of the deposited materials below the shadow mask. Fortunately, under-deposition can be almost completely avoided by a careful engineering of the shadow masks [356]. An illustration of the working principle of shadow-mask patterning is proposed in Figure 3.28 (right).

Interestingly, other additive patterning techniques commonly used for the processing of PV devices, such as screen-printing, have already been demonstrated for non-CSPC IBC devices [358], but are still to be proven for CSPC-based IBC devices.

3.5.3   *Processing of IBC solar cells*

3.5.3.1   High-temperature IBC solar cells

The variety of high-temperature back contact solar cells is tremendous – reaching from MWA (metallization wrap around), MWT (metallization wrapped through), EWT (emitter wrap through) to the so-called IBC solar cells – today also called sometimes TBC (TOPCon Back Contact or Tunneling Back Contact) solar cells, as depicted in cross sections in Figure 3.29.

Such a variety of concepts have been developed in the past by different companies with the aim of bringing back-contact solar cells to the PV market. Sunpower, for example, has been producing high-priced IBC solar cells for several decades. However, a number of companies – such as BP Solar and Solland – wanted to establish lower-cost backcontact technologies on the market in early 2000, such as MWA, MWT, and EWT. The only concept besides IBC that has been implemented successfully and launched on the PV market, and which is still in PV production, is MWT from, for example, Sunrise, Risen Energy, and Sunport Power. However, similar as in the case of PERC, the average efficiency is limited to values of below 23%, when passivating contacts are not used and that is why we will concentrate on high-temperature IBC technology in the following. High temperature

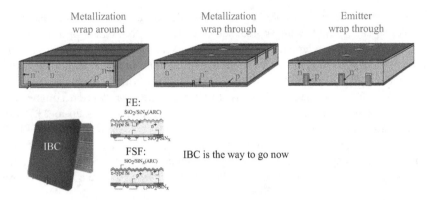

Figure 3.29   Different geometries of various back-contact solar cell concepts [359]

is not a disadvantage only as compared to low-temperature IBC, even though higher thermal budget is used. High temperatures during diffusions are gettering (purifying) the Si material making the use of lower cost Si-materials possible as well as allowing lower purity during the process itself. In addition, the module process can be run standardly at temperatures where the cell layers are not harmed. That is why the processing of standard high-temperature IBC process can be done with standard equipment.

3.5.3.1.1   Comparison of high-temperature IBC solar cells

IBC solar cell architectures are characterized by an interdigitated p/n pattern on the rear side. Accordingly, no frontside metallization is required, allowing for highest energy conversion efficiencies. In addition, the devices are ideal for the implementation of polysilicon-based passivating contacts for both polarities, as the layers with their high parasitic absorption are applied to the rear side of the device. However, the interdigitated pattern of such cells complicates the process sequence substantially, as both polarities are moved to the same surface of the silicon wafer. This task is most "easily" solved by using cost-intensive patterning technology known from semiconductor manufacturing like photolithography, ion-implantation, plating, PVD processes, etc. Accordingly, although IBC solar cell technology has been on the market for many years, with major contenders being Sunpower [360] and LG electronics [361], it has been serving exclusively high price niche markets such as residential rooftops or device integrated applications.

In order to reduce the production costs and environmental impact, ISC Konstanz has developed the ZEBRA solar cell technology [362] without the need of any passivating contact so far reaching 700 mV. The key feature of this IBC solar cell is that it can be processed on any wafer size using standard solar cell production equipment, such as tube diffusion furnaces and screen-printing metallization. The screen-printing pattern on the rear side consists of several layers (3D approach) with busbars which are printed on fingers that are partly covered by insulation paste and which reach the whole length of the cell. This approach makes module interconnection simple, e.g. string-soldering is possible, but also very flexible, so-called multi-wire approaches or conductive backsheet optimized patterns can be easily implemented. Similar to standard technology, the cells can be cut into stripes to reduce the series resistance of the solar module. The ZEBRA cell was successfully transferred to SPIC (Xining, China) with a production start in 2020 [363] and to ValoeCell (Vilnius, Lithuania) where production commenced in 2021. The solar cell efficiency achieved in mass production is up to 23.7% in average. Modules can be purchased from SPIC [364] and from Futura Sun [365] in monofacial and since 2022 in a bifacial configuration. The bifaciality factor (rear-side efficiency/front-side efficiency) of standard ZEBRA prototype modules in glass-glass or glass-transparent backsheet configuration is known to be 0.75 (on module level).

An alternative IBC cell design has been developed by ISFH in the past years. The POLO IBC solar cell (also known as PERPoly IBC) design employs carrier selective polysilicon on oxide (POLO) contacts. The POLO IBC cell design is based on today's industry typical PERC+ solar cells as it also uses Ga-doped Cz wafers, an AlO_x/SiN rear passivation and Al finger base contacts. However, it replaces the efficiency limiting phosphorus emitter and Ag contacts by SiO_x/n-poly-Si (POLO) [366] contacts which are moved to the rear side of the device. Device simulations demonstrate an efficiency potential of POLO IBC of 25.5%. ISFH has been able to achieve solar cell efficiencies up to 23.7% on an area of 4 cm^2 by applying LPCVD poly-Si deposition and a rather complex poly-Si structuring process using laser ablation. ISFH has developed a PECVD SiON/n-a-Si deposition process [367] using a lab-type deposition tool resulting in an excellent passivation quality after firing with iV_{oc} up to 744 mV. LPKF developed an industrially manufacturable glass-based shadow mask which enables local deposition of the SiO_x/n-a-Si layer stack onto the silicon wafer without the need of further structuring the poly-Si layer. This local PECVD SiO_x/n-a-Si deposition through shadow mask enables a very lean POLO IBC manufacturing process flow with very high cell iV_{oc} up to 738 mV [368]. Very recently, the first POLO IBC with PECVD shadow masks including screen-printed metal contacts have been processed at ISFH using mostly industry-typical tools demonstrating conversion efficiencies up to 21.5% [369]. The solar cell analysis shows a clear path to obtain much higher efficiencies in the future.

The third possibility is to apply carrier selective contacts (passivating contacts) on each polarity. This then leads to the highest voltages above 730 mV. There are at the moment few producers that are developing this technology as a consortium is doing in an EU Project HighLight. The question there is still how to realize and contact the poly-Si B-doped layer, as well as how to easily pattern the p+/n+ regions. For the first challenge, APCVD, PECVD, and PVD depositions are tested to create in-site and one-sided doped B-rich poly-Si layers. The patterning is realized either via laser scribing and laser doping or use of a shadow mask.

Figure 3.30 depicts all the discussed concepts starting from 0 passivating contacts in the ZEBRA cell on the left, POLO IBC in the middle and TBC cell on the right. The blue and red areas are representing the doped layers and the yellow

Figure 3.30 High-efficiency processes for different types of IBC solar cells on p- and n-type materials. The structures include no passivated contact (PC), one PC or two PCs for both polarities

Table 3.5 High-efficiency processes for different types of IBC solar cells on p- and n-type material. The structures include no passivated contact (PC), one PC or two PCs for both polarities

	ZEBRA/Jolywood IBC (without PC)		POLO IBC (one PC)		TBC (both PC)	
Company/ Institute	SPIC/ISC	Jolywood	ISFH/ISC	Sunpower	LG electro- nics	TRINA
Brand-named	ZEBRA/ Andromeda2.0	–	POLO IBC	Maxeon 6	Neon R	–
Type	n-type	n-type	p-type	n-type	n-type	p-type
In production	Yes	Yes	No	Yes	Yes	No
Efficiency	24% [370]	23%	21.5% [369]	25%	25%	25.04% [372]
PERC related	80%	70%	90%	0%	0%	20%
Module efficiency	22.3% [371]					
Can be bifacial	Unknown	Unknown	22.8% [371]	22.3%, not listed any more [371]	Unknown	
COO	<2 × PERC	<2 × PERC	<2 × PERC	>4 × PERC	>4 × PERC	Unknown

areas represent the thin 1–1.5 nm SiO_x layer which passivates the surface, however allowing the necessary charge carriers transport via tunneling through this layer. The complexity of the processes is increased from left to right, however the voltage is improving, respectively. When a simple and low-cost cell fabrication process will be developed, it will make sense to switch to the TBC concept in the future. Table 3.5 summarizes the companies that are involved in these developments, the properties, the reached efficiencies, and the costs compared to PERC of the solar cells.

On the very left, you see the two solar cell concepts without any passivating contact from SPIC/ISC and Jolywood. The ZEBRA IBC cell is reaching 24% in production and is based to about 80% on standard PERC production using the same equipment. The details will be described in the next paragraph. There is not much known about the IBC process from Jolywood – only that it is also not based on passivated contact technology and that the P-diffusion is done by implantation. The costs are in both cases below 2 × PERC costs. The POLO IBC technology by ISFH is using one passivating contact and still Al-screen printing also for the emitter formation. The efficiency for large area devices is at 21.8% and the projected costs in production also close to PERC. The IBC technologies with both passivating contacts have the highest efficiency but also the highest costs – namely four times the costs for PERC which makes them only attractive for high-power density markets such as roof-tops.

3.5.3.1.2   *Diffused solar cell without passivating contacts: ZEBRA technology*

The only IBC solar cell without passivating contacts on the market is the ZEBRA technology. It is leaned on the PERC process including only few additional steps and not needing AlO_x passivation as it is based on phosphorus silicon glass (PSG) and boron silicon glass (BSG) passivation technology (Figure 3.31).

ZEBRA is composed of $POCl_3$- and BBr_3-tube furnace diffusions, BSG/PSG passivation, PECVD SiN_x ARC depositions and screen printing in five steps including printing of an isolation paste.

Figure 3.32 shows a picture of the metallization pattern of the rear side as well as a schematical drawing of the busbar implementation contacting every second finger of the same polarity. The screen printing sequence consists of five subsequent printing steps: two printing steps for finger formation, two for isolation paste, and one for the busbar creation. In future, it will be reduced to four steps only. Figure 3.33 summarizes the production data in one week of April 2022.

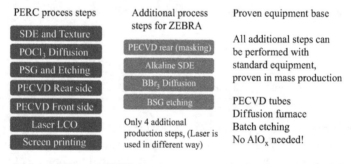

Figure 3.31   Additional ZEBRA process steps compared to PERC technology

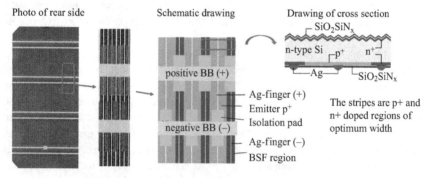

Figure 3.32   Rear-side metallization geometry and cross section of ZEBRA

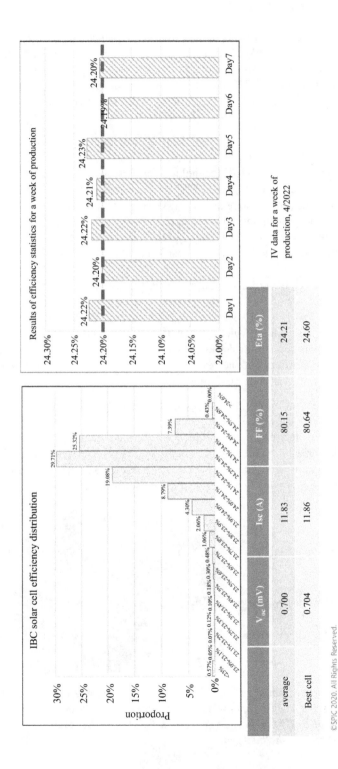

Figure 3.33 ZEBRA production data from SPIC's production in Xining from 1 week in April 2022

The average efficiency is 24.2% with a maximum efficiency of 24.6% with a respective voltage of 704 mV. This is, to our knowledge, the highest voltage reached for a standardly diffused c-Si solar cell without the use of passivating contacts. In order to overcome 710 mV TBC, cell technology is being developed by ISC Konstanz, SPIC, and an EU consortium in EU's Horizon 2020 HighLite project.

3.5.3.1.3 IBC solar cells with passivating contacts: PERPoly IBC/POLO IBC technology

The industrial version of POLO IBC technology was first proposed and patented by ECN (now TNO), called PERPoly IBC [373], as an upgrade for PERC technology. It consists of the same process steps as for PERC expect that before the $POCl_3$ tube diffusion, thin oxide growth and poly-Si deposition were done or the $POCl_3$ tube diffusion being replaced by in situ P-doped poly-Si deposition. Here again, as for TOPCon cell process, one-sided poly-Si deposition by APVCD, PECVD, or PVD is preferred.

Figure 3.34 shows the cross-section of the POLO IBC cell on the left and the glass mask and rear side of the IBC cell depicted on the right. The advantage of this cell technology is that it is very close to PERC, p-type substrates can be used, as well as the Al screen printing which creates also the Al-emitter. The potential of this cell technology is between 24% and 25% with V_{oc} limited to 720 mV by the not passivated emitter contact. The newest results including the process COO are published by Dullweber at WCPEC-8 in Milano [369].

3.5.3.1.4 TBC solar cells: poly-Si IBC technology as the holy grail of c-Si PV

Sunpower and LG electronics are producing IBC technology with passivating contacts already for a long time, having many patents on their technology. Even though the devices are very similar as depicted in Figure 3.35, the processes are differing in terms of complexity as well as process sequences.

Figure 3.34   *Cross-section of industrial POLO IBC and photograph of a high precision glass shadow mask (orange) manufactured by LPKF for local PECVD n-poly-Si deposition for POLO IBC solar cells (shown in blue in the background)*

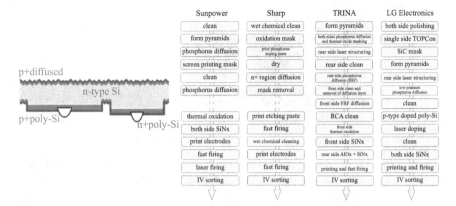

Sunpower	Sharp	TRINA	LG Electronics
clean	wet chemical clean	form pyramids	both side polishing
form pyramids	oxidation mask	both sides phosphorus diffusion and thermal oxide masking	single side TOPCon
phosphorus diffusion	print phosphorus doping paste	rear side laser structuring	SiC mask
screen printing mask	dry	rear side clean	form pyramids
clean	n+ region diffusion	rear side phosphorus diffusion (BSF)	rear side laser structuring
phosphorus diffusion	mask removal	front side clean and removal of diffusion layer	low pressure phosphorus diffusion
		front side FSF diffusion	clean
thermal oxidation	print etching paste	RCA clean	p-type doped poly-Si
both side SiNx	fast firing	front side thermal oxidation	laser doping
print electrodes	wet chemical cleaning	front side SiNx	clean
fast firing	print electrodes	rear side AlOx + SiNx	both side SiNx
laser firing	fast firing	printing and fast firing	printing and firing
IV sorting	IV sorting	IV sorting	IV sorting

Figure 3.35   Process sequences of passivated contacts IBC solar cells from Sunpower, Sharp, Trina, and LG

Figure 3.35 shows process sequences for Sunpower, LG electronics, and TRINA. All three processes are very complex and the COO are far away from standard PERC. That is why new developments try to combine standard IBC technology with TOPCon technology calling it TBC. There are many institutes and companies that are developing this technology with a high focus using shadow mask and laser processes for the patterning of poly-Si depositions by LPVCD, PECVD, APCVD, and PVD doped layers. An additional development of huge interest is the replacement of Ag-metallization by Cu and/or Al.

3.5.3.1.5   Summary of high-temperature IBC technology

All want it – few have it. The new n-type technologies are being implemented so fast, that few large Tier 1 manufacturers, such as AIKO and LONGi, already want to go the next step after TOPCon – to implement IBC technology on the PV market. LONGi is going the POLO IBC way, whereas AIKO is moving toward TBC, calling it all back contact (ABC), and are recently claiming an IBC module efficiency of 23.5% [374] – which is however not confirmed by a third party yet.

The only two IBC players currently being visible in the PV market are Sunpower with a module efficiency of 22.8% and SPIC with 22.3%, respectively. LG electronics stepped out in June 2022 [375] (a final module efficiency of 22.3%).

3.5.3.2   SHJ-IBC solar cells

The fabrication of SHJ-IBC solar cells is in essence similar to the one of two side-contacted HJT devices, with the addition of new materials (e.g. the SiN_x ARC at the front) and additional patterning steps. It starts with the usual cleaning and texturing wet chemical steps. Wafers are then briefly dipped in a diluted HF bath to remove the native oxide. They are then promptly loaded into a PECVD reactor where the a-Si:H and other dielectric layers are deposited. At the front-side, the passivation is ensured either with a single intrinsic a-Si:H layer or with a bilayer composed of

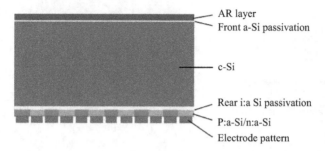

Figure 3.36   Illustration of the cross-section of the record-holder IBC-SHJ solar cell from Kaneka, Japan. Reprinted with Macmillan Publishers Limited permission from [319]

Table 3.6   Characteristics of the 26.3%-efficient IBC-SHJ device from Kaneka. Data gathered from [319]

V_{oc} (mV)	J_{sc} (mA/cm^2)	FF (%)	CE (%)	R_s ($\Omega \cdot$cm^2)	Area (cm^2)	Thickness (μm)	Doping ($\Omega \cdot$cm^2)
744	42.3	83.8	26.3	0.32	180 (da)	165	3

intrinsic a-Si:H capped with n-type a-Si:H or similar, all deposited full area. The anti-reflective properties are usually performed using a SiN$_x$ layer or similar. At the rear side, a-Si:H, TCO, and metal layers are deposited and patterned using a combination of the methods presented in Section 3.5.2.2.

The record-holder IBC-SHJ solar cell from Kaneka, Japan, with conversion efficiency reaching up to 26.7%, is arguably the most accomplished realization among IBC-SHJ devices. Its cross-sectional schematics is illustrated in Figure 3.36. Its actual process flow has still not been officially disclosed to date; it is often assumed to be based on photolithography patterning. As already stated in Table 3.4, note that no insulation gap is required between the n- and the p-type a-Si:H fingers on the rear-side, owing to their poor lateral conductivity.

The IV characteristics of such IBC-SHJ solar cell are given in Table 3.6, here in the case of the device with "only" 26.3% conversion efficiency. Impressively, the V_{oc}, the J_{sc}, and the FF were shown to fall short by only 1.1%, 3.4%, and 5.3%, respectively, from their theoretical limit [319]. This illustrates the almost perfect passivation, light management, and carrier transport and collection achieved in this device. Would the few remaining losses be further reduced, Kaneka hypothesized that a conversion efficiency of 27.1% could be practically achieved [319,320], a mark they approached even closer with their 26.7%-efficient device demonstrated later on [320].

With the conversion efficiency potential of IBC-SHJ devices been unambiguously demonstrated by the above-discussed realizations of Kaneka, a successful route for the cost-effective mass-production of IBC-SHJ was still elusive.

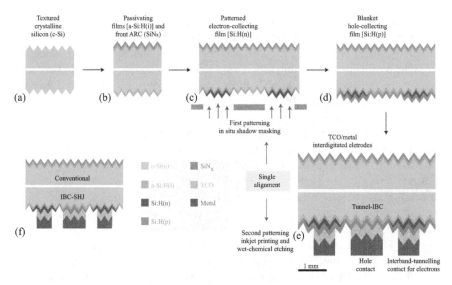

Figure 3.37   Cross-sectional schematics and process flow of tunnel-IBC solar cells (a to e), and comparison with a conventional IBC-SHJ devices. Reprinted with Macmillan Publishers Limited permission from [323]

Unsurprisingly, and as for any IBC architecture, the associated complexity in patterning the hole- and electron-collecting areas at the back side of such devices induces important additional processing costs. This Gordian Knot was cut by the "tunnel-IBC" concept [323, 376] pioneered by EPFL and CSEM, Switzerland, whose functioning and processing are now discussed.

Figure 3.37 compares the architecture of a conventional IBC-SHJ device (Figure 3.37(f)) and the tunnel-IBC concept (Figure 3.37(e)). In a conventional IBC-SHJ device, such as the one of Kaneka depicted in Figure 3.36, both the electron and the hole collectors are patterned and must be aligned one with respect to the other. In contrast, in the tunnel-IBC device, only the electron collector is patterned, whereas the hole collector covers the entire rear surface, including the electron-collecting fingers, hence forming a tunnel junction at these locations.

Compared to conventional IBC-SHJ devices, the tunnel-IBC relies on a drastically simplified process flow, illustrated in Figure 3.37(a–e). Indeed, it simultaneously eliminates any step that would otherwise be required to pattern the hole-collecting regions, as well as any alignment between the hole- and the electron-collecting regions. Combined with in-situ shadow masking for the patterning of the n-type a-Si:H fingers (see Section 3.5.2.2.1), the total number of process steps is reduced to 10 only [339], which, as discussed in more details in Section 3.5.3.3, is arguably the leanest process flow to date for IBC devices. This is a major leap toward the successful spread of the IBC-SHJ technology at industrial level.

However, to work efficiently, the tunnel-IBC device must overcome some challenges. First, the tunnel junction located at the electron-collecting regions must not impede the carrier transport to the back electrode. To do so, highly doped materials are required in order to narrow the tunnel junction depletion width (requirement (i)). Second, as the hole collector covers the entire rear surface, the hole- and the electron-collecting regions are electrically connected. To prevent short circuits, the hole collector materials must hence feature a low lateral conductance (requirement (ii)). Importantly, this second requirement is in apparent contrast with the first one, where a high doping – and hence a high conductivity – is required.

To fulfill these two competing requirements, a key enabler was to exploit the specificities of the anisotropic growth mechanism of hydrogenated nano-crystalline silicon (*nc*-Si:H). Indeed, the crystallinity of *nc*-Si:H layers strongly depends on the substrate on which they are deposited [377]. If grown on an amorphous substrate, an amorphous nucleation layer will form before the crystalline growth sets in. In contrast, if the substrate has already some crystallinity, the *nc*-Si:H layer will present large crystallites spanning over its entire thickness.

The TEM pictures presented in Figure 3.38 illustrate this spatial differentiation. At the tunnel junction location (see Figure 3.38(a–c)), large crystallites are seen at the *nc*-Si:H(n)/*nc*-Si:H(p) interface as well as over the entire thickness of the *nc*-Si:H(p) layer. This is due to the fact that this latter is grown on top of the patterned *nc*-Si:H(n) fingers, which are themselves crystalline in their upper part and hence act as a nucleation layer. Consequently, the materials forming the tunnel junction can be highly doped, which is a prerequisite for an efficient tunnel junction. In contrast, between the electron-collecting fingers, the *nc*-Si:H(p) layer is directly grown on the fully amorphous *a*-Si:H(i) buffer layer (see Figure 3.38(d–f)). At these locations, the *nc*-Si:H(p) layer features a 5-nm-thick amorphous portion in its lower part originating from the protocrystalline growth regime. Owing to this amorphous region, the *nc*-Si:H(p) layer features a low lateral conductance, hence preventing any shunt to the electron-collecting fingers. Remarkably, with a single layer, it is thus possible to fulfill the requirements (i) and (ii) of the tunnel-IBC by exploiting the influence of the underlying layer on the growth mechanism of the *nc*-Si:H(p) layer. The tunnel-IBC concept was applied to the processing of IBC-SHJ solar cells with conversion efficiencies up to 25.4% on 200-cm^2 sizes [324]. Note that a variation of the tunnel-IBC concept, based on the laser crystallization of doped a-Si:H stacks, was also proposed [349].

3.5.3.3   Comparison of IBC technologies

As an overview, Figure 3.39 plots the number of process steps and the highest conversion efficiency reported to date for a variety of IBC technologies. Interestingly, the IBC concepts featuring the highest conversion efficiencies to date, such as the HJ-IBC (Kaneka) and the POLO^2-IBC (ISFH), are plagued by a high number of process steps, up to 32 in the case of the POLO^2-IBC. Conversely, IBC concepts with a lower number of process steps, such as the ZEBRA (ISC) or the simplified POLO-IBC (ISFH), did not yet reach the 25% conversion efficiency landmark. The only exception to this trend is the tunnel-IBC concept (been acknowledged that the complete process flow for SunPower devices is undisclosed), which demonstrated up to 25.4% conversion efficiency in only ten process steps.

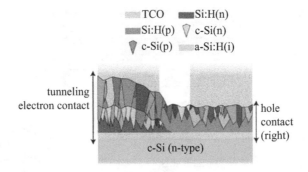

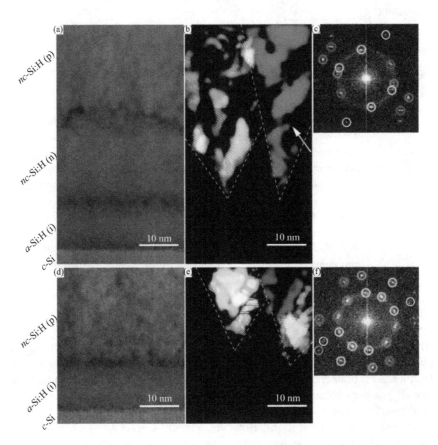

Figure 3.38 *Top: Schematic closeup view of the tunneling electron contact and*
the hole contact, evidencing the crystallites inside the nc-Si:H layers.
Bottom: TEM micrographs of the nc-Si:H layers microstructure: at
the tunnel junction location (a) and in the middle of the hole collector
(d), with the corresponding colored inverse Fourier transform of
selected reflections (b and e) obtained from the Fourier transform (c
and f). Pictures adapted from [323]

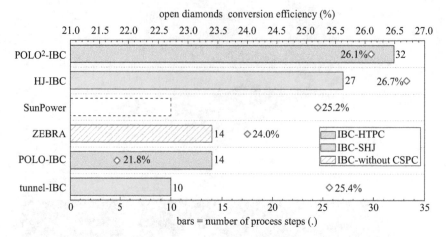

Figure 3.39   *Comparison of the number of process steps and the highest reported conversion efficiency for a variety of IBC technologies. The names of the different technologies are taken as reported in the references. The references used to build this figure are: POLO²-IBC: process flow and conversion efficiency from Ref. [322]; HJ-IBC: author's own estimation for the process flow, Ref. [320] for the conversion efficiency; SunPower: as no official data are available for the process flow of SunPower's devices, the author assumed the smallest number of process steps published for IBC devices, namely the ten steps of the tunnel-IBC technology; Ref. [378] was used for the conversion efficiency; ZEBRA: Refs. [379–381] for the process flow, Ref. [311] for the conversion efficiency; POLO-IBC: process flow and conversion efficiency from Ref. [350]; tunnel-IBC: Ref. [339] for the process flow, Ref. [382] for the conversion efficiency*

3.5.4   Conclusion & outlook

IBC devices are often seen as the "last frontier" of single-junction c-Si solar cells owing to their unrivalled conversion efficiency potential and pleasing aesthetics. Despite the commercial success of SunPower's and ZEBRA's products, the overall production volume of IBC modules is still limited, with ~1.5 GW for SunPower's Maxeon and ~0.5 GW for ZEBRA, restricting them so far to "premium" market applications. However, a recent market analysis by the Dutch company DSM is forecasting the market for back-contacted devices to grow above the average, with a compound annual growth rate of 15% between 2020 and 2030 [383]. As an outlook, IBC devices are also an interesting steppingstone for perovskite-on-silicon tandems in three-terminal wiring. Indeed, it was recently demonstrated that in contrast to two-terminal tandems, three-terminals do not require to be current-matched and can therefore provide up to 10%$_{rel}$ higher energy yield than standard two-terminal tandems [384]. Convincing proof-of-concepts of such perovskite-on-IBC three-terminal tandems were already demonstrated [385]. The already long-lasting story of back-contacted devices is therefore poised to continue further.

3.6 Silver usage reduction in n-type solar cells

In a solar cell, the cost of the silver metallization is the highest contributor after the wafer. The evolution toward n-type bifacial cells increases the importance to reduce the amount of silver deposited on the device.

3.6.1 Fine line screen-printing

Cells' performances are impacted by the metallization through various contributions including shadowing (Isc impact) and electrical losses (too large finger and contact resistances affect the FF). A compromise must be found between the finger width (and number) and the finger resistance.

Furthermore, the optimum strongly depends on the interconnection technology: with more interconnections, the finger length between two interconnections is smaller, reducing the finger resistance impact. This is the main reason behind the evolution of M2 cells from 3 to 5 busbars (ribbons attachment), then 9 MBB (wires connected to pads linked by a vertical line) or no Busbar at all (e.g. BB-less cells dedicated to 18 wires connected by the SmartWire Connection Technology SWCT® [386]).

The finger resistance is linked to the bulk resistivity and to the finger section. The finger aspect ratio is dependent on the printing technology and the paste rheology. Screen-printing remains the most used technology for the metallization of PV cells, due to its robustness, low cost and compatibility with high production throughputs.

Narrow fingers with no or few interruptions have to be deposited. Before 2010, it was difficult to print continuous fingers with a width lower than 100 μm: finger widths of 100 μm reported in ITRPV 3rd Edition (2011 results) with predictions of 40 μm for 2017 and 38 μm for 2020 [387]. Very impressive progresses were made by the paste suppliers and the screen makers. Nowadays, fingers widths below 40 μm are usual for the front metallization on homojunctions cells: finger width of ~36 μm reported for 2020 in the last ITRPV with a prediction of 20 μm for 2031 [388]. The paste printability was improved by the use of small spherical particles combined with excellent paste rheology. Paste transfer progresses are mainly due to the evolution of the screen mesh. In 2010, the wire diameter was above 20 μm, 5 years ago, it decreased to 16 μm and nowadays the more aggressive meshes are based on 11 μm diameter wires. Pastes and meshes progresses allow to transfer the paste at high speed (e.g. squeegee speed of 500 mm/sec) through opening widths below 25 μm. With such narrow openings, the double print (DP) process (fingers printed two times) is often used in order to avoid the risk of finger interruptions. Another investigated option concerned the use of stencils in order to avoid the wire crossing, which limits the paste transfer. However, this technology was recently

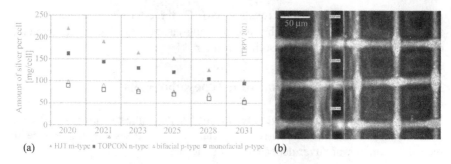

(a)   ▲ HJT m-type ■ TOPCON n-type ▲ bifacial p-type □ monofacial p-type   (b)

Figure 3.40   *(a) Amount of silver deposited per cell (front & rear sides), for various cell technologies. Reprinted from [388]. (b) Picture of a knotless screen with an opening of 24 μm*

replaced by knotless screens (Figure 3.40) having the mesh oriented at 0° (wires parallel to the fingers). Good transfers through opening widths of only 20 μm can be achieved with such screens, which are more and more used in PV [389].

The "high-temperature" pastes are now very conductive with bulk resistivity values around 2 μΩ cm, rather close to the silver bulk resistivity. However, with such narrow lines, the cell performances can be limited by the metal/silicon contact resistances. Therefore, to limit this issue, in parallel of the optimization of the doping profiles, significant progresses were also conducted by the paste suppliers on the frit (glass to be mixed to the silver) added in the high temperature pastes.

The problematic is slightly different for the cell structures using "low-temperature" pastes, for example heterojunction cells. These pastes do not contain glass, just silver and resin. Only a curing around 200°C is possible. Thus, the paste is not densified by sintering and the adhesion is provided by the resin instead of the glass. The bulk resistivity is almost two times higher (around 4–5 μΩ cm for the best candidates) than the values obtained with "high-temperature" pastes. Silver flakes must generally be used to minimize the bulk resistivity and the paste rheology is adapted to get higher finger aspect ratio. Consequently, the paste transfer is more difficult, realized at lower speeds (e.g. 250 mm/s) through screens having wider openings (e.g. 30 μm). The deposit per cell is higher than the values corresponding to devices using high temperature pastes, as shown in Figure 3.40.

The cell metallization is not only composed of fingers but also of « bus areas » dedicated to the interconnections. These buses can represent a significant part of the deposit. This was particularly true in the 2010s when continuous wide buses were systematically printed in order to solder ribbons on them (e.g. 4 continuous buses with a width of 1.2 mm). An interesting approach to limit the deposit was then to use two different pastes. The first was used for the printing of the fingers (rather thick fingers required to minimize the finger resistance, silver content paste >90%). The second was used for the printing of the buses (rather thin buses just dimensioned to ensure a good soldering because the conductivity will be brought by the ribbons, silver content paste <70%). Nowadays, the current interconnection technology is multi bus bars (MBB). The « bus area » is composed of small pads linked by a narrow vertical line, which allow a limited silver deposit. Shadowing is drastically reduced at the cell level but also at the module level due to the use of wires (shadowing of only 70%). The reduced wire shadowing allows to increase the number of interconnections, which results in shorter fingers with a higher tolerance regarding their resistance. The pad width is around 1 mm, this value depending on the stringer alignment accuracy. The pad height depends on the paste adhesion. This parameter is more critical for the cell architectures using low temperature pastes. In this case, due to silver leaching, thick pads (15–25 μm) are mandatory in order to get acceptable peel forces. A dual process is then recommended to avoid too high deposits (high thickness limited only to the small pads). Another option is the gluing of the interconnections by an electrical conductive adhesive (ECA), which allows to get good features in terms of reliability with a lower deposit due to much thinner metallization in the connection area [390]. In terms of deposit, the ultimate solution concerns the suppression of the busbar areas. This is allowed by the SmartWire Connection Technology, which leads to significant silver consumption reductions. However, the silver saving is partially compensated by some additional interconnection costs (foils, wire coating). Furthermore, this technology is now proprietary. As examples of Ag usage reductions for these new approaches, in 2018, Faes [391] reported Ag deposits of 335 mg for the soldering and 145 mg for the gluing of 6 ribbons on M2 SHJ solar cells; in comparison with 100 mg for the SmartWire Connection Technology using 18 wires. In 2021, CEA succeeded in getting reliable glued modules using M2 cells metallized with only 125 mg of Ag. A similar deposit per cell area was reported by the company HUASUN, with 140 mg of Ag for M6 cells soldered with 12 wires [392].

3.6.2 Copper metallization

3.6.2.1 Rationale for Ag replacement

The current PV production requires more than 2,000 metric tons of silver per year for its metallization, thereby consuming more than 10% of the entire worldwide annual silver supply [393]. With PV poised to be the key pillar of the future low-carbon energy mix, PV production is postulated to reach the TW-scale by 2030 already [394], hence the silver consumption by the PV industry is expected to skyrocket in the coming years [395]. Even assuming a strongly reduced silver consumption per cell, as postulated in the ITRPV roadmap projections [388], namely 50 mg of Ag for PERC and 100 mg for n-type and bifacial solar cells, between 12,000 and 21,000 metric tons of silver would be required for an annual 1.5 TW PV production. This might turn into a strong shortage risk for the PV industry in the coming years and is doubtlessly endangering the overall sustainability of PV. In addition, and strongly related to the above-discussed supply issue, the spot price of silver is very volatile, experiencing for instance a nine-fold price increase between 2000 and 2011, from USD ~5 per ounce in 2000 to USD ~45 per ounce in 2011. After falling back to USD ~15 per ounce in 2020, it again experienced a sudden increase by 50% in less than one year. Consequently, this puts the

PV industry under strong pressure, as metallization is the highest non-silicon materials cost in the current crystalline silicon-based solar cells [388].

Regardless of these supply and costs issues, alternatives to traditional silver screen-printed contacts are also eagerly sought for to further decrease the width of the metallization fingers (hence reducing the optical shadowing losses), as well as being compatible with the latest generation of high-efficiency solar cells, among which perovskite-on-silicon tandems, whose delicate materials cannot withstand curing temperatures above ~150°C [396].

In this part, it will be shown that copper-based metallization possesses serious assets to overcome the challenges presented above. Copper is indeed now identified as the best materials to replace silver for the metallization of PV devices, owing to its good bulk conductivity, lower price, and abundancy.

3.6.2.2   Cu containing screen-printable pastes

Attempting to reduce the silver consumption without disrupting the established solar cell process flow, pastes manufacturers have been introducing in the last years screen-printable pastes with a lower amount of silver. Pure copper pastes feature so far poorer bulk conductivity and printability than their standard pure silver counterparts and are subject to oxidation [397]. In contrast, silver-coated copper pastes offer good printability, bulk conductivity, are resilient to oxidation, and passed standard module degradation tests [398], yet the amount of silver is so far only reduced by 30–40%.

3.6.2.3   Cu plating

The preferred method for the copper metallization of PERC-like solar cells, and more recently of TOPCon-like ones, is the plating of nickel–copper–silver stacks. First, the front or rear dielectric is opened, usually using laser ablation potentially followed by damage annealing and surface cleaning. Then, a thin nickel layer is plated at these openings, the dielectric acting as a plating mask. Alternatively, a sintered seed-grid can be used, potentially fired through the dielectric. This thin nickel layer forms a nickel-silicide alloy, acting as an efficient barrier against copper diffusion. The nickel seed layer is then thickened by plating thick copper fingers and capped with a thin silver finish layer to prevent oxidation. Ni/Cu/Ag fingers with a width down to 12 μm and a height of 5–7 μm were recently obtained using this approach (Figure 3.41), allowing for a conversion efficiency gain of $0.4\%_{abs}$ in TOPCon solar cells compared to standard silver screen-printed references [399].

The application of copper plating to silicon heterojunction solar cells is somehow simplified by leveraging on the fact that TCOs are excellent barriers against copper diffusion [401,402], therefore no additional barrier layer is required, such as Ni in the case of PERC or TOPCon. In contrast, an additional patterning solution for the Cu-plated fingers must be provided. Here, various approaches have been demonstrated, such as the use of an inkjet-printed hotmelt resist plating mask [403], a sacrificial thin Al layer opened by inkjetting NaOH droplets [404], and more recently the use of a Ag or Cu screen-printed seed grid reinforced by subsequent Cu electroplating [405]. In 2022, the Australian start-up SunDrive jointly

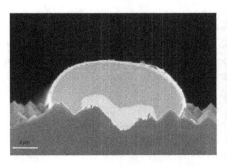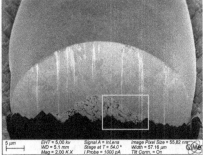

Figure 3.41 *Scanning electron microscope cross-sectional images of Cu-plated fingers. Left: Ni/Cu/Ag-plated fingers applied to a TOPCon solar cell. Picture taken from [399]. Right: Cu plated finger on a screen-printed Ag seed-grid applied to a silicon heterojunction solar cell. Picture taken from [400]*

with the Chinese equipment manufacturer Maxwell demonstrated a Cu-plated HJT solar cell with a certified efficiency of 26.07% [406], convincingly show-casing the maturity of Cu plating at the industrial scale.

3.7 General conclusions

n-Type silicon solar cells have deeply marked the story of PVs. Yet, the greatest pages are undeniably forthcoming since we are only at the beginning of a massive transition of the PV industry toward n-type solar cells. The trajectory of n-type devices can be divided in three main steps.

First, n-type cells constitute the cradle of modern PV, since the first practical solar cell developed by the Bell Labs in 1954 was processed from an n-type Si substrate. Nevertheless, despite this major achievement, n-type cells were some-how set aside in the following years. This is essentially explained by the main application of PV at that time, driving the developments, which was space. Indeed, as n-type Si is more affected by space radiations-induced defects than p-type Si, the R&D actions essentially addressed p-type devices. The first PV companies for terrestrial applications took benefit of these space-related developments and were naturally based on p-type products. This predominance of p-type technologies flourished then, benefiting from first a dynamic learning curve due to scale effects and second from the successive emergences of two iconic technologies: the cost-effective and reliable Al-BSF structure and its low CAPEX evolution toward the high-efficiency and bifacial PERC technology.

However, at the beginning of the twenty-first century, two main factors fostered a revival of n-type technologies. First, several studies pointed out the excellent electronic properties of n-type Si substrates. In parallel, Sunpower and Panasonic achieved outstanding efficiency values with their respective IBC and Si heterojunction technologies. From that point, n-type solar cells sparked a huge interest in the PV community. The developments essentially targeted two device families, the high-T processed n-PERT and the low-T processed heterojunction devices. The n-PERT devices were based on processes similar than those used for Al-BSF and PERC solar cells (e.g. thermal diffusion doping, firing through metallization), but with specific challenges to be addressed, such as the passivation of boron-emitters. As described in this chapter, the performances of n-PERT solar cells significantly improved (conversion efficiency as high as 22.8% with large surface areas), with cost-effective fabrication processes. It is worth noticing the synergetic developments between p-type PERC and n-type PERT. As an example, the passivation of the n-PERT B-emitter benefited from the development of the AlO_x layers used at the rear-side of PERC cells. On the other hand, the bifacial and stable (no B–O defects) features of n-PERT devices pushed the p-type community to find solutions to compete with these advantages (i.e. evolution toward bifacial PERC, implementation of regeneration treatments for deactivating the B–O defects and recently the transition toward Ga-doped Si). Despite considerable progresses in terms of efficiency, excellent bifaciality and first industrial adoptions, n-PERT solar cells had difficulties to compete with the PERC technology: similar efficiency values for both technologies, with a fabrication process usually slightly more complex and expensive (higher wafer price, higher amount of Ag) for n-PERT. The situation significantly changed at the end of the 2010s, with the first and successful industrial attempts of n-PERT devices integrating a poly-Si/SiO_x passivated contact on the rear surface. Such concepts, initially inspired from the IC industry, were already known in the 1980s, but experienced an intense revival in the 2010s following the outstanding efficiency results published by Sunpower and Fraunhofer. Then, many research groups and companies focused on these structures, providing a comprehensive scientific description of the transport and recombination mechanisms of charge carriers through the poly-Si/SiO_x stack, but also developing cost-effective equipment and processes for depositing and contacting the involved layers. The emergence of poly-Si/SiO_x passivated contacts enabled impressive efficiency improvements with commercial high-T processed n-PERT cells, since several industrial players reported efficiency values above 25% in 2022 (with a record value of 26.4% for a large surface area device), and various progress paths have been identified to go further: process simplifications, Ag content reduction, double-side integration of poly-Si contacts. In parallel of the development on n-PERT solar cells, the second main family of n-type devices, Si heterojunction cells, also experienced a great success in the 2010s. First, as the passivation of the contacts is inherent to this technology, it offers a huge conversion efficiency potential, demonstrated by the 26.81% efficiency value of Longi (double-side contacted configuration), which consists in the world record for a c-Si solar cell. Due to their high V_{oc}, heterojunction cells feature a favorable temperature coefficient. Furthermore, the architecture is perfectly suited for bifacial applications. Important progresses were achieved for developing cost-effective fabrication processes for heterojunction solar cells, with excellent results since major cells producers are now adopting the technology with an outstanding dynamic of progresses in 2021 and 2022. Both double-side contacted Poly-Si passivated contacts and Si hetero-junction solar cells feature efficiency values significantly higher than those of PERC and have increasing market shares. For both technologies, an important lever to further decreasing the costs is the silver reduction. As presented in this chapter, efficient solutions have already been proposed, either based on advanced screen-printing for reducing the amount of deposited Ag, or by substituting Ag by Cu via Cu-based screen-printing pastes or Cu plating. With the mass production adoption of poly-Si contacts and Si heterojunction cells, n-type technologies can be considered as fully mature at the industrial level, with beautiful promises for the years to come.

The next major transformation of the n-type solar cells landscape could come from a massive industrial deployment of IBC cells with passivated contacts (either poly-Si/SiO$_x$ structures or Si heterojunctions). Indeed, IBC devices feature an unrivalled conversion efficiency potential (record efficiency at 26.7% with a SHJ-IBC device) and pleasing aesthetics. Recently, cost effective processes emerged and key industrial players showed considerable interests for these architectures, important ingredients for a bright future of these technologies.

Last, poly-Si/SiO$_x$ and Si heterojunction structures are particularly suitable for the passivation of the front surface of c-Si bottom cells used in 2-terminal mono-lithic multi-junction devices (e.g. perovskite-on-Si tandem cells). Furthermore, IBC devices are also an interesting steppingstone for perovskite-on-silicon tandems in three-terminal wiring. Therefore, the already long-lasting and successful story of n-type solar cells is poised to continue further.

References

[1] Chapin, D. M., Fuller, C. S., and Pearson, G. L. (1954). A new silicon p-n junction photocell for converting solar radiation into electrical power. *Journal of Applied Physics*, 25(5), 676–677.

[2] Prince, M. (1955). Silicon solar energy converters. *Journal of Applied Physics*, 26(5), 534–540.

[3] Chapin, D. M., Fuller, C. S., and Pearson, G. L. (1957). U.S. Patent No. 2,780,765. Washington, DC: U.S. Patent and Trademark Office.

[4] Cooley, W. C. and Janda, R. J. (1963). *Handbook of Space-Radiation Effects on Solar-Cell Power Systems* (Vol. 3003). USA: *US National Aeronautics and Space Administration*.

[5] Taguchi, M., Kawamoto, K., Tsuge, S., *et al* . (2000). HITTM cells—high-efficiency crystalline Si cells with novel structure. *Progress in Photovoltaics: Research and Applications*, 8(5), 503–513.

[6] Mulligan, W. P., Rose, D. H., Cudzinovic, M. J., *et al* .(2004). Manufacture of solar cells with 21% efficiency. In *Proceedings of the 19th EPVSEC* (p. 387).

[7] Cheek, G. C., Mertens, R. P., Van Overstraeten, R., and Frisson, L. (1984). Thick-film metallization for solar cell applications. *IEEE Transactions on Electron Devices*, 31(5), 602–609.

[8]   Narayanan, S. (2002). Large area multicrystalline silicon solar cells in high volume production environment—history, status, new processes, technology transfer issues. *Solar Energy Materials and Solar Cells*, 74(1–4), 107–115.

[9]   Hahn, G. (2010). Status of selective emitter technology. In *25th European Photovoltaic Solar Energy Conference and Exhibition. 5th World Conference on Photovoltaic Energy Conversion* (pp. 1091–1096).

[10] Röder, T. C., Eisele, S. J., Grabitz, P., *et al.* (2010). Add-on laser tailored selective emitter solar cells. *Progress in Photovoltaics: Research and Applications*, 18(7), 505–510.

[11] Chen, N. and Ebong, A. (2016). Towards 20% efficient industrial Al-BSF silicon solar cell with multiple busbars and fine gridlines. *Solar Energy Materials and Solar Cells*, 146, 107–113.

[12] Dubois, S., Palais, O., Ribeyron, P. J., Enjalbert, N., Pasquinelli, M., and Martinuzzi, S. (2007). Effect of intentional bulk contamination with iron on multicrystalline silicon solar cell properties. *Journal of Applied Physics*, 102(8), 083525.

[13] Blakers, A. W., Wang, A., Milne, A. M., Zhao, J., and Green, M. A. (1989). 22.8% Efficient silicon solar cell. *Applied Physics Letters,* 55(13), 1363–1365.

[14] Equipment, VDMA Photovoltaic. International Technology Roadmap for Photovoltaic (ITRPV), 2022.

[15] Kranz, C., Wyczanowski, S., Baumann, U., *et al.* (2013). Wet chemical polishing for industrial type PERC solar cells. *Energy Procedia*, 38, 243–249.

[16] Schmidt, J., Werner, F., Veith, B., *et al.* (2012). Advances in the surface passivation of silicon solar cells. *Energy Procedia*, 15, 30–39.

[17] Dullweber, T. and Schmidt, J. (2016). Industrial silicon solar cells applying the passivated emitter and rear cell (PERC) concept—a review. *IEEE Journal of Photovoltaics*, 6(5), 1366–1381.

[18] Veith, B., Werner, F., Zielke, D., Brendel, R., and Schmidt, J. (2011). Comparison of the thermal stability of single Al2O3 layers and Al2O3/SiNx stacks for the surface passiviation of silicon. *Energy Procedia*, 8, 307–312.

[19] Glunz, S. W., Preu, R., Schaefer, S., *et al.* (2000). New simplified methods for patterning the rear contact of RP-PERC high-efficiency solar cells. In *Conference Record of the Twenty-Eighth IEEE Photovoltaic Specialists Conference-2000* (Cat. no. 00CH37036) (pp. 168–171). IEEE.

[20] Agostinelli, G., Szlufcick, J., Choulat, P., and Beaucarne, G. (2005). Local contact structures for industrial PERC-type solar cells. In *Proceedings of the 20th European Photovoltaic Solar Energy Conference* (pp. 942–945).

[21] Pysch, D., Schmitt, C., Latzel, B., *et al.* (2014). Implementation of an ALD-Al2O3 PERC-technology into a multi-and monocrystalline industrial pilot production. In *29th European Photovoltaic Solar Energy Conference*, Amsterdam.

[22] Dullweber, T., Kranz, C., Peibst, R., *et al.* (2016). PERC+: industrial PERC solar cells with rear Al grid enabling bifaciality and reduced Al paste

consumption. *Progress in Photovoltaics: Research and Applications,* 24(12), 1487–1498.

[23] Dullweber, T., Schulte-Huxel, H., Blankemeyer, S., *et al.* (2018). Present status and future perspectives of bifacial PERC+ solar cells and modules. *Japanese Journal of Applied Physics,* 57(8S3), 08RA01.

[24] Lee, B. G., Cieslak, J., Schwabedissen, A., *et al.* (2019). Development and mass production of bifacial Q. ANTUM p-Cz PERC cells. In *2019 IEEE 46th Photovoltaic Specialists Conference (PVSC)* (pp. 1460–1462). IEEE.

[25] Chen, W., Liu, R., Zeng, Q., and Zhou, L. (2019). Low cost multicrystalline bifacial PERC solar cells – fabrication and thermal improvement. *Solar Energy,* 184, 508–514.

[26] Green, M. A., Dunlop, E. D., Hohl-Ebinger, J., Yoshita, M., Kopidakis, N., and Hao, X. (2021). Solar cell efficiency tables (Version 58). *Progress in Photovoltaics: Research and Applications,* 29 (NREL/JA-5900-80028).

[27] Fischer, H. (1973). Investigation of photon and thermal changes in silicon solar cells. In *Conference Record of the 10th IEEE Photovoltaic Specialists Conference, 1973.*

[28] Schmidt, J. and Bothe, K. (2004). Structure and transformation of the metastable boron-and oxygen-related defect center in crystalline silicon. *Physical Review B,* 69(2), 024107.

[29] Cascant, M., Enjalbert, N., Monna, R., and Dubois, S. (2014). Influence of various p-type Czochralski silicon solar cell architectures on light-induced degradation and regeneration mechanisms. In *Proceedings of the 29th European Photovoltaic Solar Energy Conference and Exhibition.*

[30] Herguth, A., Schubert, G., Kaes, M., and Hahn, G. (2006). A new approach to prevent the negative impact of the metastable defect in boron doped Cz silicon solar cells. In *2006 IEEE 4th World Conference on Photovoltaic Energy Conference* (Vol. 1, pp. 940–943). IEEE.

[31] Herguth, A., Schubert, G., Kaes, M., and Hahn, G. (2006). Avoiding boron-oxygen related degradation in highly boron doped Cz silicon. In *21st European Photovoltaic Solar Energy Conference: 21th EU PVSEC* (pp. 530–537).

[32] Yeo, Z. Y., Ling, Z. P., Ho, J. W., Lim, Q. X., So, Y. H., and Wang, S. (2022). Status review and future perspectives on mitigating light-induced degradation on silicon-based solar cells. *Renewable and Sustainable Energy Reviews,* 159, 112223.

[33] Lauermann, T., Herguth, A., Scholz, S., *et al.* (2010). Large area solar cells made from degradation-free, low resistivity gallium doped Cz wafers. In *25th European Photovoltaic Solar Energy Conference and Exhibition. 5th World Conference on photovoltaic Energy Conversion* (pp. 2002–2005).

[34] Kersten, F., Engelhart, P., Ploigt, H. C., *et al.* (2015). A new mc-Si degradation effect called LeTID. In *2015 IEEE 42nd Photovoltaic Specialist Conference (PVSC)* (pp. 1–5). IEEE.

[35] Chen, D., Kim, M., Stefani, B. V., *et al.* (2017). Evidence of an identical firing-activated carrier-induced defect in monocrystalline and multicrystalline silicon. *Solar Energy Materials and Solar Cells*, 172, 293–300.

[36] Grant, N. E., Scowcroft, J. R., Pointon, A. I., Al-Amin, M., Altermatt, P. P., and Murphy, J. D. (2020). Lifetime instabilities in gallium doped monocrystalline PERC silicon solar cells. *Solar Energy Materials and Solar Cells*, 206, 110299.

[37] née Wenham, A. C., Wenham, S., Chen, R., *et al.* (2018). Hydrogen-induced degradation. In *2018 IEEE 7th World Conference on Photovoltaic Energy Conversion (WCPEC) (A Joint Conference of 45th IEEE PVSC, 28th PVSEC & 34th EU PVSEC)* (pp. 0001–0008). IEEE.

[38] Jensen, M. A., Zuschlag, A., Wieghold, S., *et al.* (2018). Evaluating root cause: the distinct roles of hydrogen and firing in activating light-and elevated temperature-induced degradation. *Journal of Applied Physics*, 124(8), 085701.

[39] Sharma, R., Aberle, A. G., and Li, J. B. (2018). Optimization of belt furnace anneal to reduce light and elevated temperature induced degradation of effective carrier lifetime of p-type multicrystalline silicon wafers. *Solar RRL*, 2(9), 1800070.

[40] Zuschlag, A., Skorka, D., and Hahn, G. (2017). Degradation and regeneration in mc-Si after different gettering steps. *Progress in Photovoltaics: Research and Applications*, 25(7), 545–552.

[41] Fertig, F., Lantzsch, R., Mohr, A., *et al.* (2017). Mass production of p-type Cz silicon solar cells approaching average stable conversion efficiencies of 22%. *Energy Procedia*, 124, 338–345.

[42] Naber, R. (2015). Doping techniques and process integration solutions for n-type solar cells. In *nPV Workshop* 2015.

[43] Oliver, C. (2011). Thèse de doctorat, Dopage bore du silicium multicristallin de type n: application à la fabrication de cellules photovoltaïques par un procédé industriel, Univ. de Montpellier II, Sept 2011.

[44] Lanterne, A. (2014). Étude, Ph.D. Thesis, Université de Grenoble, November 2014.

[45] Kessler, M. A., Ohrdes, T., Wolpensinger, B., and Harder, N. (2009). Characterisation and implications of the boron rich layer resulting from open-tube liquid source BBr3 boron diffusion processes. In 2009 *34th IEEE Photovoltaic Specialists Conference (PVSC)* (pp. 001556–001561). doi:10.1109/PVSC.2009.5411365.

[46] Cuevas, A., Samundsett, C., Kerr, M. J., Macdonald, D. H., Mäckel, H., and Altermatt, P. P. (2003). In *Proceedings of the 3rd WCPEC Osaka, Japan* (pp. 963–966).

[47] Schmiga, C., Froitzheim, A., Ghosh, M., *et al.* (2005). Solar cells on N-type silicon materials with screen-printed rear Al-p+ emitter. In *20th European Photovoltaic Solar Energy Conference*, Barcelona, Spain (p. 918).

[48] Kopecek, R., Buck, T., Libal, J., and Rover, I. (2006). Large area screen printed N-type silicon solar cells with rear aluminium emitter:

efficiencies exceeding 16%. In *2006 IEEE 4th World Conference on Photovoltaic Energy Conference* (pp. 1044–1047). doi:10.1109/WCPEC. 2006.279319

[49]   Schmiga, C., Horteis, M., Rauer, M., *et al.* (2009). In Large-area n-type silicon solar cells with printed contacts and aluminium-alloyed rear emitter. In *Proceedings of the 24th European Photovoltaic Solar Energy Conference*, Hamburg, Germany (pp. 1167–1170).

[50]   Schmiga, C., Rauer, M., Rudiger, M., *et al.* (2010). In aluminium-doped p+ silicon for rear emitters and back surface fields: results and potentials of industrial n-and p-type solar cells. In *Proceedings of 25th European Photovoltaic Solar Energy Conference*, Valencia, Spain (pp. 1163–1168).

[51]   Book, F., Wiedenmann, T., Schubert, G., Plagwitz, H., and Hahn, G. (2011). Influence of the front surface passivation quality on large area n-type silicon solar cells with Al-alloyed rear emitter. *Energy Procedia*, 8, 487–492.

[52]   Xi Xi, X. C., Song Zhang, W. L., and Zhengrong Shi, G. L. (2015). The investigation on the front surface field of aluminum rear emitter N-type silicon solar cells. *Solar Energy*, 114, 198–205. ISSN 0038-092X, https://doi.org/10.1016/j.solener.2015.01.019.

[53]   Schmiga, C., Rauer, M., Rudiger, M., Glatthaar, M., and Glunz, S. W. (2012). Status and perspectives of n-type silicon solar cells with aluminium alloyed rear emitter. In *27th European Photovoltaic Solar Energy Conference and Exhibition* (pp. 915–918).

[54]   Bock, R., Schmidt, J., Mau, S., *et al.* (2010). The ALU+ concept: n-type silicon solar cells with surface-passivated screen-printed aluminium-alloyed rear emitter. *IEEE Transactions on Electron Devices*, 57, 1966.

[55]   Dauwe, S., Mittelstädt, L., Metz, A., and Hezel, R. (2002). Experimental evidence of parasitic shunting in silicon nitride rear surface passivated solar cells. *Progress in Photovoltaics: Research and Applications*, 10, 271–278. https://doi.org/10.1002/pip.420

[56]   Cesar, I., Kay, A., and Grätzel, M. (2009). All-side SiNx passivated mc-Si solar cells evaluated with respect to parasitic shunting. In *Photovoltaic Specialists Conference (PVSC), 2009 34th IEEE* (pp. 001386–001391).

[57]   Lauermann, T., Zuschlag, A., Scholz, S., *et al.* (2011). In *26th European Photovoltaic Solar Energy Conference and Exhibition*, Hamburg, Germany (pp.1137–1143).

[58]   Rauer, M., Schmiga, C., Woehl, R., *et al.* (2011). Investigation of aluminum-alloyed local contacts for rear surface-passivated silicon solar cells. *IEEE Journal of Photovoltaics*, 1, 22–28.

[59]   Uruena, A., Horzel, J., Singh, S., *et al.* (2012). Rear contact and BSF formation for local Al-BSF solar cells. *Energy Procedia*, 27, 561–566. https://doi.org/10.1016/j.egypro.2012.07.110.

[60]   Chen, D., Weiwei, D., Sheng, J., and Wang, W. (2013). Preventing the formation of voids in the rear local contact areas for industrial-type PERC solar cells. In *28th European Photovoltaic Solar Energy Conference and Exhibition*, doi:10.4229/28thEUPVSEC2013-2AO.3.5.

[61]   Mertens, V., Ballmann, T., Cieslak, J., *et al.* (2013). Large area n-type Cz double side contact back-junction boron emitter solar cell with 21.3% conversion efficiency. In *28th European Photovoltaic Solar Energy Conference and Exhibition*, France.

[62]   Uruena, A., Aleman, M., Cornagliotti, E., *et al.* (2015). Beyond 22% large area n-type silicon solar cells with front laser doping and a rear emitter. In *Proceedings of the 30th EUPVSEC*, Hamburg, Germany.

[63]   Chen, J., Duerinckx, F., Cornagliotti, E., *et al.* (2016). 21.7% Large area n-PERT silicon solar cells using screen-printed aluminium with open circuit voltage above 690 mV. In *Proceedings of the 32nd EUPVSEC*, Munich, Germany (pp. 439–442).

[64]   Stefan Bordihn, Mertens, V., Cieslak, J., *et al.* (2016). Status of industrial back junction n-type Si solar cell development. *Energy Procedia*, 92, 678–683. https://doi.org/10.1016/j.egypro.2016.07.042.

[65]   Engelhart, P., Manger, D., Kloter, B., *et al.* (2011). Q.ANTUM – Q-cells next generation high-power silicon cell & module concept. In *Proceedings of the 26th EUPVSEC*, Hamburg, Germany (pp. 821–826).

[66]   Dullweber, T., Schulte-Huxel, H., Blankemeyer, S., *et al.* (2018). Present status and future perspectives of bifacial PERC+ solar cells and modules. *Japanese Journal of Applied Physics*, 57, 08RA01.

[67]   Wehmeier, N., Nowack, A., Lim, B., *et al.* (2016). 21.0%-Efficient screen-printed n-PERT back-junction silicon solar cell with plasma-deposited boron diffusion source. *Solar Energy Materials and Solar Cells*, 158, 50.

[68]   Peng, Z., Nakahara, M., Buck, T., and Kopecek, R. (2019). Investigation on industrial screen-printed aluminum point contact and its application in n-PERT rear junction solar cells. *IEEE Journal of Photovoltaics*, 9(6), 1554–1562. doi:10.1109/JPHOTOV.2019.2940142

[69]   Kopecek, R., Peng, Z. W., Buck, T., *et al.* Low-cost standard nPERT solar cells towards 23% efficiency and 700 mV voltage using Al paste technology. *PV International*, 42, 74–84.

[70]   Tous, L., Chen, J., Choulat, P., *et al.* (2018). Large area monofacial screen-printed rear emitter nPERT cells approaching 23% efficiency. In *Proceedings of the 35th EU PVSEC*, Brussels, Belgium.

[71]   Zhao, A., Wang, P. P., Altermatt, and Green, M. A. (2001). *Proceedings of the 12th International Photovoltaic Science and Engineering Conference* (p. 2001).

[72]   Zhao, J., Wang, A., Altermatt, P. P., Green, M. A., Rakotoniaina, J. P., and Breitenstein, O. (2002). In *Proceedings of the 29th IEEE Photovoltaic Specialist Conference* (p. 218).

[73]   Zhao, J., Wang, A., and Green, M. A. (1999). 24.5% Efficiency silicon PERT cells on MCZ substrates and 24.7% efficiency PERL cells on FZ substrates. *Progress in Photovoltaics: Research and Applications*, 7, 471–474.

[74]   Bernd Steinhauser, A. B., Henning Nagel, S. G., Sven Kluska, J. B., Pierre Saint-Cast, J. B., and Martin, H. (2017). Advances in PassDop technology: recombination and optics. *Energy Procedia*, 124, 313–320.

[75] Kopecek, R., Buck, T., Libal, J., *et al.* Large area N-type multicrystalline silicon solar cells with B-emitter: efficiencies exceeding 14%. In *15th International Photoboltaic Science and Engineering Conference & 2005 Renewable Energy Expo of Shanghai* (PVSEC 15th).

[76] Mihailetchi, V. D., Coletti, G., Komatsu, Y., *et al.* (2008). Large area and screen printed N-type silicon solar cells with efficiency exceeding 18%. In *Proceedings of 23rd EU PVSEC*, Valencia (pp. 1036–1039).

[77] Romijn, I. G., *et al.* (2012). Industrial n-type solar cells: towards 20% efficiency. *Photovoltaics International (PV-Tech)*, 15, 63–69.

[78] Buchholz, F., Preisa, P., Chua, H., Lossena, J., and Wefringhausa, E. (2017). Progress in the development of industrial nPERT cells. *Energy Procedia*, 124, 649–656.

[79] Rodriguez, J., Wang, E.-C., Chen, N., *et al.* (2018). Towards 22% efficient screen-printed bifacial n-type silicon solar cells. *Solar Energy Materials and Solar Cells*, 187, 91–96. https://doi.org/10.1016/J.SOLMAT.2018.07.02

[80] Cornagliotti, E., Tous, L., Russell, R., *et al.* (2017). High efficiency bifacial n-PERT cells with co-plated Ni/Ag contacts for multi-wire interconnection. *Energy Procedia*, 130, 50–54.

[81] Yong, L. (2018). nPERT and nIBC solar cell technology in mass production. Presented at the *nPV Workshop*, Lausanne, Switzerland.

[82] Bawa, M. S., Bell, W. J., Grimes, H. M., and Shaffner, T. J. (1989). Temperature dependence of oxidation induced stacking faults and oxygen incorporation in dislocation-free Czochralski silicon. *Journal of Crystal Growth*, 94(3), 803–806.

[83] Veirman, J., Martel, B., Letty, E., *et al.* (2016). Thermal History Index as a bulk quality indicator for Czochralski solar wafers. *Solar Energy Materials and Solar Cells*, 158, 55–59.

[84] Dominguez, E., Lora-Tamayo, E., Blanchard, B., and Bellanato, J. (1978). Analytic study of the Si-B phase when B2O3 is diffused in Si. *Journal of the Electrochemical Society,* 125(9), 1521.

[85] Brown, D. M. and Kennicott, P. R. (1971). Glass source B diffusion in Si and SiO$_2$. *Journal of Electrochemical Society: Solid State Science*, 118, 293–300.

[86] Aselage, T. L. (1997). The coexistence of silicon borides with boron-saturated silicon: metastability of SiB3. *Journal of Material Research*, 13, 1786–1794.

[87] Kyungsun, R., Chel-Jong, C., Hyomin, P., Donghwan, K., Ajeet, R., and Young-Woo, O. K. (2016). Fundamental understanding, impact, and removal of boron-rich layer on n-type silicon solar cells. *Solar Energy Materials and Solar Cells*, 146, 58–62.

[88] Phang, S. P., Liang, W., Wolpensinger, B., Kessler, M. A., and Macdonald, D. (2013). Tradeoffs between impurity gettering, bulk degradation, and surface passivation of boron-rich layers on silicon solar cells. *IEEE Journals of Photovoltaics*, 3(1), 261–266.

[89]   Kessler, M. A., Ohrdes, T., Wolpensinger, B., and Harder, N. P. (2010). Charge carrier lifetime degradation in Cz silicon through the formation of a boron rich layer during BBr3 diffusion processes. *Semiconductor Science and Technology*, 25, 055001-1–055001-9.

[90]   Libal, J., Petres, R., Buck, T., *et al.* (2005). N-type multicrystalline silicon solar cells: BBr3 -diffusion and passivation of p+-diffused silicon surfaces. In *Proceedings of the 20th European Photovoltaic Solar Energy Conference*, Barcelona, Spain (pp. 793–796).

[91]   Veschetti, Y., *et al.* (2010). High efficiency N-type silicon solar cells with novel diffusion technique for emitter formation. In *25th European Photovoltaic Solar Energy Conference and Exhibition*, Valencia, Spain (pp. 6–10).

[92]   Mihailetchi, V. D., Komatsu, Y., and Geerligs, L. J. (2008). Nitric acid pretreatment for the passivation of boron emitters for n-type base silicon solar cells. *Applied Physics Letters*, 92, 063510.

[93]   Burgers, A. R., Geerligs, L. J., Carr, A. J., *et al.* (2011). 19.5% Efficient n-type Si solar cells made in production. In *Proceedings of the 26th European Photovoltaic Solar Energy Conference and Exhibition*, Hamburg.

[94]   Benick, J., Schultz, O., Schön, J., and Glunz, S. (2008). Passivation of boron emitters by local overcompensation with phosphorus. In *Proceedings of the 23rd European Photovoltaic Solar Energy Conference and Exhibition*, Valencia.

[95]   Edler, A. (2014). Development of Bifacial n-Type Solar Cells for Industrial Application. Ph.D. Thesis, Library of the University of Konstanz, Konstanz, Germany.

[96]   Brand, P., Veschetti, Y., Sanzone, V., *et al.* (2011). Integration of aluminum oxide as a passivation layer in a high efficiency process on n-type silicon solar cells. In *Proceedings of the 37th IEEE Photovoltaics Specialist Conference*, Seattle.

[97]   Hoex, B., Schmidt, J., Bock, R., Altermatt, P. P., van de Sanden, M. C. M., and Kessels, W. M. M. (2007). Excellent passivation of highly doped p-type Si surfaces by the negative-charge-dielectric Al2O3. *Applied Physics Letters,* 91, 112107.

[98]   Duttagupta, S., Lin, F., Shetty, K. D., *et al.* (2012). State-of-the-art surface passivation of boron emitters using inline PECVD AlOx/SiNx stacks for industrial high-efficiency silicon wafer solar cells. In *2012 38th IEEE Photovoltaic Specialists Conference*, IEEE (pp. 001036–001039).

[99]   Edler, A., Mihailetchi, V. D., Koduvelikulathu, L. J., Comparotto, C., Kopecek, R., and Harney, R. (2015). Metallization-induced recombination losses of bifacial silicon solar cells. *Progress in Photovoltaics: Research and Applications*, 23(5), 620–627.

[100]   Veschetti, Y., Lanterne, A., Sanzone, V., *et al.* (2013). High efficiency on large area n-type Si bifacial solar cells. In *N-PV Workshop*, Chambery.

[101]   Liu, J., Janssen, G. J. M., Koppes, M., *et al.* (2015). Selective emitters in n-type c-Si solar cells. In *Proceedings of the 31st European Photovoltaic Solar Energy Conference and Exhibition,* Hamburg.

[102]   Engelhardt, J., Frey, A., Gloger, S., Hahn, G., and Terheiden, T. (2015). Passivating boron silicate glasses for co-diffused high-efficiency n-type silicon solar cell application. *Applied Physics Letters*, 107, 042102.

[103]   Krügener, J., Peibst, R., Bugiel, E., *et al.* (2015). Ion implantation of boric molecules for silicon solar cells. *Solar Energy Materials and Solar Cells*, 142, 12–17.

[104]   Kiefer, F., Krügener, J., Heinemeyer, F., *et al.* (2016). Bifacial, fully screen-printed n-PERT solar cells with BF2 and B implanted emitters. *Solar Energy Materials and Solar Cells*, 157, 326–330.

[105]   Müller, R., Benick, J., Bateman, N., *et al.* (2014). Evaluation of implantation annealing for highly-doped selective boron emitters suitable for screenprinted contacts. *Solar Energy Materials and Solar Cells*, 120, 431–435.

[106]   Lanterne, A., Desrues, T., Lorfeuvre, C, *et al.* (2019). Plasma-immersion ion implantation: a path to lower the annealing temperature of implanted boron emitters and simplify PERT solar cell processing. *Progress in Photovoltaics: Research and Applications*, 27(12), 1081–1091.

[107]   Das, A., Ryu, K., and Rohatgi, A. (2011). 20% Efficient screen-printed n-type solar cells using a spin-on source and thermal oxide/silicon nitride passivation. *IEEE Journal of Photovoltaics*, 1(2), 146–152.

[108]   Ryu, K., Cho, E., Rohatgi, A,, and Ok, Y.-W. (2016). Process development and comparison of various boron emitter technologies for high-efficiency (∼21%) ntype silicon solar cells. *Progress in Photovoltaics: Research and Applications*, 24(8), 1109–1115.

[109]   Lim, B., Brendemühl, T., Dullweber, T., and Brendel, R. (2016). Loss analysis of n-type passivated emitter rear totally diffused back-junction silicon solar cells with efficiencies up to 21.2%. *IEEE Journal of Photovoltaics*, 6(2), 447–453.

[110]   Ryu, K., Madani, K., Rohatgi, A., and Ok, Y. W. (2018). High efficiency screen-printed ntype silicon solar cell using co-diffusion of APCVD boron emitter and POCl3 back surface field. *Current Applied Physics*, 18(2), 231–235.

[111]   Rothhardt, P., Demberger, C., Wolf, A., and Biro, D. (2014). Codiffused bifacial n-type solar cells (CoBiN). *Energy Procedia*, 55, 287–294.

[112]   Dullweber, T., Wehmeier, N., Nowack, A., *et al.* (2016). Industrial bifacial n-type silicon solar cells applying a boron co-diffused rear emitter and an aluminum rear finger grid. *Physica Status Solidi A*, 213(11), 3046–3052.

[113]   Blévin, T., Lanterne, A., Grange, B., Cabal, R., Vilcot, J. P., and Veschetti, Y. (2014). Development of industrial processes for the fabrication of high efficiency n-type PERT cells. *Solar Energy Materials and Solar Cells*, 131, 24–29.

[114]   Cabal, R., Blévin, T., Monna, R., and Veschetti, Y. (2016). *PV International*, 31.

[115]   Peng, Z. W., Buck, T., Koduvelikulathu, L. J., Mihailetchi, V. D., and Kopecek, R. (2019). Industrial screen-printed n-PERT-RJ solar cells:

efficiencies beyond 22% and open-circuit voltages approaching 700 mV. *IEEE Journal of Photovoltaics*, 9(5), 1166–1174.

[116]   Shanmugam, V., Khanna, A., Basu, P. K., Aberle, A. G., Mueller, T., and Wong, J. (2016). Impact of the phosphorus emitter doping profile on metal contact recombination of silicon wafer solar cells. *Solar Energy Materials and Solar Cells*, 147, 171–176.

[117]   Hoenig, R., Kalio, A., Sigwarth, J., *et al.* (2012). Impact of screen printing silver paste components on the space charge region recombination losses of industrial silicon solar cells. *Solar Energy Materials and Solar Cells*, 106, 7–10.

[118]   Mihailetchi, V. D., Chu, H., and Kopecek, R. (2018). Insight into metal induced recombination losses and contact resistance in industrial silicon solar cells. In *2018 IEEE 7th World Conference on Photovoltaic Energy Conversion (WCPEC)(A Joint Conference of 45th IEEE PVSC, 28th PVSEC & 34th EU PVSEC)* (pp. 2673–2677). IEEE.

[119]   Herrmann, D., Fell, A., Lohmüller, S., *et al.* (2021). Microstructure beneath screen-printed silver contacts and its correlation to metallization-induced recombination parameters. *Solar Energy Materials and Solar Cells*, 230, 111182.

[120]   Kamioka, H., Takeda, H., and Takagi, M. (1972). A new method of emitter formation for microwave transistors and high speed monolithic integrated circuits. In *1972 International Electron Devices Meeting*. IEEE.

[121]   Post, I. R. C., Ashburn, P., and Wolstenholme, G. H. (1992). Polysilicon emitters for bipolar transistors: a review and re-evaluation of theory and experiment. *IEEE Transactions on Electron Devices,* 39(7), 1717–1731.

[122]   De Graaff, H. C. and Gerard De Groot, J. (1979). The SIS tunnel emitter: a theory for emitters with thin interface layers. *IEEE Transactions on Electron Devices*, 26(11), 1771–1776.

[123]   Mochizuki, H., Aoki, T., Yamoto, H., Okayama, M., Abe, M., and Ando, T. (1976). Semi-insulating polycrystalline-silicon (SIPOS) films applied to MOS integrated circuits. *Japanese Journal of Applied Physics,* 15(S1), 41.

[124]   Matsushita, T., Oh-uchi, N., Hayashi, H., and Yamoto, H. (1979). A silicon heterojunction transistor. *Applied Physics Letters,* 35(7), 549–550.

[125]   Van Overstraeten, R. J. (1981). Advances in silicon solar cell processing. In *Photovoltaic Solar Energy Conference*. Springer, Dordrecht.

[126]   Green, M. A., and Blakers, A. W. (1983). Advantages of metal-insulator-semiconductor structures for silicon solar cells. *Solar Cells*, 8(1), 3–16.

[127]   Tarr, N. G. (1985). A polysilicon emitter solar cell. *IEEE Electron Device Letters*, 6(12), 655–658.

[128]   Yablonovitch, E., Gmitter, T., Swanson, R. M., and Kwark, Y. H. (1985). A 720 mV open circuit voltage SiO_x:c-Si:SiO_x double heterostructure solar cell. *Applied Physics Letters*, 47(11), 1211–1213.

[129]   Kwark, Y. H. and Swanson, R. M. (1987). N-type SIPOS and poly-silicon emitters. *Solid-State Electronics*, 30(11), 1121–1125.

[130] Gan, J. Y. (1990). *Polysilicon Emitters for Silicon Concentrator Solar Cells* (Doctoral dissertation, Stanford University).

[131] Cousins, P. J., Smith, D. D., Luan, H. C., *et al.* (2010). Generation 3: improved performance at lower cost. In *2010 35th IEEE Photovoltaic Specialists Conference* (pp. 000275–000278). IEEE.

[132] Manning, J. (2012). Method of forming contacts for a back-contact solar cell. Unites States Patent Application Publication. Pub. No: US 20120138135 A1.

[133] Feldmann, F., Bivour, M., Reichel, C., Hermle, M., and Glunz, S. W. (2013). A passivated rear contact for high-efficiency n-type silicon solar cells enabling high Vocs and FF> 82%. In *28th European PV Solar Energy Conference and Exhibition*.

[134] Römer, U., Peibst, R., Ohrdes, T., *et al.* (2014). Recombination behavior and contact resistance of n+ and p+ poly-crystalline Si/mono-crystalline Si junctions. *Solar Energy Materials and Solar Cells*, 131, 85–91.

[135] Richter, A., Benick, J., Feldmanna, F., Fella, A., Hermlea, M., and Stefan W. (2017). n-Type Si solar cells with passivating electron contact: identifying sources for efficiency limitations by wafer thickness and resistivity variation. *Solar Energy Materials and Solar Cells,* 173, 96–105.

[136] Glunz, S. W., Bivour, M., Messmer, C., *et al.* (2017). Passivating and carrier-selective contacts-basic requirements and implementation. In *2017 IEEE 44th Photovoltaic Specialist Conference (PVSC)* (pp. 2064–2069). IEEE.

[137] Glunz, S. W. and Feldmann, F. (2018). SiO_2 surface passivation layers–a key technology for silicon solar cells. *Solar Energy Materials and Solar Cells*, 185, 260–269.

[138] van de Loo, B. W., Macco, B., Schnabel, M., *et al.* (2020). On the hydrogenation of Poly-Si passivating contacts by Al2O3 and SiN_x thin films. *Solar Energy Materials and Solar Cells*, 215, 110592.

[139] Morisset, A., Cabal, R., Giglia, V., *et al.* (2021). Evolution of the surface passivation mechanism during the fabrication of ex-situ doped poly-Si (B)/SiO_x passivating contacts for high-efficiency c-Si solar cells. *Solar Energy Materials and Solar Cells*, 221, 110899.

[140] Steinkemper, H., Feldmann, F., Bivour, M., and Hermle, M. (2015). Numerical simulation of carrier-selective electron contacts featuring tunnel oxides. *IEEE Journal of Photovoltaics*, 5(5), 1348–1356.

[141] Lanterne, A., Yang, J., Loretz, J. C., *et al.* (2020). Phosphorus doped poly-Si passivated contacts by LPCVD and PECVD for industrial large-area solar cells. In *37th European PV Solar Energy Conference and Exhibition*.

[142] Feldmann, F., Reichel, C., Müller, R., and Hermle, M. (2017). The application of poly-Si/SiO_x contacts as passivated top/rear contacts in Si solar cells. *Solar Energy Materials and Solar Cells*, 159, 265–271.

[143] Rienäcker, M., Bossmeyer, M., Merkle, A., *et al.* (2016). Junction resistivity of carrier-selective polysilicon on oxide junctions and its impact on solar cell performance. *IEEE Journal of Photovoltaics*, 7(1), 11–18.

[144]   Shewchun, J., Singh, R., and Green, M. A. (1977). Theory of metal-insulator-semiconductor solar cells. *Journal of Applied Physics*, 48(2), 765–770.

[145]   Peibst, R., Römer, U., Larionova, Y., *et al.* (2016). Working principle of carrier selective poly-Si/c-Si junctions: Is tunnelling the whole story? *Solar Energy Materials and Solar Cells*, 158, 60–67.

[146]   Zhang, Z., Zeng, Y., Jiang, C. S., *et al.* (2018). Carrier transport through the ultrathin silicon-oxide layer in tunnel oxide passivated contact (TOPCon) c-Si solar cells. *Solar Energy Materials and Solar Cells*, 187, 113–122.

[147]   Kale, A. S., Nemeth, W., Guthrey, H., *et al.* (2019). Understanding the charge transport mechanisms through ultrathin SiOx layers in passivated contacts for high-efficiency silicon solar cells. *Applied Physics Letters*, 114(8), 083902.

[148]   Liu, A., Yan, D., Phang, S. P., Cuevas, A., and Macdonald, D. (2018). Effective impurity gettering by phosphorus-and boron-diffused polysilicon passivating contacts for silicon solar cells. *Solar Energy Materials and Solar Cells*, 179, 136–141.

[149]   Hayes, M., Martel, B., Alam, G. W., *et al.* (2019). Impurity gettering by boron-and phosphorus-doped polysilicon passivating contacts for high-efficiency multicrystalline silicon solar cells. *Physica Status Solidi (A)*, 216 (17), 1900321.

[150]   Liu, A., Yan, D., Wong-Leung, J., *et al.* (2018). Direct observation of the impurity gettering layers in polysilicon-based passivating contacts for silicon solar cells. *ACS Applied Energy Materials*, 1(5), 2275–2282.

[151]   Liu, A., Yang, Z., Feldmann, F., *et al.* (2021). Understanding the impurity gettering effect of polysilicon/oxide passivating contact structures through experiment and simulation. *Solar Energy Materials and Solar Cells*, 230, 111254.

[152]   Oliveau, C., Desrues, T., Lanterne, A., Seron, C., Rousseau, S., and Dubois, S. (2020). Double-side integration of high temperature passivated contacts: application to Cast-Mono Si. In *2020 47th IEEE Photovoltaic Specialists Conference (PVSC)* (pp. 1163–1166). IEEE.

[153]   Wang, Q., Wu, W., Yuan, N., Li, Y., Zhang, Y., and Ding, J. (2020). Influence of SiO$_x$ film thickness on electrical performance and efficiency of TOPCon solar cells. *Solar Energy Materials and Solar Cells*, 208, 110423.

[154]   Moldovan, A., Feldmann, F., Zimmer, M., Rentsch, J., Benick, J., and Hermle, M. (2015). Tunnel oxide passivated carrier-selective contacts based on ultra-thin SiO2 layers. *Solar Energy Materials and Solar Cells*, 142, 123.

[155]   Yang, G., Ingenito, A., Isabella, O., and Zeman, M. (2016). IBC c-Si solar cells based on ion-implanted poly-silicon passivating contacts. *Solar Energy Materials and Solar Cells*, 158, 84–90.

[156]   Morisset, A., Cabal, R., Grange, B., *et al.* (2019). Highly passivating and blister-free hole selective poly-silicon based contact for large area

crystalline silicon solar cells. *Solar Energy Materials and Solar Cells*, 200, 109912.

[157] Tong, H., Liao, M., Zhang, Z., *et al.* (2018). A strong-oxidizing mixed acid derived high-quality silicon oxide tunneling layer for polysilicon passivated contact silicon solar cell. *Solar Energy Materials and Solar Cells*, 188, 149–155.

[158] Tong, J., Ma, F. J., Hoex, B., and Lennon, A. (2018). Tunnel oxides formed by field-induced anodization for silicon solar cell passivation. In *2018 IEEE 7th World Conference on Photovoltaic Energy Conversion (WCPEC) (A Joint Conference of 45th IEEE PVSC, 28th PVSEC & 34th EU PVSEC)* (pp. 2078–2081). IEEE.

[159] Yan, D., Cuevas, A., Bullock, J., Wan, Y., and Samundsett, C. (2015). Phosphorus-diffused polysilicon contacts for solar cells. *Solar Energy Materials and Solar Cells*, 142, 75–82.

[160] Jeon, M., Kang, J., Shim, G., *et al.* (2017). Passivation effect of tunnel oxide grown by N2O plasma for c-Si solar cell applications. *Vacuum*, 141, 152–156.

[161] Huang, Y., Liao, M., Wang, Z., *et al.* (2020). Ultrathin silicon oxide prepared by in-line plasma-assisted N_2O oxidation (PANO) and the application for n-type polysilicon passivated contact. *Solar Energy Materials and Solar Cells*, 208, 110389.

[162] Alzahrani, A., Allen, T. G., De Bastiani, M., *et al.* (2020). In situ plasma-grown silicon-oxide for polysilicon passivating contacts. *Advanced Materials Interfaces*, 7(21), 2000589.

[163] Lozac'h, M., Nunomura, S., and Matsubara, K. (2020). Double-sided TOPCon solar cells on textured wafer with ALD SiOx layer. *Solar Energy Materials and Solar Cells*, 207, 110357.

[164] Lanterne, A., Yang, J., Loretz, J. C., *et al.* (2020). LPCV-deposited poly-Si passivated contacts: surface passivation, gettering and integration in high efficiency devices. In *2020 47th IEEE Photovoltaic Specialists Conference (PVSC)* (pp. 2675–2678). IEEE.

[165] Gao, T., Yang, Q., Guo, X., *et al.* (2019). An industrially viable TOPCon structure with both ultra-thin SiOx and n+-poly-Si processed by PECVD for p-type c-Si solar cells. *Solar Energy Materials and Solar Cells*, 200, 109926.

[166] Kafle, B., Goraya, B. S., Mack, S., Feldmann, F., Nold, S., and Rentsch, J. (2021). TOPCon – technology options for cost efficient industrial manufacturing. *Solar Energy Materials and Solar Cells*, 227, 111100.

[167] Sze, S. M. (1988). *VLSI Technology.* London: McGraw-hill.

[168] Naber, R. C., Van de Loo, B. W. H., and Luchies, J. M. (2019). LPCVD in-situ n-type doped polysilicon process throughput optimization and implementation into an industrial solar cell process flow. In *Proceeding of the 34th European Photovoltaics Solar Energy Conference* (Vol. 53).

[169] Kopecek, R., Hoß, J., and Lossen, J. (2020). TOPCon technology: what exactly is it and how mature is it in production? *Photovoltaic International*, 44, 76–84.

[170] Zhou, Y., Tao, K., Liu, A., *et al.* (2020). The impacts of LPCVD wrap-around on the performance of n-type tunnel oxide passivated contact c-Si solar cell. *Current Applied Physics*, 20(7), 911–916.

[171] Lee, E. G. and Rha, S. K. (1993). A study of the morphology and micro-structure of LPCVD polysilicon. *Journal of Materials Science*, 28(23), 6279–6284.

[172] Chen, W., Stuckelberger, J., Wang, W., *et al.* (2020). Influence of PECVD deposition power and pressure on phosphorus-doped polysilicon passivating contacts. *IEEE Journal of Photovoltaics*, 10(5), 1239–1245.

[173] Feldmann, F., Fellmeth, T., Steinhauser, B., *et al.* (2019). Large area TOPCon cells realized by a PECVD tube process. In *Proceedings of the 36th European Photovoltaic Solar Energy Conference and Exhibition* (pp. 304–308).

[174] Li, S., Pomaska, M., Hoß, J., *et al.* (2019). Poly-Si/SiOx/c-Si passivating contact with 738 mV implied open circuit voltage fabricated by hot-wire chemical vapor deposition. *Applied Physics Letters*, 114(15), 153901.

[175] Mousumi, J. F., Ali, H., Gregory, G., *et al.* (2021). Phosphorus-doped polysilicon passivating contacts deposited by atmospheric pressure chemi-cal vapor deposition. *Journal of Physics D: Applied Physics*, 54(38), 384003.

[176] Lossen, J., Hoß, Jan., Eisert, S., *et al.* (2018). Electron beam evaporation of silicon for poly-silicon/SiO2 passivated contacts. In *Proceedings of the 35th European Photovoltaic Solar Energy Conference and Exhibition*.

[177] Hoß, J., Baumann, J., Berendt, M., *et al.* (2019). Sputtering of silicon thin films for passivated contacts. In *AIP Conference Proceedings* (Vol. 2147, No. 1, p. 040007). AIP Publishing LLC.

[178] Yan, D., Cuevas, A., Phang, S. P., Wan, Y., and Macdonald, D. (2018). 23% Efficient p-type crystalline silicon solar cells with hole-selective passivating contacts based on physical vapor deposition of doped silicon films. *Applied Physics Letters*, 113(6), 061603.

[179] Taiyang News Report. High Efficiency Solar Cells, 2019. http://taiyang-news.info/reports/high-efficiency-solar-cells-2019/

[180] Yang, X., Kang, J., Liu, W., Zhang, X., and De Wolf, S. (2021). Solution-doped polysilicon passivating contacts for silicon solar cells. *ACS Applied Materials & Interfaces*, 13(7), 8455–8460.

[181] Feldmann, F., Müller, R., Reichel, C., and Hermle, M. (2014). Ion implantation into amorphous Si layers to form carrier-selective contacts for Si solar cells. *Physica Status Solidi (RRL) – Rapid Research Letters*, 8(09), 767–770.

[182] Young, D. L., Nemeth, W., LaSalvia, V., *et al.* (2016). Low-cost plasma immersion ion implantation doping for interdigitated back passivated

contact (IBPC) solar cells. *Solar Energy Materials and Solar Cells*, 158, 68–76.

[183] Veau, A., Desrues, T., Lanterne, A., *et al.* (2018). Plasma immersion ion implantation (PIII): new path for optimizing doping profiles of advanced phosphorus emitters. In *AIP Conference Proceedings* (Vol. 1999, No. 1, p. 110007). AIP Publishing LLC.

[184] Yang, Q., Liao, M., Wang, Z., *et al.* (2020). In-situ phosphorus-doped polysilicon prepared using rapid-thermal anneal (RTA) and its application for polysilicon passivated-contact solar cells. *Solar Energy Materials and Solar Cells*, 210, 110518.

[185] Ingenito, A., Nogay, G., Jeangros, Q., *et al.* (2018). A passivating contact for silicon solar cells formed during a single firing thermal annealing. *Nature Energy*, 3(9), 800–808.

[186] Jiang, L., Song, L., Yan, L., Becht, G., Zhang, Y., and Hoerteis, M. (2017). The challenge of screen printed Ag metallization on nano-scale poly-silicon passivated contacts for silicon solar cells. In *Nanoengineering: Fabrication, Properties, Optics, and Devices XIV* (Vol. 10354, p. 103541D). San Diego, CA: International Society for Optics and Photonics.

[187] Çiftpınar, H. E., Stodolny, M. K., Wu, Y., *et al.* (2017). Study of screen printed metallization for polysilicon based passivating contacts. *Energy Procedia*, 124, 851–861.

[188] Feldmann, F., Bivour, M., Reichel, C., *et al.* (2017). The application of poly-Si/SiOx contacts as passivated top/rear contacts in Si solar cells. *Solar Energy Materials and Solar Cells*, *159*, 265–271.

[189] Feldmann, F., Nicolai, M., Müller, R., Reichel, C., and Hermle, M. (2017). Optical and electrical characterization of poly-Si/SiOx contacts and their implications on solar cell design. *Energy Procedia*, 124, 31–37.

[190] Stodolny, M. K., Lenes, M., Wu, Y., *et al.* (2016). n-Type polysilicon passivating contact for industrial bifacial n-type solar cells. *Solar Energy Materials and Solar Cells*, 158, 24–28.

[191] Duttagupta, S., Nandakumar, N., Padhamnath, P., *et al.* (2018). monoPoly^{TM} cells: large-area crystalline silicon solar cells with fire-through screen printed contact to doped polysilicon surfaces. *Solar Energy Materials and Solar Cells,* 187, 76–81.

[192] Ok, Y.-W., Upadhyaya, A. D., Rounsaville, B., *et al.* (2018). Screen printed, large area bifacial N-type back junction silicon solar cells with selective phosphorus front surface field and boron doped poly-Si/SiOx passivated rear emitter. *Applied Physics Letters* 113(26), 263901.

[193] Ling, Z. P., Ge, J., Stangl, R., Aberle, A., and Mueller, T. (2019). Double-sided passivated contacts for solar cell applications: an industrially viable approach toward 24% efficient large area silicon solar cells. In *Silicon Materials* (Vol. 89). London: IntechOpen.

[194] Padhamnath, P., Wong, J., Nagarajan, B., *et al.* (2019). Metal contact recombination in monoPoly^{TM} solar cells with screen-printed & fire-through contacts. *Solar Energy Materials and Solar Cells*, 192, 109–116.

[195] Wu, W., Bao, J., Ma, L., *et al.* (2019). Development of industrial n-type bifacial topcon solar cells and modules. In *Proceedings of the 36th European Photovoltaic Solar Energy Conference and Exhibition* (pp. 100–102).

[196] Chen, Y., Chen, D., Liu, C., *et al.* (2019). Mass production of industrial tunnel oxide passivated contacts (i-TOPCon) silicon solar cells with average efficiency over 23% and modules over 345 W. *Progress in Photovoltaics: Research and Applications,* 27(10), 827–834.

[197] Chen, R., Wright, M., Chen, D., *et al.* (2021). 24.58% Efficient commercial n-type silicon solar cells with hydrogenation. *Progress in Photovoltaics: Research and Applications*, 29(11), 1213–1218.

[198] https://www.pv-magazine.com/2021/06/02/longi-achieves-25-21-efficiency-for-topcon-solar-cell-announces-two-more-records/

[199] https://www.trinasolar.com/fr/resources/newsroom/thu-03242022-1124

[200] https://www.pv-magazine.com/2022/12/09/chinese-pv-industry-brief-jinkosolar-achieves-26-1-efficiency-for-n-type-topcon-solar-cell/

[201] Nandakumar, N., Rodriguez, J., Kluge, T., *et al.* (2019). Investigation of 23% monopoly screen-printed silicon solar cells with an industrial rear passivated contact. In *2019 IEEE 46th Photovoltaic Specialists Conference (PVSC)* (pp. 1463–1465). New York, NY: IEEE.

[202] Feldmann, F., Fellmeth, T., Steinhauser, B., *et al.* (2019). Large area TOPCon cells realized by a PECVD tube process. In *Proceedings of the 36th European Photovoltaic Solar Energy Conference and Exhibition* (pp. 304–308).

[203] https://www.pv-magazine.com/press-releases/jolywood-solar-promotes-the-standardization-of-25-topcon-mass-production-development/

[204] Ingenito, A., Limodio, G., Procel, P., *et al.* (2017). Silicon solar cell architecture with front selective and rear full area ion-implanted passivating contacts. *Solar RRL*, 1(7), 1700040.

[205] Peibst, R., Kruse, C., Schäfer, S., *et al.* (2020). For none, one, or two polarities—how do POLO junctions fit best into industrial Si solar cells? *Progress in Photovoltaics: Research and Applications*, 28(6), 503–516.

[206] Dullweber, T., Stöhr, M., Kruse, C., *et al.* (2020). Evolutionary PERC+ solar cell efficiency projection towards 24% evaluating shadow-mask-deposited poly-Si fingers below the Ag front contact as next improvement step. *Solar Energy Materials and Solar Cells*, 212, 110586.

[207] Young, D. L., Chen, K., Theingi, S., *et al.* (2020). Reactive ion etched, self-aligned, selective area poly-Si/SiO2 passivated contacts. *Solar Energy Materials and Solar Cells*, 217, 110621.

[208] Yuan, S., Cui, Y., Zhuang, Y., *et al.* (2021). Passivated emitter and rear cell silicon solar cells with a front polysilicon passivating contacted selective emitter. *Physica Status Solidi (RRL) – Rapid Research Letters*, 15, 2100057.

[209] Yu, B., Shi, J., Li, F., *et al.* (2021). Selective tunnel oxide passivated contact on the emitter of large-size n-type TOPCon bifacial solar cells. *Journal of Alloys and Compounds*, 870, 159679.

[210] Larionova, Y., Schulte-Huxel, H., Min, B., *et al.* (2020). Ultra-thin poly-Si layers: passivation quality, utilization of charge carriers generated in the poly-Si and application on screen-printed double-side contacted poly-crystalline Si on oxide cells. *Solar RRL*, 4(10), 2000177.

[211] Desrues, T., Lanterne, A., Seron, C., *et al.* (2021). Poly-Si/SiO$_x$ passivating contacts on both sides: a versatile technology for high efficiency solar cells. In *2021 IEEE 48th Photovoltaic Specialists Conference (PVSC)* (pp. 1069–1072). New York NY: IEEE.

[212] Meyer, F., Savoy, A., Leon, J. J. D., *et al.* (2021). Optimization of front SiNx/ITO stacks for high-efficiency two-side contacted c-Si solar cells with co-annealed front and rear passivating contacts. *Solar Energy Materials and Solar Cells*, 219, 110815.

[213] Yan, X., Suhaimi, F. B., Xu, M., *et al.* (2021). Process development and integration of double-side Poly-Si passivated solar cells with printed contacts via LPCVD and ex-situ tube diffusion. *Solar Energy Materials and Solar Cells*, 230, 111249.

[214] Shen, H., Omelchenko, S. T., Jacobs, D. A., *et al.* (2018). In situ recombination junction between p-Si and TiO2 enables high-efficiency monolithic perovskite/Si tandem cells. *Science Advances*, 4(12), eaau9711.

[215] Green, M. A., Dunlop, E. D., Siefer, G., *et al.* (2023). Solar cell efficiency tables (Version 61). *Progress in Photovoltaics: Research and Applications*, 31, 3–16.

[216] Zhang, Y., Kim, M., Wang, L., Verlinden, P., and Hallam, B. (2021). Design considerations for multi-terawatt scale manufacturing of existing and future photovoltaic technologies: challenges and opportunities related to silver, indium and bismuth consumption. *Energy and Environmental Sciences*, 14, 5587–5610. DOI:10.1039/D1EE01814K

[217] Razzaq, A., Allen, T. G., Liu, W. Z., *et al.* (2022). Silicon heterojunction solar cells: Techno-economic assessment and opportunities. *Joule*, 6, 514–542.

[218] Grigorovici, R. Amorphous germanium and silicon. *Materials Research Bulletin*, 3, 13–24, 196.

[219] Fuhs, W., Niemann, K., and Stuke, J. (1974). Heterojunctions of amorphous silicon and silicon single crystals. *AIP Conference Proceedings*, 20, 345. https://doi.org/10.1063/1.2945985

[220] Taguchi, M. (2021). Review—Development history of high efficiency silicon heterojunction solar cell: from discovery to practical use. *ECS Journal of Solid State Science and Technology*, 10, 025002.

[221] Chunduri, S. K. (2010). A HIT for all? *Photon International*, 130–140.

[222] Kekelidze, G., Bubnov, G., Matsapey, N. J., *et al.* (2016). A solar module prototype assembled from silicon heterojunction solar cells manufactured in Gen5 KAI PECVD reactors. In *32nd European Photovoltaic Solar Energy Conference and Exhibition* (pp. 1–1).

[223]   Li, L., Zang, L., Xu, Z., *et al.* (2016). Mass production of high efficiency silicon heterojunction solar cells: a low-cost approach by upgrading GEN8.5 thin film solar line. In *32nd European Photovoltaic Solar Energy Conference and Exhibition* (2BO.9).

[224]   Condorelli, G., Favre, W., Battaglia, A., *et al.* (2018). High efficiency hetero-junction: From pilot line to industrial production. In *2018 IEEE 7th World Conference on Photovoltaic Energy Conversion (WCPEC)* (pp. 1970–1973), doi:10.1109/PVSC.2018.8548197.

[225]   De Wolf, S., Descoeudres, A., Holman, Zachary, C., and Ballif, C. (2012). High-efficiency silicon heterojunction solar cells: a review. *Green*, 2(1), 7–24. https://doi.org/10.1515/green-2011-0018

[226]   van Sark, W. G. J. H. M., Korte, L., Roca, F. (eds.) (2012). Physics and technology of amorphous-crystalline heterostructure silicon solar cells. In *Engineering Materials*. New York, NY: Springer.

[227]   Haschke, J., Dupré, O., Boccard, M., and Ballif, C. (2018). Silicon heterojunction solar cells: recent technological development and practical aspects – from lab to industry. *Solar Energy Materials and Solar Cells*, 187, 140–153. https://doi.org/10.1016/j.solmat.2018.07.018.

[228]   Wenjing, W. Towards over 25% average efficiency of HJT solar cell for mass production. In *Huasun, Taiyangnews Virtual Conference*, 15 December 2021. https://www.youtube.com/watch?v=rcyEb1Riz7A

[229]   Fujiwara, H., Kaneko, T., and Kondo, M. (2007). Application of hydrogenated amorphous silicon oxide layers to c-Si heterojunction solar cells. *Applied Physics Letters*, 91, 133508. https://doi.org/10.1063/1.2790815

[230]   Pysch, D., Bivour, M., Hermle, M., and Glunz, S. W. (2011). Amorphous silicon carbide heterojunction solar cells on p-type substrates. *Thin Solid Films*, 519(8), 2550–2554. ISSN 0040-6090, https://doi.org/10.1016/j.tsf.2010.12.028

[231]   Mazzarella, L., Kirner, S., Stannowski, B., Korte, L., Rech, B., and Schlatmann, R. (2015). p-Type microcrystalline silicon oxide emitter for silicon heterojunction solar cells allowing current densities above 40 mA/cm^2. *Applied Physics Letters,* 106, 023902. https://doi.org/10.1063/1.4905906

[232]   Mazzarella, L., Kirner, S., Stannowski, B., Korte, L., Rech, B., and Schlatmann, R. (2018). Nanocrystalline n-type silicon oxide front contacts for silicon heterojunction solar cells: photocurrent enhancement on planar and textured substrates. *IEEE Journal of Photovoltaics*, 8(1), 70–78. doi:10.1109/JPHOTOV.2017.2770164

[233]   Zhao, Y., Mazzarella, L., Procel, P., *et al.* (2020). Doped hydrogenated nanocrystalline silicon oxide layers for high-efficiency c-Si heterojunction solar cells. *Progress in Photovoltaics: Research and Applications,* 28, 425–435. https://doi.org/10.1002/pip.3256

[234]   Sharma, M., Panigrahi, J., and Komarala, V. K. (2021). Nanocrystalline silicon thin film growth and application for silicon heterojunction solar cells: a short review. *Nanoscale Advances*, 3, 3373–3383.

[235] Koida, T., Fujiwara, H., and Kondo, M. (2009). High-mobility hydrogen-doped In2O3 transparent conductive oxide for a-Si:H/c-Si heterojunction solar cells. *Solar Energy Materials and Solar Cells*, 93(6–7), 851–854. ISSN 0927-0248, https://doi.org/10.1016/j.solmat.2008.09.047

[236] Tohsophon, T., Dabirian, A., De Wolf, S., *et al.* (2015). Environmental stability of high-mobility indium-oxide based transparent electrodes. *APL Materials,* 3, 116105. https://doi.org/10.1063/1.4935125

[237] Senaud, L. L., Christmann, G., Descoeudres, A., *et al.* (2019). Aluminium-doped zinc oxide rear reflectors for high-efficiency silicon heterojunction solar cells. *IEEE Journal of Photovoltaics*, 9(5), 1217–1224. doi:10.1109/JPHOTOV.2019.2926860

[238] Cruz, A., Wang, E. C., Morales-Vilches, A. B., *et al.* (2019). Effect of front TCO on the performance of rear-junction silicon heterojunction solar cells: insights from simulations and experiments. *Solar Energy Materials and Solar Cells*, 195, 339–345. https://doi.org/10.1016/j.solmat.2019.01.047

[239] Wu, Z., Duan, W., Lambertz, A., *et al.* (2021). Low-resistivity p-type a-Si: H/AZO hole contact in high-efficiency silicon heterojunction solar cells. *Applied Surface Science*, 542, 148749. ISSN 0169-4332, https://doi.org/10.1016/j.apsusc.2020.148749

[240] Han, C., Santbergen, R., van Duffelen, M., *et al.* (2022). Towards bifacial silicon heterojunction solar cells with reduced TCO use. *Progress in Photovoltaics: Research and Applications*, 30, 1–13.

[241] Yu, J., Li, J., Zhao, Y., *et al.* (2021). Copper metallization of electrodes for silicon heterojunction solar cells: process, reliability and challenges. *Solar Energy Materials and Solar Cells*, 224, 110993. ISSN 0927-0248, https://doi.org/10.1016/j.solmat.2021.110993

[242] Hatt, T., Bartsch, J., Schellinger, S., *et al.* (2021). Stable copper plated metallization on SHJ solar cells & investigation of selective Al/AlOx laser patterning. In *Proceedings of 38th European PV Solar Energy Conference and Exhibition, 2DV3.*

[243] Stüwe, D., Mager, D., Biro, D., Korvink, J. G. (2015). Inkjet technology for crystalline silicon photovoltaics. *Advanced Materials*, 27, 599–626.

[244] Lossen, J., Matusovsky, M., Noy, A., Maier, C., and Bähr, M. (2015). Pattern transfer printing (PTPTM) for c-Si solar cell metallization. *Energy Procedia*, 67, 156–162. ISSN 1876-6102, https://doi.org/10.1016/j.egypro.2015.03.299

[245] Nos, O., Favre, W., and Jay, F. (2016). Quality control method based on photoluminescence imaging for the performance prediction of c-Si/a-Si:H heterojunction solar cells in industrial production lines. *Solar Energy Materials and Solar Cells*, 144, 210–220.

[246] Cattin, J., Haschke, J., Ballif, C., and Boccard, M. (2020). Influence of local surface defects on the minority-carrier lifetime of passivating-contact solar cells. *Applied Physics Letters*, 116, 113901. https://doi.org/10.1063/1.5145351

[247]   Giglia, V., Varache, R., Veirman, J., and Fourmond, E. (2020). Understanding of the influence of localized surface defectivity properties on the performances of silicon heterojunction cells. *Progress in Photovoltaics: Research and Applications*, 28, 1333–1344.

[248]   Varache, R., Nguyen, N., and Muñoz, D. (2014). 2D p-FF simulations for the interpretation of junction isolation's influence on silicon heterojunction solar cells. *Energy Procedia*, 55, 149–154.

[249]   Giglia, V., Varache, R., Veirman, J., and Fourmond, E. (2022). Influence of cell edges on the performance of silicon heterojunction solar cells. *Solar Energy Materials and Solar Cells*, 238, 111605. ISSN 0927-0248, https:// doi.org/10.1016/j.solmat.2022.111605

[250]   https://www.longi.com/en/news/propelling-the-transformation/

[251]   https://www.longi.com/en/news/new-hjt-world-record/

[252]   Richter, A., Hermle, M., and Glunz, S. W. (2013). Reassessment of the limiting efficiency for crystalline silicon solar cells. *In IEEE Journal of Photovoltaics*, 3(4), 1184–1191. doi:10.1109/JPHOTOV.2013.2270351

[253]   Veith-Wolf, B. A., Schäfer, S., Brendel, R., and Schmidt, J. (2018). Reassessment of intrinsic lifetime limit in n-type crystalline silicon and implication on maximum solar cell efficiency. *Solar Energy Materials and Solar Cells*, 186, 194–199. ISSN 0927-0248, https://doi.org/10.1016/j. solmat.2018.06.029

[254]   https://www.pv-magazine-australia.com/2022/09/05/sundrive-hits-efficiency-high-with-copper-based-solar-cell-technology/

[255]   Wenjing, W. Towards over 25% average efficiency of HJT solar cell for mass production. In *Huasun, Taiyangnews Virtual Conference*, 15 December 2021, https://www.youtube.com/watch?v=rcyEb1Riz7A

[256]   Haschke, J., Christmann, G., Messmer, C., *et al.* (2020). Lateral transport in silicon solar cells. *Journal of Applied Physics,* 127, 114501. https://doi.org/ 10.1063/1.5139416

[257]   Paviet-Salomon, B., Descoeudres, A., Christmann, G., *et al.* (2019). High performance back-contacted silicon heterojunction solar cells (p. 70). CSEM Sci., Tech. Rep..

[258]   Wolf, S. D. (2012). Intrinsic and doped a-Si:H/c-Si interface passivation. *Engineering Materials*, 105, 223–259.

[259]   Reusch, M., Bivour, M., Hermle, M., and Glunz, S. W. (2013). Fill factor limitation of silicon heterojunction solar cells by junction recombination. *Energy Procedia*, 38, 297–304. ISSN 1876-6102, https://doi.org/10.1016/j. egypro.2013.07.281

[260]   Adachi, D., Hernández, J. L., and Yamamoto, K. (2015). Impact of carrier recombination on fill factor for large area heterojunction crystalline silicon solar cell with 25.1% efficiency. *Applied Physics Letters,* 107(23), 233506. https://doi.org/10.1063/1.4937224

[261]   Veirman, J., Varache, R., Albaric, M., *et al.* (2021). Silicon wafers for industrial n-type SHJ solar cells: bulk quality requirements, large-scale

availability and guidelines for future developments. *Solar Energy Materials and Solar Cells*, 22815, 111128. 10.1016/j.solmat.2021.111128

[262] Garin, Rau, U., Brendle, W., *et al.* (2005). Characterization of a-Si:H/c-Si interfaces by effective-lifetime measurements. *Journal of Applied Physics*, 98, 093711.

[263] Olibet, S., Vallat-Sauvain, E., and Ballif, C. (2007). Model for a-Si:H/c-Si interface recombination based on the amphoteric nature of silicon dangling bonds. *Physical Review B – Condensed Matter and Materials Physics*, 76(3), 1–14.

[264] Leendertz, C., Stangl, R., Schulze, T. F., *et al.* (2010). A recombination model for a-Si:H/c-Si heterostructures. *Physica Status Solidi C*, 7(3–4), 1005–1010. doi: 10.1002/pssc.200982698

[265] Schulze, T. F., Korte, L., and Rech, B. (2010). Interplay of amorphous silicon disorder and hydrogen content with interface defects in amorphous/crystalline silicon heterojunctions. *Applied Physics Letters*, 96, 252102. https://doi.org/10.1063/1.3455900

[266] Ru, X., Quab, M. Wang, J., *et al.* (2020). 25.11% efficiency silicon heterojunction solar cell with low deposition rate intrinsic amorphous silicon buffer layers. *Solar Energy Materials & Solar Cells*, 215, 110643.

[267] De Wolf, S., Demaurex, B., Descoeudres, A., and Ballif, C. (2011). Very fast light-induced degradation of a-Si:H/c-Si(100) interfaces. *Physical Review B*, 83, 233301.

[268] Demaurex, B., Wolf, S. D., Descoeudres, A., Holman, C. Z., and Ballif, C. (2012). Damage at hydrogenated amorphous/crystalline silicon interfaces by indium tin oxide overlayer sputtering. *Applied Physics Letters,* 101, 171604. https://doi.org/10.1063/1.4764529

[269] Favre, W., Coignus, J., Nguyen, N., Lachaume, R., Cabal, R., and Muñoz, D. (2013). Influence of the transparent conductive oxide layer deposition step on electrical properties of silicon heterojunction solar cells. *Applied Physics Letters*, 102(18), 181118.

[270] Maslova, O., Brézard-Oudot, A., Gueunier-Farret, M.-E., *et al.* (2010). Explicit analytical modeling of the low frequency a-Si:H/c-Si heterojunction capacitance: analysis and application to silicon heterojunction solar cells. *Applied Physics Letters,* 97, 252110. https://doi.org/10.1063/1.3525166

[271] Kleider, J.-P., Alvarez, J., Ankudinov, A. V., *et al.* (2011). Characterization of silicon heterojunctions for solar cells. *Nanoscale Research Letters*, 6, 152.

[272] Kleider, J.-P., Marchat, C., Morisset, A., *et al.* (2019). Electrical scanning probe microscopy approaches to investigate solar cell junctions and devices. *Physica Status Solidi A*, 216, 1800877.

[273] Lachaume, R., Favre, W., Scheiblin, P., *et al.* (2013). Influence of a-Si:H/ITO interface properties on performance of heterojunction solar cells. *Energy Procedia*, 38, 770–776.

[274] Varache, R., Aguila, O. N., Valla, A., Nguyen, N., and Munoz, D. (2015). Role of the front electron collector in rear emitter silicon heterojunction solar cells. *IEEE Journal of Photovoltaics*, 5(3), 711–717.

[275] Kleider, J. P., Soro, Y. M., Chouffot, R., *et al.* (2008). High interfacial conductivity at amorphous silicon/crystalline silicon heterojunctions. *Journal of Non-Crystalline Solids*, 354, 2641.

[276] Halliop, B., Salaun, M. F., Favre, W., *et al.* (2012). Interface properties of amorphous-crystalline silicon heterojunctions prepared using DC saddle-field PECVD. *Journal of Non-Crystalline Solids*, 358, 2227.

[277] Favre, W., Labrune, M., Dadouche, F., *et al.* (2010). Study of the interfacial properties of amorphous silicon/n-type crystalline silicon heterojunction through static planar conductance measurements. *Physica Status Solidi C*, 7, 1037.

[278] Varache, R., Kleider, J. P., Favre, W., and Korte, L. (2012). Band bending and determination of band offsets in amorphous/crystalline silicon hetero-structures from planar conductance measurements. *Journal of Applied Physics*, 112(12), 123717.

[279] Nogay, G., Seif, J. P., Riesen, Y., *et al.* (2016). Nanocrystalline silicon carrier collectors for silicon heterojunction solar cells and impact on low-temperature device characteristics. *IEEE Journal of Photovoltaics*, 6(6), 1654–1662.

[280] Crandall, R. S., Iwaniczko, E., Li, J. V., and Page, M. R. (2012). A comprehensive study of hole collection in heterojunction solar cells. *Journal of Applied Physics*, 112(9), 093713.

[281] Kanevce, A. and Metzger, W. K. (2009). The role of amorphous silicon and tunneling in heterojunction with intrinsic thin layer (HIT) solar cells. *Journal of Applied Physics*, 105(9), 094507.

[282] Procel, P., Xu, H., Saez, A., *et al.* (2020). The role of heterointerfaces and subgap energy states on transport mechanisms in silicon heterojunction solar cells. *Progress in Photovoltaics: Research and Applications*, 28, 935–945.

[283] Filipič, M., Holman, Z. C., Smole, F., De Wolf, S., Ballif, C., and Topič, M. (2013). Analysis of lateral transport through the inversion layer in amorphous silicon/crystalline silicon heterojunction solar cells. *Journal of Applied Physics*, 114(7), 074504.

[284] Basset, L. (2020). Ph.D. Thesis. Contact electrodes for heterojunction silicon solar cells: Evaluations and optimizations of the electron contact. Physics [physics]. Université de Lille, English ⟨tel-03159905⟩.

[285] Bivour, M., Schröer, S., Hermle, M., and Glunz, S. W. (2014). Passivated rear contacts for high-efficiency n-type Si solar cells providing high interface passivation quality and excellent transport characteristics. *Solar Energy Materials & Solar Cells*, 122, 120–129.

[286] Lachenal, D., Baetzner, D., Frammelsberger, W., *et al.* (2016). Heterojunction and passivated contacts: a simple method to extract both n/tco and p/tco contacts resistivity. *Energy Procedia*, 92, 932–938.

[287] Shenghao, L., Andreas Lambertz, M. P., Karsten Bittkau, W. D., *et al.* (2021). Transparent-conductive-oxide-free front contacts for high-efficiency silicon heterojunction solar cells. *Joule*, 5(6), 1535–1547. ISSN 2542-4351, https://doi.org/10.1016/j.joule.2021.04.004

[288] Basset, L., Favre, W., Bonino, O., Sudre, J., Ménard, G., and Vilcot, J. P. (2021). In depth analysis of transfer length method application on passivated contacts under illumination. *Solar Energy Materials and Solar Cells*, 230, 111255. ISSN 0927-0248, https://doi.org/10.1016/j.solmat.2021.111255

[289] Senaud, L.-L., Procel, P., Christmann, G., *et al.* (2021). Advanced method for electrical characterization of carrier-selective passivating contacts using transfer-length-method measurements under variable illumination. *Journal of Applied Physics,* 129, 195707. doi:10.1063/5.0042854

[290] Holman, Z. C., Descoeudres, A., Barraud, L., *et al.* (2012). Current losses at the front of silicon heterojunction solar cells. *IEEE Journal of PV*, 2(1), 7–15.

[291] Danel, A., Eymard, J., Pernoud, F., *et al.* Bifaciality optimization of silicon heterojunction solar cells. In *Proceedings of the 36th EUPVSEC*, 2CO.10.6, 2019, Marseille.

[292] Kobayashi, E., de Wolf, S., Levrat, J., *et al.* (2016). Light-induced performance increase of silicon heterojunction solar cells. *Applied Physics Letters*, 109(15), 153503.

[293] Letty, E., Veirman, J., Favre, W., and Lemiti, M. (2017). Bulk defect formation under light soaking in seed-end n-type Czochralski silicon wafers – effect on silicon heterojunction solar cells. *Solar Energy Materials & Solar Cells*, 166, 147–156.

[294] Kobayashi, E., De Wolf, S., Levrat, J., *et al.* (2017). Increasing the efficiency of silicon heterojunction solar cells and modules by light soaking. *Solar Energy Materials and Solar Cells,* 173, 43–49.

[295] Cattin, J., Senaud, L. -L., Haschke, J., *et al.* (2021). Influence of light soaking on silicon heterojunction solar cells with various architectures. *IEEE Journal of Photovoltaics*, 11(3), 575–583.

[296] Veirman, J., Leoga, A. J. K., Basset, L., *et al.* Understanding the improvement of silicon heterojunction solar cells under light soaking. In *AIP Conference Proceedings, SiliconPV2021, the 11th International Conference on Crystalline Silicon Photovoltaics.*

[297] Bao, S., Yang, L., Huang, J., *et al.* (2021). The rapidly reversible processes of activation and deactivation in amorphous silicon heterojunction solar cell under extensive light soaking. *Journal of Materials Science: Materials in Electronics*, 32, 4045–4052. https://doi.org/10.1007/s10854-020-05146-0

[298] Wright, M., Soeriyadi, A., Wright, B., *et al.* (2022). High-intensity illuminated annealing of industrial SHJ solar cells: a pilot study. *IEEE Journal of Photovoltaics*, 12(1), 267–273.

[299] Zhang, Y., Kim, M., Wang, L., Verlinden, P., and Hallam, B. (2021). Design considerations for multi-terawatt scale manufacturing of existing

and future photovoltaic technologies: challenges and opportunities related to silver. *Energy Environmental Science*, 14, 5587–5610. doi:10.1039/D1EE01814K

[300]   Chang, N. L., Wright, M., Egan, R., and Hallam, B. (2020). The technical and economic viability of replacing n-type with p-type wafers for silicon heterojunction solar cells. *Cell Reports Physical Science*, 1, 100069.

[301]   Danel, A., Chaugier, N., Veirman, J., Varache, R., Albaric, M., and Pihan, E. (2022). Closing the gap between n- and p-type silicon heterojunction solar cells: 24.47% efficiency on lightly doped Ga wafers. Prog. Photovolt. Res. Appl., doi: 10.1002/pip.3635

[302]   https://www.longi.com/en/news/p-type-hjt-record/

[303]   Harrison, S., Nos, O., Danel, A., *et al.* (2016). How to deal with thin wafers in a heterojunction solar cells industrial pilot line: first analysis of the integration of cells down to 70 μm thick in production mode, 2.B.9.2. In *Proceedings of the EUPVSEC*, Amsterdam.

[304]   Danel, A., Harrison, S., Gerenton, F., *et al.* (2018). Silicon heterojunction solar cells with open-circuit voltage above 750 mV, 2.DO.1.2. In *EUPVSEC Proceedings*, Bruxelles.

[305]   Long, W., Yin, S., Peng, F., *et al.* (2021). On the limiting efficiency for silicon heterojunction solar cells. *Solar Energy Materials and Solar Cells*, 231, 111291. ISSN 0927-0248, https://doi.org/10.1016/j.solmat.2021.111291

[306]   Sahli, F., Werner, J., Kamino, B. A., *et al.* (2018). Fully textured monolithic perovskite/silicon tandem solar cells with 25.2% power conversion efficiency. *Nature Materials,* 17, 820–826. https://doi.org/10.1038/s41563-018-0115-4

[307]   Al-Ashouri, A., Kohnen, E., Li, B., *et al.* (2020). Monolithic perovskite/silicon tandem solar cell with >29% efficiency by enhanced hole extraction. *Science*, 370(6522), 1300–1309. doi:10.1126/science.abd4016

[308]   Schwartz, R. J. and Lammert, M. D. (1975). Silicon solar cells for high concentration applications. In *International Electron Devices Meeting 1975*, 21, 350–352.

[309]   Lammert, M. D. and Schwartz, R. J. (1977). The interdigitated back contact solar cell: a silicon solar cell for use in concentrated sunlight. *IEEE Transactions on Electron Devices*, 24(4), 337–342.

[310]   SunPower, "History." [Online]. Available: https://us.sunpower.com/company/history [Accessed: 11-May-2020].

[311]   Hutchins, M. (2020). Zebra IBC cell with 24% efficiency moves to large-scale production. PV Magazine, 2020. [Online]. Available: https://www.pv-magazine.com/2020/09/04/zebra-ibc-cell-with-24-efficiency-moves-to-large-scale-production/.

[312]   Lu, M., Bowden, S., Das, U., and Birkmire, R. (2007). Interdigitated back contact silicon heterojunction solar cell and the effect of front surface passivation. *Applied Physics Letters*, 91(6), 063507.

[313]   Cousins, P. (2007). Solar cell having doped semiconductor heterojunction contacts. WO 2007/130188 A2.

[314]   Swanson, R. M. (2008). Backside contact solar cell with doped polysilicon regions. US007468485B1.

[315]   Masuko, K., Shigematsu, M., Hashiguchi, T., *et al.* (2014). Achievement of more than 25 % conversion efficiency with crystalline silicon heterojunction solar cell. *IEEE Journal of Photovoltaics*, 4(6), 1433–1435.

[316]   Nakamura, J., Asano, N., Hieda, T., Okamoto, C., Katayama, H., and Nakamura, K. (2014). Development of heterojunction back contact Si solar cells. *IEEE Journal of Photovoltaics*, 4(6), 1491–1495.

[317]   Smith, D. D., Cousins, P., Westerberg, S., De Jesus-Tabajonda, R., Aniero, G., and Shen, Y.-C. (2014). Toward the practical limits of silicon solar cells. *IEEE Journal of Photovoltaics*, 46, 1–5.

[318]   Zhao, J., Wang, A., Green, M. A., and Ferrazza, F. (1998). 19.8% efficient 'honeycomb' textured multicrystalline and 24.4% monocrystalline silicon solar cells. *Applied Physics Express*, 73(14), 1991–1993.

[319]   Yoshikawa, K., Kawasaki, H., Yoshida, W., *et al.* (2017). Silicon heterojunction solar cell with interdigitated back contacts for a photoconversion efficiency over 26%. *Nature Energy*, 2, 17032.

[320]   Green, M. A., Hishikawa, Y., Warta, W., *et al.* (2017). Solar cell efficiency tables (version 50). *Progress in Photovoltaics: Research and Applications*, 25(7), 668–676.

[321]   Osborne, M. SunPower hits average cell conversion efficiencies of 25% at Fab 4, 2017. [Online]. Available: https://www.pv-tech.org/news/sunpower-hits-average-cell-conversion-efficiencies-of-25-at-fab-4.

[322]   Haase, F., Hollemann, C., Schäfer, S., *et al.* (2018). Laser contact openings for local poly-Si-metal contacts enabling 26.1%-efficient POLO-IBC solar cells. *Solar Energy Materials and Solar Cells*, 186, 184–193.

[323]   Tomasi, A., Salomon, B. P., Jeangros, Q., *et al.* (2017). Simple processing of back-contacted silicon heterojunction solar cells using selective-area crystalline growth. *Nature Energy*, 2, 17062.

[324]   Boccard, M., Paratte, V., Antognini, L., *et al.* (2021). Perfecting silicon. In *Proceedings of the 38th European Photovoltaic Solar Energy Conference and Exhibition (WCPEC)*, Milan, Italy.

[325]   Peibst, R., Haase, F., Hollemann, C., *et al.* (2021). Exploiting single-junction efficiency potentials and going beyond. In *EUPVSEC* 2021.

[326]   PRNewswire. LONGi breaks world record for HJT solar cell efficiency twice in one week, 2021. [Online]. Available: https://www.prnewswire.com/news-releases/longi-breaks-world-record-for-hjt-solar-cell-efficiency-twice-in-one-week-301412741.html.

[327]   M. of S. Energy, "SunPower Maxeon II." [Online]. Available: https://solarmuseum.org/cells/sunpower-maxeon-ii/.

[328]   Stangl, R., Haschke, J., Bivour, M., Schmidt, M., Lips, K., and Rech, B. (2008). Planar rear emitter back contact amorphous/crystalline silicon heterojunction solar cells (RECASH/PRECASH). In *Conference Record Of the IEEE Photovoltaic Specialists Conference* (pp. 8–13).

[329]   Van Kerschaver, E., Einhaus, R., and Szlufcik, J. (1998). A novel silicon solar cell structure with both external polarity contacts on the back surface. In *2nd World Conference on Photovoltaic Energy Conversion* (pp. 1479–1482).

[330]   Gee, J. M., Schubert, W. K., and Basore, P. A. (1993). Emitter wrap-through solar cell. In *Conference Record of the Twenty Third IEEE Photovoltaic Specialists Conference* (pp. 265–270).

[331]   Thaidigsmann, B., Spribille, A., and Plagwitz, H. (2011). HIP-MWT – a new cell concept for industrial processing of high-performance metal wrap through silicon solar cells. In *26th EUPVSEC* (pp. 817–820).

[332]   Hermann, S., Merkle, A., Ulzhöfer, C., *et al.* (2011). Progress in emitter wrap-through solar cell fabrication on boron doped Czochralski-grown silicon. *Solar Energy Materials and Solar Cells*, 95(4), 1069–1075.

[333]   Van Kerschaver, E. and Beaucarne, G. (2006). Back-contact solar cells: a review. *Progress in Photovoltaics: Research and Applications*, 14(2), 107–123.

[334]   Tomasi, A. (2016). Back-contacted silicon heterojunction solar cells. In *EPFL*.

[335]   Hendrichs, M., Thaidigsmann, B., Fellmeth, T., *et al.* (2013). Cost-optimized metallization layout for metal wrap through (MWT) solar cells and modules. In *28th EUPVSEC*.

[336]   Hermle, M., Granek, F., Schultz-wittmann, O., and Glunz, S. W. (2008). Shading effects in back-junction back-contacted silicon solar cells. In *Proceedings of the 33rd IEEE Photovoltaic Specialists Conference* (pp. 10–13).

[337]   Reichel, C., Granek, F., Hermle, M., and Glunz, S. W. (2011). Investigation of electrical shading effects in back-contacted back-junction silicon solar cells using the two-dimensional charge collection probability and the reciprocity theorem. *Journal of Applied Physics*, 109(2), 024507.

[338]   Procel, P., Yang, G., Isabella, O., and Zeman, M. (2018). Theoretical evaluation of contact stack for high efficiency IBC-SHJ solar cells *Solar Energy Materials and Solar Cells*, 186, 66–77.

[339]   Lachenal, D., Papet, P., Legradic, B., *et al.* (2019). Optimization of tunnel-junction IBC solar cells based on a series resistance model. *Solar Energy Materials and Solar Cells*, 200, 110036.

[340]   Procel, P., Xu, H., Mazzarella, L., *et al.* (2019). On the correlation between contact resistivity and high efficiency in (IBC-) SHJ solar cells. In *Proceedings of the 36th European Photovoltaic Solar Energy Conference and Exhibition (2019) 2CO.12.6* (pp. 251–254).

[341]   Rienäcker, M., Bossmeyer, M., Merkle, A., *et al.* (2017). Junction resistivity of carrier selective polysilicon on oxide junctions and its impact on the solar cell performance. *IEEE Journal of Photovoltaics*, 7(1), 11–18.

[342]   Procel, P., Xu, H., Saez, A., *et al.* (2020). The role of heterointerfaces and subgap energy states on transport mechanisms in silicon heterojunction solar cells. *Progress in Photovoltaics: Research and Applications,* 28(9), 935–945.

[343]   Stang, J.-C., Franssen, T., Haschke,J., *et al.* (2017). Optimized metallization for interdigitated back contact silicon heterojunction solar cells. *Solar RRL*, 1(3–4), 1700021.

[344] Xu, M., Bearda, T., Radhakrishnan, H. S., and Jonnak, S. K. (2016). Process development of silicon heterojunction interdigitated back-contacted (SHJ-IBC) solar cells bonded to glass. In *Proceedings of the 32nd European Photovoltaic Solar Energy Conference.*

[345] Haase, F., Schäfer, S., Kiefer, F., Krügener, J., Brendel, R., and Peibst, R. (2017). Perimeter recombination in 25%-efficient IBC solar cells with passivating POLO contacts for both polarities. In *Proceedings of the 44th IEEE Photovoltaic Specialists Conference IEEE Photovoltaic Specialists Conference.*

[346] Desrues, T., De Vecchi, S., Souche, F., Munoz, D., and Ribeyron, P. J. (2012). SLASH concept: a novel approach for simplified interdigitated back contact solar cells fabrication. In *Conference Record of the IEEE Photovoltaic Specialists Conference* (pp. 1602–1605).

[347] Xu, M., Bearda, T., Filipič, M., *et al.* (2018). Simple emitter patterning of silicon heterojunction interdigitated back-contact solar cells using damage-free laser ablation. *Solar Energy Materials and Solar Cells*, 186, 78–83.

[348] Krause, S., Miclea, P. T., Schweizer, S., and Seifert, G. (2013). Optimized scribing of TCO layers on glass by selective femtosecond laser ablation. In *2013 IEEE 39th Photovoltaic Specialists Conference (PVSC)* (pp. 2432–2435).

[349] Vasudevan, R., Harrison, S., D'Alonzo, G., *et al.* (2018). Laser-induced BSF: a new approach to simplify IBC-SHJ solar cell fabrication. In *AIP Conference Proceedings*, 1999.

[350] Haase, F., Hollemann, C., Schafer, S., Krugener, J., Brendel, R., and Peibst, R. (2019). Transferring the record p-type Si POLO-IBC cell technology towards an industrial level. In *Conference Record of the IEEE Photovoltaic Specialists Conference* (pp. 2200–2206).

[351] Tucci, M., Serenelli, L., Salza, E., and Pirozzi, L. (2008). Back enhanced heterostructure with INterDigitated contact – BEHIND – solar cell. In *Optoelectronics and Microelectronic Materials and Devices* (pp. 242–245).

[352] Tomasi, A., Salomon, B. P., Lachenal, D., *et al.* (2014). Back-contacted silicon heterojunction solar cells with efficiency > 21%. *IEEE Journal of Photovoltaics*, 4(4), 1046–1054.

[353] Tomasi, A., Salomon, B. P., Lachenal, D., *et al.* (2014). Photolithography-free interdigitated back-contacted silicon heterojunction solar cells with efficiency > 21%. In *Proceedings of the 40th IEEE Photovoltaic Specialists Conference.*

[354] Desrues, T. (2009). Développement de cellules photovoltaïques à hétérojonctions silicium et contacts en face arrière. *Institut National des Sciences Appliquées de Lyon.*

[355] Guittienne, P., Grange, D., Hollenstein, C., and Gindrat, M. (2012). Plasma jet-substrate interaction in low pressure plasma spray-CVD processes. *Journal of Thermal Spray Technology*, 21(2), 202–210.

[356] Ledinský, M., Salomon, B. P., Vetushka, A., *et al.* (2016). Profilometry of thin films on rough substrates by Raman spectroscopy. *Scientific Reports*, 6, 37859.

[357]   Spee, D. A. (2008). *Preparations for Making Back Contacted Heterojunction Solar Cells.* Utrecht University.

[358]   Zhang, X., Yang, Y., Liu, W., *et al.* Development of high efficiency interdigitated back contact silicon solar cells and modules with industrial processing. In *WCPEC-6.*

[359]   Jooß, W. (2002). Multicrystalline and back contact buried contact silicon solar cells. Dissertation [https://kops.uni-konstanz.de/handle/123456789/9116].

[360]   https://us.sunpower.com/solar-resources/what-ibc-solar-cell-technology

[361]   https://www.lg.com/de/business/neon-r

[362]   Halm, A., Mihailetchi, V. D., Galbiati, G., *et al.* (2012). The zebra cell concept-large area n-type interdigitated back contact solar cells and one-cell modules fabricated using standard industrial processing equipment. In *27nd EU-PVSEC*, Frankfurt, Germany.

[363]   Kopecek, R., Libal, J., Lossen, J., *et al.* (2020). ZEBRA technology: low cost bifacial IBC solar cells in mass production with efficiency exceeding 23.5%. In *2020 47th IEEE Photovoltaic Specialists Conference (PVSC)* (pp. 1008–1012). doi:10.1109/PVSC45281.2020.9300503.

[364]   http://www.spicsolar.com/en/product.asp?id=431

[365]   https://www.futurasun.com/de/produkte/monokristalline-photovoltaikmodule/zebra-back-contact-120-half-cut-ibc-zellen-all-black/

[366]   Kruse, C. N., Schäfer, S., Haase, F., *et al.* (2021). Simulation-based road-map for the integration of poly-silicon on oxide contacts into screen-printed crystalline silicon solar cells. *Scientific Reports*, 11, 996.

[367]   Stöhr, M., Aprojanz, J., Brendel, R., and Dullweber, T. (2021). Firing-stable PECVD SiOxNy/n-poly-Si surface passivation for silicon solar cells. *ACS Applied Energy Materials*, 4(5), 4646–4653.

[368]   Mertens, V., Schäfer, S., Stöhr, M., *et al.* (2021). Local PECVD SiOxNy/n-Poly-Si deposition through a shadow mask for POLO IBC solar cells. In *Proceedings of the 38th European Photovoltaic Solar Energy Conference and Exhibition (WCPEC)*, Milan, Italy; pp. 135–139.

[369]   Dullweber, T., Mertens, V., Stöhr, M., *et al.* (2022). Towards cost-competitive high-efficiency POLO IBC solar cells with minimal conversion invest for existing PERC+ production lines. In *Proceedings of the 8th World Conference on Photovoltaic Energy Conversion (WCPEC)*, Milan, Italy; pp. 35–39.

[370]   ZEBRA 24%: Plenary WCPEC-8

[371]   Clean Energy Review, 2022 July. https://www.cleanenergyreviews.info/blog/most-efficient-solar-panels

[372]   TRINA 25% IBC. https://www.trinasolar.com/us/resources/newsroom/trina-solar-announces-new-efficiency-record-2504-large-area-ibc-mono-crystalline

[373]   Stodolny, M. K., Anker, J., Geerligs, L. J., *et al.* (2017). Material properties of LPCVD processed n-type polysilicon passivating contacts and its application in PERPoly industrial bifacial solar cells, ECN-L–17-039 September 2017; 19 pp, Presented at: EUPVSEC, Amsterdam, Netherlands, 28 September 2017.

[374] https://taiyangnews.info/technology/solar-grids-launches-all-back-contact-modules/

[375] https://www.lgnewsroom.com/2022/02/lg-to-close-solar-panel-business/

[376] Paviet-Salomon, B., Tomasi, A., Lachenal, D., *et al.* (2018). Interdigitated back contact silicon heterojunction solar cells featuring an interband tunnel junction enabling simplified processing. *Solar Energy*, 175, 60–67.

[377] Seif, J.-P. (2015). *Window Layers for Silicon Heterojunction Solar Cells: Properties and Impact on Device Performance*. École polytechnique fédérale de Lausanne.

[378] Smith, D. D., Reich, G., Baldrias, M., Reich, M., Boitnott, N., and Bunea, G. (2016). Silicon solar cells with total area efficiency above 25%. In *Conference Record of the IEEE Photovoltaic Specialists Conference* (pp. 3351–3355).

[379] Galbiati, G., Mihailetchi, V. D., Roescu, R., *et al.* (2013). Large-area back-contact back-junction solar cell with efficiency exceeding 21%. *IEEE Journal of Photovoltaics*, 3(1), 560–565.

[380] Mihailetchi, V. D., Chu, H., Galbiati, G., Comparroto, C., Halm, A., and Kopecek, R. (2015). A comparison study of n-type PERT and IBC cell concepts with screen printed contacts. *Energy Procedia*, 77, 534–539.

[381] Galbiati, G., Mihailetchi, V. D., Halm, A., Roescu, R., and Kopecek, R. (2011). Results on n-type IBC solar cells using industrial optimized techniques in the fabrication processing. *Energy Procedia*, 8, 421–426.

[382] NextBase. NextBase deliverable report, IBC-SHJ device with efficiency > 26.0 % on 6-in wafer.

[383] Schoot, H. Introduction to back-contact cell/module technology and DSM conductive backsheets. In *PV Magazine* webinar "Back-contact's move to the front," 2020, no. September.

[384] Gota, F., Langenhorst, M., Schmager, R., Lehr, J., and Paetzold, U. W. (2020). Energy yield advantages of three-terminal perovskite-silicon tandem photovoltaics. *Joule*, 4(11), 2387–2403.

[385] Tockhorn, P., Wagner, P., Kegelmann, L., *et al.* (2020). Three-terminal perovskite/silicon tandem solar cells with top and interdigitated rear contacts. *ACS Applied Energy Materials*, 3(2), 1381–1392.

[386] Faes, A., Despeisse, M., Levrat, J. (2014). SmartWire solar cell interconnection technology. In *Proceedings of the 29th European Photovoltaic Solar Energy Conference* (pp. 2555–2561).

[387] ITRPV 2012, International Technology Roadmap for Photovoltaics Results 2012, Third Edition, Berlin 2012

[388] VDMA Photovoltaic. International Technology Roadmap for Photovoltaic (ITRPV), 2021.

[389] Tepner, S., Ney, L., Linse, M., Lorenz, A., Pospischil, M., and Clement, F. (2020). Studying knotless screen patterns for fine-line screen printing of Si-solar cells. *IEEE Journal of Photovoltaics*, 10(2), 319–325.

[390] Bettinelli, A., Debourdeau, M., Diaz, J., *et al.* (2019). SHJ cell metallization for various interconnection technologies. In *5th Solar Cell Paste and Metallization Forum*, Changzhou.

[391]   Faes, A., Lachowicz, A., Bettinelli, A., *et al.* (2018). Metallization and interconnection for high-efficiency bifacial silicon heterojunction solar cells and modules. *Photovoltaics International*, 41, 65–76.

[392]   HJT Fully made in China – TaiyangNews April, 27th 2021. https://taiyangnews.info/technology/hjt-fully-made-in-china/

[393]   World Silver Survey 2021. [Online]. Available: chrome-extension://efaidnbmnnnibpcajpcglclefindmkaj/viewer.html?pdfurl=https%3A%2F%2Fwww.silverinstitute.org%2Fwp-content%2Fuploads%2F2021%2F04%2FWorld-Silver-Survey-2021.pdf&clen=14979104&chunk=true.

[394]   Haegel, N. M., Atwater H. Jr, Barnes, T., *et al.* (2019). Terawatt-scale photovoltaics: transform global energy. *Science*, 364(6443), 836–838.

[395]   Verlinden, P. J. (2020). Future challenges for photovoltaic manufacturing at the terawatt level. *Journal of Renewable and Sustainable Energy*, 12(5), 053505.

[396]   Kamino, B. A., Salomon, B. P., Moon, S.-J., *et al.* (2019). Low-temperature screen-printed metallization for the scale-up of two-terminal perovskite-silicon tandems. *ACS Applied Energy Materials*, 2(5), 3815–3821.

[397]   Galiazzo, M. and Frasson, N. (2021). Evaluation of different approaches for HJT cells metallization based on low temperature Cu paste. In *MIW 2021*.

[398]   Wang, W. (2022). HJT mass production in Huasun. In *nPV Workshop* 2022.

[399]   Schmiga, C. (2021). 23.8% Efficient bifacial i-TOPCon silicon solar cells with < 20 μm wide Ni/Cu/Ag-plated contact fingers. In *MIW2021*.

[400]   Lachowicz, A. (2021). Reinforcement of screen-printed copper paste by electroplated copper for metallization of silicon heterojunction solar cells. In *MIW2021*.

[401]   Liu, C. M., Liu, W. L., Chen, W. J., Hsieh, S. H., Tsai, T. K., and Yang, L. C. (2005). ITO as a diffusion barrier between Si and Cu. *Journal of the Electrochemical Society*, 152(3), G234.

[402]   Yu, J., Bian, J., Duan, W., *et al.* (2016). Tungsten doped indium oxide film: ready for bifacial copper metallization of silicon heterojunction solar cell. *Solar Energy Materials and Solar Cells*, 144, 359–363.

[403]   Lachowicz, A., Christmann, G., Descoeudres, A., Nicolay, S., and Ballif, C. (2020). Silver- and indium-free silicon heterojunction solar cell. In *Proceedings of the 37th European Photovoltaic Solar Energy Conference and Exhibition*.

[404]   Hatt, T., Bartsch, J., Schellinger, S., Jahn, M., and Tutsch, L. (2021). Copper electroplating for SHJ solar cells – adequate contact by electrolyte tuning. In *MIW2021*.

[405]   Lachowicz, A., Andreatta, G., Blondiaux, N., *et al.* (2021). Patterning techniques for copper electroplated metallization of silicon heterojunction cells. In *Conference Record of the IEEE Photovoltaic Specialists Conference* (pp. 1530–1533).

[406]   PV Magazine . (2022). SunDrive sets 26.07% efficiency record for heterojunction PV cell in mass production, 2022. [Online]. Available: https://www.pv-magazine.com/2022/03/17/sundrive-sets-26-07-efficiency-record-for-heterojunction-pv-cell-in-mass-production/.

Chapter 4

n-type silicon modules

*Andreas Schneider[1], Emilio Muñoz-Cerón[2],
Jorge Rabanal-Arabach[3], Eszter Voroshazi[4],
Vincent Barth[4] and Rubén Contreras Lisperguer[5]*

4.1 General module requirements

4.1.1 Potential energy generation increase

The photovoltaic industry is facing an exponential growth in the recent years fostered by a dramatic decrease in installation prices. This cost reduction is achieved by means of several mechanisms. First, because of the optimization of the design and installation process of current PV projects, and second, by the optimization, in terms of performance, in the manufacturing techniques and material combinations within the modules, which also has an impact on both, the installation process, and the levelized cost of electricity (LCOE).

One popular trend is to increase the power delivered by photovoltaic modules, either by using larger wafer sizes or by combining more cells within the module unit. This solution means a significant increase in the size of these devices, but it implies an optimization in the design of photovoltaic plants. This results in an installation cost reduction which turns into a decrease in the LCOE.

However, this solution does not represent a breakthrough in addressing the real challenge of the technology which affects the module requirements. The innovation efforts must be focused on improving the modules capability to produce energy without enlarging the harvesting area. This challenge can be faced by approaching some of the module characteristics which are summarized below.

[1]University of Applied Sciences Gelsenkirchen, Department of Electrical Engineering and Applied Natural Sciences, Gelsenkirchen, Germany
[2]University of Jaén, IDEA PV Research Group, Engineering Projects Area, Jaén, Spain
[3]University of Antofagasta, Department of Electrical Engineering, Antofagasta, Chile
[4]University of Grenoble Alpes, CEA, Liten, INES, France
[5]University of Jaén, IDEA PV Research Group, Jaén, Spain

4.1.1.1   Bifacial modules

One of the main mechanisms to increase the electrical output of a PV module without enlarging the surface area is through the development of bifacial modules. The main improvement of these devices is that they are photo-electrically active on both sides, which means that besides the front radiation, they can take advantage of the reflected or albedo radiation that falls on their rear surface. Therefore, if a module can use both sides for energy conversion without enlarging the size itself, it results in an energy gain with respect to a monofacial photovoltaic module. For this reason, bifacial modules are currently having an increasing penetration rate in the market of around 30%, and it is expected to be the predominant technology in the short-term market share [1].

The characteristic to generate energy on both sides does not automatically mean that front and rear surfaces have the same conversion efficiency. To this end, the recently published IEC TS 60904-1-2 standard defines the bifaciality factor as the ratio between the main electric parameters of both sides of the module (rear and front). For this reason, a short-circuit current bifaciality, an open-circuit bifaciality, and a maximum power bifaciality is defined, which can be calculated as given in the following equation [2]:

$$\varphi_{P_{max}} = \frac{P_{max,rear}}{P_{max,front}} \tag{4.1}$$

For example, a 400 W module (rated at its front side only) will generate 464 W at 1,000 W/m^2 if the bifaciality is 80% and the available irradiance at the rear surface is 200 W/m^2. This results in a power increase of 64 W if compared to a 400 W rated monofacial module [2].

Therefore, the closer the bifaciality gets to 100% and the higher the available rear side irradiation is the more power the module will generate. However, in bifacial PERC (passivated emitter and rear cell) modules (p-type c-Si modules) the bifaciality factor is currently around 70%.

Nevertheless, n-type modules, as an emerging technology with very high conversion efficiency potential, have reported bifaciality factors above 80%, and some manufacturers claiming bifacialities of 85%. This percentage difference of up to 15% in the bifaciality factor leads to significant power gains. Furthermore, certain n-type module technologies even show bifaciality factors of 90%. There are some reports that state energy yield increases of up to 3% if n-type bifacial modules are compared to p-type bifacial ones with 70% bifaciality [3].

In addition to the inner layers and materials that embed the n-type cells, one of the mechanisms to increase the bifaciality factor is by reducing the shading effect of the rear side silver metallization, which is possible by using advanced printing techniques.

If compared to PV modules manufactured with p-type wafer materials, n-type material usually has a higher power ratio in terms of outdoor energy generation because of the better electrical characteristics and the increased bifaciality factor.

4.1.1.2 Temperature performance

PV module performance is directly proportional to the operating temperature, which results in decreasing module power as the temperature increases. As modules are intended to operate in different climatic conditions, developing mechanisms that reduce both, the operating temperature of a module and the temperature coefficients immediately result in a power gain. In this sense, n-type modules are reported to come with both improvements that enable a better temperature performance in comparison to p-type technology.

Under the same specifications, n-type modules have a lower operating temperature of approximately 1–2°C if compared to p-type modules. This lower operating temperature will increase the expected power output by 2%. Whereas p-type modules have a temperature coefficient for power of around -0.41 %/K to -0.48 %/K, n-type can lower this coefficient down to the range from -0.20 %/K to -0.40 %/K [4,5].

Therefore, under high ambient temperature conditions, the use of n-type modules is far more appropriate. This is an advantage for the application in hotter climates, like desert areas, as the combination of both features improve energy production by up to 3% [6].

4.1.1.3 Energy losses and annual degradation

Another feature that enables higher electricity production capacity throughout the module's lifetime, while maintaining the surface area, is the reduction of the annual power degradation that the panel suffers, as well as other losses that appear during the operation of the module. In this aspect, the use of n-type module technology is an improvement over other p-type technologies such as PERC.

According to the characteristics published by the main manufacturers, the use of n-type modules can mean an improvement in reduced degradation of around 1% in the first year, while the annual degradation only ranges between 0.05% and 0.10%. Under these terms, the power warranty can be expanded if compared to p-type modules.

Among the degradation mechanisms, that PV modules in general undergo, are the so-called power-induced degradation (PID), light-induced degradation (LID), and light and elevated temperature-induced degradation (LeTID), among others. Regarding the PID, this is a loss mechanism caused by sodium ions which origin from the glass sheet. However, although PID is a loss mechanism affecting both technologies, the degradation rates for n-type modules are smaller and reach a stable-state faster than p-type modules [7].

LID is a loss mechanism produced by either boron–oxygen activation or iron–boron pair dissociation which degrades the charge carrier lifetime, representing an efficiency loss of $\sim$1.2–2.5%. On the other hand, LeTID appears after hundreds of hours of illumination and is active at temperatures above 65°C.

Based on the performance test results offered by several companies, LID test results for n-type modules are close to zero, therefore it leads to almost no degradation for this material, caused by the low boron content. There are some manufacturers that even claim a slight power gain [5].

LeTID losses also exist on n-type, but significantly less prominent than on p-type. While modules based on p-type cells experience losses between 0.9% and 1.2%, modules based on n-type cells only show losses of approximately 0.4%.

Therefore, since n-type technology comes with a lower annual degradation, it can generate more power and therefore higher energy throughout its service life-time, without the need to increase the size of its surface, if compared to a similar module of p-type technology.

4.1.1.4   Efficiency

The efficiency of a PV module with a defined surface is given as its capacity to convert irradiation into electric power under a well-defined irradiance level, based on the values of the standard test conditions (STCs). Therefore, when comparing two PV modules with the same area, a difference in efficiency also means a dif-ference in power output, or in other words assuming the same power output, a more efficient module requires less light active area.

This increase in efficiency means that a system with a given power rating is less expensive in terms of the required ground area, structural mounting material, and wiring.

In the case of PERC modules (based on p-type technology), manufacturers are coming closer to the physical efficiency limit, so there is less room for improve-ments in this aspect [8].

However, in the case of modules based on n-type cell technology, another advantage arises. Such modules come with higher efficiencies [9] and, more importantly, there is still room for efficiency improvement, since the theoretical limit lies at 28.7%, compared to PERC, which comes with a limit of 24.7%.

Based on the efficiency values currently provided by module manufacturers, the efficiency of n-type modules steadily increases.

Although the current trend is using larger wafers, as seen with p-type PERC modules. This is based on the demand for PV modules with higher output power to drive down LCOE at utility scale.

Therefore, in general terms, the use of n-type PV modules represents a quali-tative leap in improving the potential of PV technology to generate energy, due to their improved characteristics compared to p-type modules in terms of bifaciality, temperature coefficient, annual degradation, lower LID or LeTID losses, and higher efficiencies, and what is more important, with a significantly better tech-nological roadmap for improvements [1]. These improvements result in a higher energy yield and a cost reduction as power increases greater than 3.5%, leading to a 3% in cost saving for the BOS.

As a summary, Table 4.1 shows a comparison of the datasheets of two modules from the same manufacturer, one built using p-type cells, whereas the other one is the same type of module but made of n-type cells. If seen from an aesthetical view, there are no noticeable differences between both module types, as shown in Figure 4.1.

Table 4.1   Comparison of the data sheet of two modules from the same manufacturer

Characteristics	P-type version	N-type version
P_{max} (of the series) (W)	545	570
V_{max} (V)	41.8	43.85
I_{max} (A)	13.04	13
Voc (V)	49.65	51.6
I_{sc} (A)	13.92	13.81
Efficiency (%)	21.3	22.3
P_{max} temperature coefficient (%/K)	−0.35	−0.31
Dimensions (mm × mm × mm)	2,256 × 1,133 × 35	2,256 × 1,133 × 35
Weight (kg)	32.3	32.3
Bifaciality (%)	70 ± 5%	80 ± 5%

Figure 4.1   Which is the installation with n-type modules?

4.1.2   State of the art

Regardless of the specific characteristics mentioned in the previous section, several technological approaches are currently coexisting within the n-type technology [1,8]. The most extended one is the tunnel oxide passivating-contact (TOPCon) approach, which is an evolution of the PERT technology. The construction is based on thin layers of oxide on poly-Silicon as tunneling oxide passivated contact to replace n+ BSF on the back side of bifacial cells.

Another cell design based on n-type is the HJT solar cell, which adds an a-Si:H layer to create a tandem solar cell (dual junction). Although the process is more advanced and expensive than PERT based with passivated contacts, the temperature coefficient for power is extremely low and comes with a higher bifaciality factor, hence modules with HJT cells are better suited for summertime or hotter climates. HJT-based modules currently possess the highest efficiency rates of n-type technology.

The third module approach is based on interdigitated back contact (IBC) solar cells which is used by some manufacturers in production. Different technological approaches for manufacturing IBC solar cells exist, for example the Zebra-technology from ISC-Konstanz. The larger photovoltaic module manufacturing companies are betting heavily on IBC technology, as it is the case for example for Longi, JASolar, Jolywood, Jinko, LG, REC, Hanwha Q CELLS, and Sunpower at the date this chapter is written.

Based on the data published, the record efficiencies reported for several module technologies, n-type based technology is positioned among the most efficient one [10]. In addition, according to some publications related to best operating solar panels in 2021, among the top 10 module manufacturers, three n-type modules are positioned, with efficiencies ranging from 21.9% to 22.8%, while the efficiency of the best p-type modules on this list reaches efficiencies up to 21.3% [11].

In relation to the annual degradation shown in the previous section, it is worth noting that these three top efficiency modules come with a power warranty of 90.8–92.0% of their initial output power after 25 years, while the power warranty of the other module products do not exceed 86.0%, with 84.8% being the most widespread value [11].

This corroborates not only the achievement of higher efficiencies for this type of modules but also states a greater potential for larger electricity production during the service lifetime of the modules.

It is estimated that by 2022, the proportion of n-type modules on this list is expected to increase significantly with a promising roadmap in the coming years [1].

4.2   Characterization and performance monitoring

The characterization of any device is usually understood as the description of features or quality of the device under test (DUT). In a PV module, production line electrical features are essential for product labeling and classification. The key electrical parameters of a PV module can be determined by measuring its current–voltage (I–V) characteristic curve. Nevertheless, to assess the PV module quality its I–V curve is not sufficient and further inspections need to be performed. Non-contacted ribbons, cracks due to soldering, lamination, or handling; and inactive cell areas can be detected by using the electroluminescence (EL) technique. Other typical inline inspection tools are high potential testing, insulation testing, and frame continuity testing, which are performed to ensure a minimal leakage current and to avoid any dangerous and undesirable electrical discharge during normal operation.

Characterization techniques that complement the I–V curve and EL measurements for inline module manufacture are the insulation resistance test, the wet leakage current test, the spectral response (SR), and the gel content test; whilst for field or outdoor characterization I–V and EL are complemented by the infrared thermography (IRT) and the ultraviolet fluorescence (UVF).

The insulation resistance test and wet leakage current test are meant to determine the electrical safeness of the module. The SR is mainly used to determine the wavelength dependency as in-depth inspection after material failure or for finding the most suitable material combination allowing to obtain a low optical loss after the encapsulation. Besides, the gel content test is a chemical test that gives the percentage of gel that the encapsulant material contains after lamination, allowing the manufacturer to determine if the lamination process was conducted with material matching time and temperature regimes and if the raw material is of a suitable quality. It is important to remark that such tools are not restricted to quantify production quality but recommended to determine the module performance within the time of warranty. These tools are also of great importance for the development of new materials or new applications.

For field (outdoor) inspection, I–V measurements are highly recommended to be carried out in combination with IRT and EL imaging. These measurements are desirable to be applied during commissioning and later during PV plant operations, at frequent intervals, as for example, every 2–3 years, as part of a preventive maintenance schedule. The frequency of such inspection depends on the knowledge of the components' reliability for the specific environment where the PV power plant is installed. The IRT is also an important tool to test and validate the components and materials under real field conditions, as the power electronic devices for example. In the field, EL and IRT are often used to detect hot spots, bypass diode failures and defects caused usually by PID, LeTID, and unproper transportation, mounting or manipulation of the PV modules, within other actions that might cause the solar cells or the electric conductors to break or crack.

4.2.1 Electrical characterization

4.2.1.1 I–V characterization

As for any semiconductor device, the I–V and power–voltage (P–V) curves are used to determine the main electrical parameters of a PV module or a solar cell. The I–V curve measurement is the most fundamental characterization technique for photovoltaic devices. Values such as open circuit voltage (V_{OC}), short-circuit current (I_{SC}), and voltage, current and the power at maximum power point (V_{MP}, I_{MP} and P_{Max}, respectively) are extracted. Based on these main parameters, other important variables can be calculated, among them are the fill factor (FF) and the efficiency (η). The series and shunt resistance (R_S, R_{Sh}) and the photocurrent or light-generated current (I_{Ph}) can also be determined from the I–V curves under a constant illumination condition [12,13].

An I–V curve is usually obtained from indoor measurements in a laboratory by using a solar simulator and an electronic load according to the IEC 60904-9 or ASTM E927-10 standards. The light source of the solar simulator is classified based on three parameters: spectral content, spatial uniformity, and temporal stability. The classification leads to the designation of one out of three classes: A, B, or C, where the first one indicates the better performance. These classifications are specified in Table 4.2. Regardless of their class, there are three types of solar

Table 4.2   Classifications of solar simulators according to the IEC 60904-9 and ASTM E927-10 standards

Performance parameter	Class		
	A	B	C
Spectral match	0.75 to 1.25	0.6 to 1.4	0.4 to 2.0
Irradiation non-uniformity in the measurement plane	<2%	<5%	<10%
Temporal instability	<2%	<5%	<10%

simulator systems: flasher, pulsed, and steady state. The first two are often used to measure the performance of PV modules in manufacturing lines, at least for labeling. Nonetheless, steady state is mostly used to determine the nominal module operating temperature (NMOT), for degradation tests and light soaking. Whichever solar simulator is used, a homogeneous light source with a certain intensity is required.

The spectrum specification for the solar simulator is further defined via the integrated irradiance across several wavelength intervals for three different air mass (AM) conditions. This is done since the solar spectrum has not the same intensity for different wavelength intervals. The AM is assigned by a number corresponding to "the relative length of the direct-beam path through the atmosphere compared with a vertical path directly to sea level" [14,15]—the longer the path, the larger the attenuation. The ASTM defines then three main solar spectra according to a certain AM: the AM1.5D, AM1.5G, and AM0. The former is defined as the direct solar spectrum, while the latter stands for the extraterrestrial solar spectra. The AM1.5G is defined as the global solar spectra, and it is of great interest for solar application developers.

Furthermore, STC are defined [16,17] to have a reproducible base line condition:

- AM1.5G spectral distribution
- Total irradiance in plane of measurement of 1,000 W m^{-2} (also called one sun)
- Device temperature of 25°C

To obtain the I–V curve, voltage sweep is necessary to be applied, which is usually done when the device is illuminated but not restricted to it. For non-illuminated conditions, the so-called dark I–V curve allows one to extract parameters such as the equivalent diode saturation current (I_0) and the diode ideality factor (n_D). Either for illuminated or dark I–V curve, an electronic load is used to do the voltage sweep while the current and voltage are measured by the Kelvin sensing method for electrical impedance measure (4T or four-wire sensing). This method allows to avoid the voltage drop in the wires. However, even using such method does not secure to measure the true I_{SC}. The voltage drop in the wires needs to be compensated by using an active electronic load. Most of the indoor I–V measurement devices make use of active loads, while handheld and outdoor devices, commonly used for commissioning, use passive loads (e.g., capacitor and/or resistor networks).

The voltage sweep is usually performed in less than 60 ms from V_{OC} to I_{SC} operative points. For such sweep velocity, the n-type solar cells show a certain distortion in the measurement due to their capacitive characteristics [18]. Therefore, the hysteresis method must be applied to avoid under- or over estimation of the electrical parameters. This method consists of performing the voltage sweep from V_{OC} to I_{SC}, and later from I_{SC} to V_{OC}, in a flash step with longer duration.

For medium- to long-term outdoor measurements of the I–V curve, it is recommended to store the data as 5 min average with minimums and maximums of each data batch. The currents and voltages shall be measured using sensors with accuracy of $\pm0.2\%$ for I_{SC} and for V_{OC}, according to the IEC 60904-1 standard [16]. The data logger should be set to perform the voltage sweep in between 1 and 2 sec. This should allow one to partly minimize the hysteresis effects and to minimize data scatter due to cloud variations.

For n-type bifacial modules (nPV), the IEC TS 60904-1-2 is the standard recommended to follow. In terms of output power characterization for bifacial devices, three main steps are required to be assessed [19]: (1) the bifaciality factor at STCs; (2) the rear-irradiance driven power gain yield, BiFi (also considered as the bifacial power gain); and (3) the determination of the output power at rear irradiances of 100 W/m^2 and 200 W/m^2.

The bifaciality factor of the DUT can be determined by means of a single- or by a double-sided solar simulator, depending on the capabilities of the apparatus (e.g., light source, space, mirrors, filters) [20]. The first option is the recommended approach for the already established facilities. With this method, the I–V characteristics at STC of each side, front and rear, needs to be individually measured. To do so, a non-reflecting background must be used at the non-exposed side. It is considered that "such side is dark" if the irradiance reaching on at least two points is lower than 3 W/m^2. This way, two main bifacial factors can be calculated: (1) the short-circuit current bifaciality coefficient, $\varphi_{I_{SC}}$ (Eq. (4.2)); and (2) the maximum power bifaciality coefficient, $\varphi_{P_{max}}$ (Eq. (4.3)). The former accounts for the ratio between the short-circuit current generated by both sides of the DUT, rear over front side. In a similar way, the $\varphi_{P_{max}}$ is calculated but this time from the output power of each side separately. The spectral mismatch correction needs to be applied according to IEC 60904-7, unless it is known that the front and back of the bifacial DUT have identical SR:

$$\varphi_{I_{SC}} = \frac{I_{SC,r}}{I_{SC,f}} \tag{4.2}$$

$$\varphi_{P_{max}} = \frac{P_{max,r}}{P_{max,f}} \tag{4.3}$$

$$\Phi = min\left(\varphi_{I_{SC}}; \varphi_{P_{max}}\right) \tag{4.4}$$

In best situation a solar simulator can illuminate both sides of the bifacial PV module at once. To achieve this, two variables in terms of setup are distinguished: double light source, and single light source with mirrors and filters. No matter which setup is utilized, the non-uniformity of the irradiance must be lower than 5% on both sides. It is important to ensure that the front side is always at STC, while the read side irradiance can vary either by controlling the light intensity or by means of different gray filters.

4.2.1.2   Electrical safety tests

For a safe operation of a PV module, the generated electric potential and current must flow in the circuit, within the devices, in the cables, and not leak to the outer elements. Electrical insulation and current leakage are of a great concern for a safe operation of a PV module. Such leakage could occur through the encapsulant-glass layer or through the encapsulant-frame path [21,22]. The severity of such leakage is linked not only to the dielectric properties of the PV module materials but also to the electric potential to which the module is subjected. In PV power plants, the modules operate in-series connected in so-called strings at more than 600 V and up to 1.5 kV. For such high voltages, the electrical insulation of the module packing materials might be affected and diminish specially under wet conditions. This way, the electrical insulation of the PV module must ensure it can withstand moisture or wetness from outdoor conditions. To test such requirements, according to IEC 61730-2 and to IEC 61215, insulation resistance test and wet leakage current test are the recommended ones for electrical safety examination. This test should be combined with the IEC 62790 for safety requirements and tests of junction boxes, which might also create a path for current to leak.

It is important to mention that a PV module shows a capacitive behavior in which its encapsulant material constitute the dielectric of such capacitor, and the metallic frame and the electric leads (e.g. ribbons) represent the metallic plates. This way, the measured current can be divided into three components: (1) the capacitance charging current; (2) the absorption current; and (3) the leakage current. The latter is the one usually measured since it rapidly increases reaching a steady-state value.

The setup required for a leakage current test includes (1) a power supply able to energize the DUT with at least 500 V_{DC}, or the voltage of the system; (2) an insulation tester (usually a Mega-Ohmmeter); and (3) an apparatus where the module is watered in a flat horizontal position with water or wetting agent solution. The latter solution should be of a resistivity no larger than 3.5 kΩ/cm, and applied at a temperature of (22 ± 2) °C. Furthermore, the insulation tester should be sensible to currents levels in the order of nA to μA, so the insulation resistance should be higher than 40 MΩ m^2 for modules of area larger than 0.1 m^2.

4.2.1.3   PID

The current leakage not only affects the safeness of the operation, but it also potentially leads to a decrease in the PV module efficiency. The significance of this degradation depends on the solar cell technology and on the design and packing materials of the module. This degradation is today known as PID. The PID mechanism for nPV modules has been reported to differ with the solar cell architecture [23]. In some cases, PID appears for positive high potential, while in others for negative high potential [24].

For n-type c-Si cells, the PID mechanism is linked to an increase of surface recombination, which is probably caused by the accumulation of negative electric charges on the anti-reflection coating of the cell. Nevertheless, at the time, this book is written the PID mechanism for nPV modules continues incomplete, but three PID types are identified, which depends on the cell structure: (1) polarization, (2) Na-penetration, and (3) corrosion-type [25].

Despite the nature of the PID effect, solar cells should be resistant to it. If this is not possible to achieve without compromising the cell efficiency, it is necessary to mitigate it at module or system level. To test if a module is prone to PID, the IEC TS 62804-1 standard is recommended to follow. According to this standard, the DUT should not lose more than 5% of its initial power when subject to a potential equal but opposite to the system potential. Besides module power, the dry and wet module insulation resistance should remain higher than 40 $M\Omega m^2$. Two methods are differentiated: (1) climate chamber and (2) conductive electrode. The latter is named because a conductive foil or water layer on the module glass (or front cover) is used to increase the surface conductivity. In this method, the module temperature must be regulated to (25 ± 1) °C, (50 ± 1) °C, or (60 ± 1) °C. The RH should be lower than 60% and the test duration larger than 168 h. For the climate chamber method, both, the module temperature, and chamber humidity shall be controlled, the latter to (85 ± 3) % and the former to (60 ± 2) °C, (62 ± 2) °C, or (85 ± 2) °C. This test should run for at least 96 h.

4.2.2 Relevant module characterization in the field

It is well known that PV modules, once exposed to the field, will produce energy according to the light intensity and to the ambient temperature. The output power will increase as the solar irradiance increase, but it will decrease with the increase of the device temperature.

4.2.2.1 Irradiance dependency

Depending on the SR of the solar cell and on the transmittance of the PV module packing materials, high intensities in the wavelengths within the solar cell SR could imply high output current and hence high energy harvesting. Nevertheless, despite a high energy generation, the solar cell and module packing materials can also be negatively affected by the large irradiation and therefore reduce the lifetime of the PV module. For instance, most of the PV packing materials degrade at higher rates when exposed to large doses of UV wavelengths.

The SR of a solar cell or module corresponds to the amount of generated output current due to a certain amount of irradiated power on the solar cell. It permits an examination of how photons of different wavelengths (energy) contribute to the output current (usually I_{SC}). The SR is defined as the $I_{SC}(\lambda)$ resulting from a single wavelength of light normalized by the maximum possible current. An experimental setup allows the measurement of the external SR, SR_{ext}, while the internal SR (SR_{int}) is determined from the latter plus the knowledge of the metal grid shadowing on the solar cell, its reflectance and optical thickness. The SR_{ext} is a perfect indicator for the optical quality of the encapsulation material. This can be done by comparing the SR_{ext} of the solar cell with the SR_{ext} of module. In this way, it is possible to identify undesirable absorption in certain ranges of wavelength due to different bill of materials (BOMs) in the fabrication of the PV module.

To perform SR measurements, a spectrophotometer is used, in which the wavelength of light passing through a DUT can be selected and the amount of light absorbed by the device is measured. This technique is very expensive and perhaps not very cost-effective, since the instruments to perform such measurements only allow small areas (a few cm^2). For industrial scale PV modules, it is more effective to measure the broadband irradiance at different wavelength range, along with the electrical output of the module. For a module manufacturing line, reference cells are the proper devices to use.

For field measurements, to monitor the irradiance doses nearby PV modules the use of a set of pyranometers plus a reference cell is recommended [26]. For the latter, the short-circuit current is measured but it is important to apply a temperature correction, and hence the sensor must include a temperature sensor. This reference cell (small module) should match the SR of the PV module under test and must be mounted under the same angular elevation (same plane-of-array). This way, the measurements give a more accurate figure of the DUT performance under low-light conditions which should be carefully assessed for n-type devices.

It is also recommended to use pyranometers of second reference type or at least of first class, according to the ISO 9060 standard. Depending on the evaluation goal, the SR of these devices shall be of broadband within the UV, visible, and/or IR range of the electromagnetic spectrum. For module performance, the pyranometer should be able to measure up to at least 1,500 W/m^2 within the wavelength range of 300–2,500 nm. Commercial first class pyranometers can easily measure up to 4,000 W/m^2 within a wavelength range of 285–2,800 nm.

If the degradation of the module materials is the focus of the investigation, then UV radiometers should be used. Such devices can measure either UVB (280–315 nm) or UVA (315–400 nm) or combined both ranges. The selection depends on the specific research target.

For PV plants with tracking systems, the measure of the direct normal irradiance shall be included. To measure this variable, the use of a first class pyrheliometer mounted in an automated two-axis tracker is recommended. This tracker can also be mounted in combination with two pyranometers, one to measure the global horizontal irradiance, and one equipped with a shading ball to measure the horizontal diffuse irradiance.

To obtain a proper characterization of bifacial modules, the irradiance reaching the back of the module must be measured. This can be done by means of the same instrument used to measure the "front irradiance", i.e., a pyranometer or a reference module. An albedometer can also be added to measure the environmental variables.

Such instrument usually comprises of two pyranometers that measure both, global and ground reflected solar irradiance in the horizontal plane. Its main characteristic is the housing shape of the downside pyranometer, which should refuse the sky content radiation. This usually entails a smaller field of view if compared to the upper-side one. It is important to note that the addition of an albedometer does not imply the avoidance of installing an irradiance sensor at the back plane of the module since both come with complementary rather than exclusive measurements.

Whichever the research target is, a proper installation and maintenance of the solar radiation sensors shall be followed. The pyranometer and reference cell mounting angle and orientation must match that of the DUT within a fraction of a degree. No matter if the DUT is of monofacial or bifacial technology, the instruments should receive the same amount of ground-reflected radiation as the modules. This is usually achieved by mounting these in proximity by and at the same height, usually in the same mounting structure.

The instruments should be complemented with ventilators and heaters, especially if there is any risk of dew, frost, or snow accumulation. Such ventilators and heaters help to reduce any thermal offset which is one of several sources of error and uncertainty in pyranometers and reference cells.

Although the use of ventilators usually reduces the dust accumulation over the pyranometers, the optics of the sensors must be regularly cleaned and inspected. In highly polluted places, the cleaning schedule might be necessary to be arranged on daily basis. In some instruments, a desiccant compound is incorporated to prevent condensation inside the instrument. Such desiccant must also be checked and periodically replaced.

Finally, but of great importance for the operation and maintenance, the calibration of the instrument must be regularly verified. If possible, multiple instruments should be checked against each other and sent to a calibration laboratory on an alternating basis.

4.2.2.2 Module temperature dependency (or thermal behavior)

A commercial PV module will convert only 17–23% of the incident solar radiation into electricity, if operating at its MPP. A large part of the solar energy is transformed into heat. Five main factors are considered to influence the temperature of a PV module: (1) energy absorbed by the module parts other than the active area of the solar cells. The amount of absorbed and reflected energy is linked to the module materials and its color. The larger the energy is, the more heating is contributed to the module. (2) Light reflected from the cover surface of the module does not contribute to the module power nor its heating. The reflectivity of a typical cover of a PV module (solar glass) is at least $\sim$4%. (3) The generated electrical current creates a voltage drop along the electrical interconnectors. This electrical energy is dissipated as heat (Joule-Lenz law). (4) Since the main heat source of a PV module is the solar cell and the bonds of the electrical connections, a higher packing factor increases the generated heat per unit area. (5) The thermal conductivity of the packing materials contributes to the module capacity to dissipate thermal energy; a low value diminishes such property hence it might increase the module temperature at steady state.

The thermal loss is directly linked to the semiconductor properties and its voltage and current dependence to temperature. For this reason, the temperature coefficients α, β, and γ for short-circuit, open voltage, and maximum power point conditions must be assessed. Such temperature dependency varies with the fabrication technology. For example, a module made of p-type cells might have a power temperature coefficient at MPP ranging from −0.41 %/K to −0.48 %/K, whilst for the n-type solar cells, the γ ranges from −0.20 %/K to −0.40 %/K, per each Kelvin increase above 25°C for rated power [27–29]. This difference in temperature dependency leads to nPV modules not necessarily operate at lower NMOT but with higher energy output due to the improved temperature coefficients.

The module operating temperature can be obtained from field measurements by means of resistance temperature detectors (usually Pt100) or by inferring it from the open-circuit voltage method. The calculation of this equivalent cell temperature is standardized in the IEC 60904-5 and is calculated as:

$$T_{ECT} = T_{STC} + \frac{1}{\beta}\left[\frac{V_{OC}}{V_{OC,STC}} - 1 - a\ln\left(\frac{G_{POA}}{G_{STC}}\right)\right] \tag{4.5}$$

From the set of data, it is recommended to use those obtained with irradiance in the plane of the array (G_{POA}) close or equal to the STC value, hence the logarithm term can be obviated.

To investigate the temperature dependency of a DUT, a similar setup is recommended as suggested in the irradiance dependency section, plus at least four Pt100 to obtain the temperature of the module. Such RTDs should be mounted at four different positions within the rear side of the module. Five sensors, one on each cell at the opposite corners of the module plus a sensor in the cell at the module center is a valid setup. The registered values need to be later averaged, and the temperature should be uniform within ±2°C from the average value.

4.2.2.3   Infrared thermography

Infrared thermography is a technique that makes use of several laws of physics to create a two-dimensional image mapping of detected energy in part of the infrared (IR) range of the electromagnetic spectrum. This 2D image shows the detected intensities, which can be related to the surface temperature of the DUT by using the Stefan–Boltzmann law. IR cameras are either used as handheld or mounted in a mobile apparatus or in a vehicle (usually an unmanned aerial vehicle, UAV). An example of an IR image of a PV installation is shown in Figure 4.2. Any spots of high temperature in this picture indicate possible solar cell shunts or soldering disconnections.

IR thermography is usually used as a tool in the O&M plan of a PV power plant. It helps to determine the risk of fire due to hotspots within the module during its operation. It can also help to identify modules with potentially lower performance, which might be further analyzed for proper failure detection. This can be

Hotter

Colder

Figure 4.2   Example of IR image (image courtesy of CDEA, University of Antofagasta)

performed by using IR cameras, usually mounted in an UAV that flies over the PV plant sending IR pictures either in real-time or in batch mode to spot and mark modules with potential failure.

The IRT is usually applied to detect hot spots, caused by cell cracks, shunted cells, interrupted interconnectors, broken down or short-circuited bypass diodes, etc. Hot spots or areas hotter than the surrounding can mainly be explained by the Joule-Lenz law. During operation, when a cell of the complete module matrix is partly or fully shaded and the remaining cells are under illumination, the shaded cell will start to behave as an electrical load instead of a generator. Hence, it will consume power generated by the other cells connected to the string. Because of the balance of energy, this power will be dissipated as heat, increasing the cell temperature in the PV module. Local heating can also be triggered if the cell area or the cell-ribbon bond are broken or contain small gaps. Those defects will create local points of high resistivity that also dissipate power as heat.

To properly perform IRT tests it is recommended to follow two standards: the IEC TS 60904-12 (draft) and the IEC TS 62446-3. The former describes general methods of IRT in laboratories or production lines, with focus on the IR imaging techniques of the PV module itself. The latter describes the application of IRT for PV modules and PV power plants in operation, under sunlight.

4.2.2.4   Electroluminescence

The electroluminescence (EL) technique allows manufacturers (cells and modules) and PV plant operators to diagnose most of the failures in the active area of the module; shorted bypass diodes, disconnected cells, short-circuited cells, or cracks/fissures within the solar cells are some of the failures one can observe by studying EL images.

EL is an electro-optical phenomenon in which a material emits light in response to an electric field (or electric current) applied to the material [30]. In a semiconductor both, EL and photoluminescence (PL), are the result of radiative recombination of electrons and holes [31,32]. The difference between PL and EL is the method to excite the charge carriers. In the latter, the excitation is made by applying an electrical potential to the sample under observation (electrical excitation), while for PL the electron–hole excitation is made by bombarding the sample with photons (optical excitation). The wavelength range of emission varies depending on the material bandgap and its temperature. For PV modules, c-Si solar cells at room temperature (~300 K) will emit between 1,000 nm and 1,300 nm with a peak at 1,150 nm approximately [33].

In PVs, with the EL characterization technique, the device photoemission is measured by using a radiation detector, such as Si-CCD or InGaAs cameras. The resulting fake color picture is a 2D-map showing regions with different recombination and generation rates. An example of such images is shown in Figure 4.3, where EL pictures of two different n-type modules are exhibited. The intensity or pixel brightness in the picture is proportional to the local voltage of the DUT [33,34]. As example, for cells with a complete dark area it can be stated that these are either electrically disconnected or shunted; if only a region of a cell is dark this indicates a complete disconnection of such area. Whatever the size of the dark area is, the electrical disconnection might be caused by interruption of the front side (or rear) metallization or by broken solder bonds. Furthermore, solar cells with varying dark patterns within the same PV module might indicate PID.

To generate the luminescence effect in an EL test in most cases, a power supply is required to create a controlled current flow through the DUT. For PV modules mounted in the field, it is recommended to have these, or the string to which these modules are connected to, first disconnected from the remaining electrical interconnection of the PV plant. The IEC 60904-13 standard gives guidelines for the environmental conditions and details of the necessary set-up for a proper EL test. The EL test should be preferably performed in a darkened room, under controlled ambient temperature, otherwise the use of Lock-In EL technique and/or measurements during the night are recommended. Measurements in laboratories should be made at a module temperature between 20°C and 30°C for consistency in achieving qualitative or quantitative comparisons. Such temperature measures should be done by means of a temperature probe with at least 1°C accuracy, mounted on the back of the module in a manner that does not interfere with the image acquisition.

The quality of the picture largely depends on the measurement setup and detector employed. The use of Si-CCD camera allows the capture of high-resolution images (8 MPx or higher) but with long exposure time due to its low sensitivity at the wavelength range of interest. With the increase in exposure time, the thermal noise increases as the sensor temperature rises. The use of a sensor cooling system helps to reduce the thermal noise. However, an InGaAs detector array has a higher sensitivity in the range of interest thereby allowing shorter exposure time and lower artefacts due to thermal noise. The drawbacks are a resolution lower than one megapixel and its elevated cost, which, among other constraints, make it less widespread.

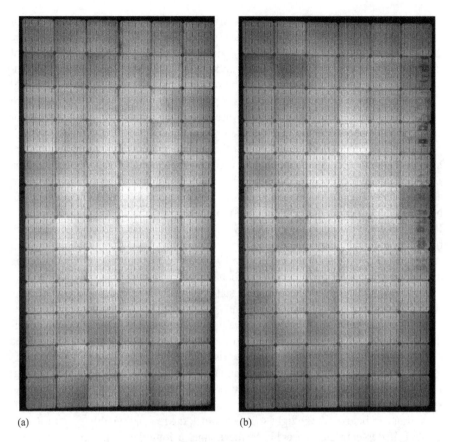

(a)   (b)

*Figure 4.3   Example of EL picture for two different 72-cells nPV modules
operating under desert climate conditions. (Left) A module showing
mostly no critical damages; (right) a module with imperfections in the
electrical path in the outer right substring (image courtesy of CDEA,
University of Antofagasta)*

Different electric potential applied to the DUT lead to varying excitation states which might result in different images and interpretations. For example, for a current close to I_{SC} the effect of the series resistance is clearly visible, while differences caused by cracks are visible mostly at levels of $\sim 10\%$ of the I_{SC} [35].

4.2.2.5   Ultraviolet fluorescence

Ultraviolet fluorescence (UVF) consists of detecting UV emissions (or reflection) of the components of the module after optical excitation. The presence of fluorophore is an indicator of the degradation of the polymers of the module materials (fluorophore is a component of a molecule that makes it fluorescent). Fluorescence is the emission of light by a material that has been activated by the absorption of light or other electromagnetic radiation [36,37]. The re-emitted light (e.g., in the visible region) has a longer wavelength than the absorbed radiation (e.g., ultraviolet light). Therefore, a fluorophore can be understood as a fluorescent chemical compound that can re-emit light after excitation of light and contains mainly several combined aromatic groups or other groups of planar or cyclic molecules with several π-bonds. Typical fluorophores are polymer degradation products and/or additives with chromophore/fluorophore groups.

Unlike the EL technique, in UVF, the radiation stems from the encapsulant material (not the cell) and is excited by a UV lamp, in the UV-A wavelength range (315–400 nm). This way, the UVF could serve as an intermediate step between aerial IRT and EL; by IRT one identifies sections that need detailed inspections and by UVF one identifies modules that need to undergo EL inspections. Together with IRT, the UVF imaging helps to identify cell cracks. The EL and UVF images can be taken in parallel in the field during inspection with little additional effort. Adding UVF imaging to the EL method, one can differentiate cell cracks in the multi-crystalline wafer from crystal defects. By itself, the UVF is usually used as a tool in an O&M plan to determine the state of the module encapsulant and back-sheet materials.

A method for UVF testing is to irradiate the entire PV module with an excitation light source and observe the fluorescence effect with a camera. Due to the small UVF conversion efficiency and the limited emission power of UV LEDs, this test is often performed in dark environments, just like in the case of EL testing. Excitation of PV modules with UV light is commonly achieved using LEDs in the spectral range from 350 nm (glass transmission) to 400 nm. An additional short-pass filter can be used at the light source to block the residual visible spectrum, thus creating better contrast. To enhance the image, a long-pass filter in front of the camera is used to exclude the excitation light from the resulting picture.

4.2.3   Performance monitoring

The recommended performance metrics are [26] (1) the module/array Performance Ratio (PR); (2) the module/array PVUSA; and (3) the module/array DC-power at MPP. When needed, such indices should be corrected to STC using measurements of irradiance, module temperature and sometimes spectrum.

The PR can be differentiated between the DC and the AC stages of a PV power plant. The PR at the final stage is defined by the IEC 61724 as:

$$PR = \frac{Y_f}{Y_R} = \frac{\frac{E_{AC}}{P_{max}}}{\frac{G_{net}}{G_{STC}}} \tag{4.6}$$

where Y_f is the generated AC-energy (E_{AC}) per kW of installed PV (P_{max}), and Y_R is the ratio of net plane-of-array (POA) irradiance G_{net} over the irradiance at STC conditions. The Y_f is also called "final yield" whilst the Y_R, reference yield. The

G_{net} corresponds to the module net irradiance and it can be calculated from the front and rear side contributions:

$$G_{net} = G_{POA,front} + G_{POA,rear} \tag{4.7}$$

For the case of the DC stage, PR is calculated based on the DC level energy per kW of installed PV, expressed as module/array yield Y_A rather than on Y_f.

$$MPR = \frac{Y_A}{Y_R} = \frac{\frac{E_{DC}}{P_{max}}}{\frac{G_{net}}{G_{STC}}} \tag{4.8}$$

The PVUSA metric stands for PV for utility scale applications, and it was developed by NREL in the 1990s. The idea is to evaluate the performance of PV modules or systems under the so-called performance test conditions (PTC), which stands for a normal irradiance of 1,000 W/m², an ambient temperature of 20°C, and a wind speed of 1 m/s. The method performs a best-fit correlation between the measured P_{max} of the DUT and the measured plane-of-array irradiance (G_{POA}), wind speed (w) and ambient temperature (T_{amb}). This metric is calculated according to Eq. (4.9), where the parameters a, b, c, d are factors used to calculate the power at PTC:

$$P_{max} = G_{net} \cdot (a + b \cdot G_{net} + c \cdot w + d \cdot T_{amb}) \tag{4.9}$$

Finally, the DC power at MPP is the simplest parameter to monitor but strongest affected by temperature, spectral effects, and soiling.

4.2.3.1   Soiling

The soiling is defined as layers of dust that deposits on the outer surface of PV modules, altering their optical properties [38,39], diminishing this way the transmission of light through the surface, reducing and attenuating the photon flux that reaches the solar cell, and hence having a direct effect on the electricity production of the PV module [40–46]. Other side effects of "soiling" accounts for an acceleration in the PID, the "hotspot" formation, and degradation of the transmittance of the external module cover (usually glass) due to the abrasive effect of the by air transported particles and unproper cleaning methodologies.

The dust deposited on the surface of the module glass is usually not homogeneous. A cell current mismatch can occur, caused by this inhomogeneity, which at its time can trigger hot spots, hence not only the module temperature and performance are affected but also the module reliability and lifetime.

To track the soiling loss, not only the module electrical current should be measured but the irradiance level and module temperature should also be considered. Hence, the variation of $I_{SC,T} : G_{POA}$ ratio is the recommended variable to track, where $I_{SC,T}$ stands for the short-circuit current compensated by the module temperature, and G_{POA} for the in-plane global irradiance. For the case of bifacial PV modules, the term G should include the contribution of the irradiance to the rear side of the PV module (G_{net}). This can be measured by means of a second pyranometer or by a reference module (monofacial) pointing in the same direction as the rear side of the bifacial device.

This way, the use of an MPP tracker is not a must, but it is mandatory to record the I_{SC}, the G_{POA}, and the module temperature. The latter can be determined either by the measurement of the resistance temperature detectors (RTD), usually by Pt100, or by inferring it from the open-circuit voltage method, which will obviously require the V_{OC} measurement along with the other two variables initially mentioned. The calculation of this equivalent cell temperature is standardized in the IEC 60904-5 (see Eq. (5)).

Several devices and setups exist that allow one to track the soiling rate of the module at determined places. The most common setup is the use of two reference modules, one of this is regularly cleaned and the second one is naturally soiled. For both modules, the $I_{SC,T} : G_{POA}$ ratio must be tracked.

4.2.3.2   Recommendations to monitor performance

The electrical performance of a PV module can be monitored not only by means of an I–V tracer but also with an inverter that uses a maximum power point tracker algorithm. The former was already discussed in the "Electrical characterization" section. One of the standards to consider for the monitoring of photovoltaic systems is the IEC 61724-1. It covers the main data acquisition requirements. It states, for example, that a measurement uncertainty of $\pm$ 2.0% at the inverter level is required to achieve the highest accuracy (class A). However, for single module measurements, an accuracy of at least $\pm$ 0.2% is recommended according to IEC 60904-1.

For accurate field monitoring, the test conditions must be guaranteed to be optimal throughout the whole test timeframe. The existence of any data point with abnormal behavior leading to data rejection will increase the uncertainty of the measurement campaign.

It is important to consider steady and appropriate mounting racks. Its geometry significantly affects the temperature distribution across the mounted PV module and the reflected albedo, which is of great importance for bifacial technologies. It is recommended to perform infrared imaging of the DUT to identify any non-uniformities in its temperatures, which shall then be addressed. The images shall be taken at solar irradiance higher than 800 Wm^{-2}. It is also recommended to install the samples under test at least 1 m above ground and with a minimum distance of 10 cm to other objects. The rear side of the mounting rack shall be as open as possible, allowing for a uniform rear irradiance contribution for bifacial modules. These guidelines reduce unwanted non-uniformity effects, for both, temperature, and albedo.

Nonetheless, the outer DUT on the left and right side of a row might operate at a lower temperature due to wind force convection larger than for the inner DUTs. Therefore, additional dummy modules shall be installed in such locations to reduce the environmental variability of the DUT. Otherwise, the samples at the inner place of the rack will see more light compared to those at the edges.

As for voltage and current data acquisition hardware requirements, it is recommended to use devices with uncertainty of 0.05% or lower. If the device or method uses shunt resistances, these should be calibrated with an uncertainty of 0.1% or lower.

4.3 Standards and quality measures

4.3.1 Why do we need standards

If calculated in terms of human lifespan solar panels come with a power warranty similar to the number of years counted for a generation. Product warranty is getting closer to this number with currently 15 years but steadily increasing. A solar panel is an electrical generator with a potentially lethal output voltage just for a bit more than one in series interconnected panels. The strong price decline for the installed kWp system costs forced the PV market into an exponential growth rate with a larger number of new PV technologies being introduced to the market, specifically over the course of the last couple of years. The beforementioned key points: long-lasting warranted product lifetime, potential hazardous output voltage and extremely large market growth emphasize the strong necessity of well-established standards which are subject to continuous improvements to cover for new materials and technologies used inside PV modules. Standards also can be seen as a security for banks guaranteeing park builders with financial loans, private investment placers, and system operators. Beside this, standards are a legal requirement in most countries and the same are often applied in dozens of countries world-wide setting global requirements in many national markets thereby reducing costs. For module makers, meeting the requirements of the testing regimes inside standards and complying with the terms of the well-defined terms of a certain standard these documents are a guideline during the early development phase of new products, module materials and technologies introduced. The most well-known standards for PV builders are the IEC 61215 and IEC61730 guidelines which define the whole value chain from raw material delivery to cell production and to the backend process of module production [47]. In the end standards are all about quality, reliability, and safety, also covering part specifications, definition of characterization routines and finally deal with material and product dimensions. The price for undergoing the long list of various test requirements according to the IEC 61215 and IEC 61730 standards is very high, specifically for small module makers or companies just entering the PV module business this adds significant costs. Beside this, a certificate stating that a product complies with all terms of a specific standard is only valid for the product and certain variation in terms of size, components and so on. Any not certified alteration on the product either it is by changing the BOM, the electrical interconnection or any other change automatically triggers a re-certification procedure the product must undergo, bringing in again costs for the manufacturer. Background of this requirement is that PV module components, if they are subject to a change, cannot be assessed in isolation from the rest of a PV module. The module must be seen in its entirely and not component by component. Any changes in the BOM require a further technical assessment to address any potential impacts associated with the change, no matter if it is a single material or a whole component, to exclude unanticipated interactions between module materials [48]. A detailed retesting guidance document is given in IEC/TS 62915 to reference any retesting requirements under IEC 61215 and IEC 61730 once changes are applied to the module.

IEC 62915 is specifying most relevant modifications of a PV module type and states retesting requirements for IEC 61215 series and IEC 61730 series. For many, these regulations seem to hinder companies to react in a fast and economic sensible way to important market changes. On the other hand, the initial statement of this chapter can now be completed with the words: what is built for generations must bring in all requirements to last for generations. A product such as solar module supposed to operate under various climatic conditions, no matter if it is close to the polar circle in Finland or close to the equator in Singapore, it must fulfill all requirements of the standards for at least one generation, as set by specialist over the globe over decades. And most of all, it must operate in a safe way. Looking at the balance of system (BOS) the year this book is written the module price is still 40% (and more, depending on the current world-wide market situation) of the total system costs and therefore plays a significant role in the total system.

4.3.2   Standards—seen from international to national level

The standardization procedures as valid for solar PV modules and its components are internationally regulated by the technical committee (TC) 82 "Solar photovoltaic energy systems" of the International Electrotechnical Commission (IEC). This committee consists out of industrial and scientific experts from all over the globe with deep experience and knowledge in PV manufacturing, testing, operation, and characterization. The main task of IEC technical committees is to prepare International Standards. Every couple of years updates (so-called standard revisions) on the IEC standards are being published to keep track with technological and material developments and where appropriate to add new or modify existing testing procedures and characterization schemes. IEC standards are valid on international level and many countries transfer these individually to national level. The national standards usually cover similar or take the same content of the original IEC standard, nevertheless certain abbreviations appear from time to time. The standard itself is the regulation framework various certification bodies as a legal entity on national level use as a strict guideline to test the product properties and specification against the standard itself.

Apart from the standardization activities of IEC other standardization bodies exist which also focus on different aspects of the value chain. To name few, the International Standardization Organization (ISO), the North American standards like ANSI/UL and the more production-focused documents of Semiconductor Equipment and Materials International (SEMI) exist. For ensuring reliable measurement procedures and test sequences as well to assess quality equally ISO standards are used. In general, most companies follow ISO/IEC 9001 for their quality management system.

Most of all the general requirements for the competence of the testing and the calibration laboratories of the certification bodies follow the ISO/IEC 17025 and

ISO/IEC 17065 standards. Furthermore, are also environmental and health subjects' part of ISO 14001, 45001, 50001 which most of the certification bodies comply with.

4.3.3 Overview on existing standards for type approval and advanced testing

Probably it is not exaggerated to state that the standard IEC 61215 is for most PV manufacturer and PV park operators the key achievement to guarantee for durability and quality. IEC 61215 defines the requirements for design qualification and type approval of terrestrial PV panels and is applied to all terrestrial flat plate module materials. The standard comes in two parts: the first part deals with the general requirements and the second part is all about defining the detailed testing procedures, also referred to MQT's: module quality tests. Modules passing the test sequences which are focused on the electrical and thermal characteristics of the panels have a high probability to withstand prolonged exposure in various climates world-wide. PV modules are often certified according to IEC61215 and IEC 61730. Latter one is a requirement by law for modules to pass in most countries and further examines the PV panels safety against electrical shock, fire hazard, mechanical and structural safety. As with IEC 61215 this standard comes in two parts: The first part describes the fundamental construction requirements which shall provide safe electrical and mechanical operation during the expected lifetime of the panels whereas the second part directly assesses the prevention of electrical shock, fire hazards, and personal injury due to mechanical and environmental stresses.

During the testing sequences of 61215, the impact of mechanical, thermal, and electrical stress on the power output and lifetime of PV modules are examined. The qualification of a given and tested module type to the defined procedures is an indication that this specific module type is suitable for long-term operation under normal environmental conditions. Emphasis must be placed here on normal environmental conditions. The individual stress test sequences were developed to prove that a solar module type is capable to withstand the environmental impacts during its lifetime: Extreme temperatures (exposure to 85°C at 85% relative humidity for 1,000 h) and humidity, extreme thermal fluctuations from 85°C down to −40°C, exposure of ultraviolet radiation and hailstone impact. The qualification testing procedures describe in detail how the modules are characterized before and after being exposed to various types of stress: Visual inspection, power measurement, testing of insulation. The standard describes the criteria to pass (or fail) the test: no visible evidence of major defects that may affect the performance of the module, a threshold value of the insulation resistance at high DC-voltages to reach and a power degradation not exceeding 5% after each single test or 8% after the whole sequence.

The BOM of a typical standard crystalline silicon solar cell module is easy to read since it just contains roughly a dozen different components. Nonetheless is the interaction of each individual component to itself and the other components crucial for the longevity of the solar module specifically if the panel is operated under non-moderate climatic conditions. For this reason, additional standards were established. IEC 62788 for example is a series of standards with the intent to pre-qualify materials for the use in a PV module which covers backsheets, encapsulants, edge sealing and front glass coatings.

IEC 63209-1:2021 aims to provide additional information for the baseline testing as defined in IEC 61215, which is a qualification test with pass and fail criteria. IEC 63209-1 on the other hand seeks to evaluate the long-term reliability of PV modules, also including different BOM´s. This is specifically important for manufacturer since it results in a higher flexibility in the choice of components used during the manufacturing process. One must consider what happens to a manufacturing plant if the supply chain breaks down in case one or two components cannot be delivered in time, for example because shipping routes or harbors are off operation. The test sequences are intended to provide further information for comparative qualitative analysis by applying stress factors which are relevant to application exposures known to become failure modes. A new scheduled IEC regulation called IEC 63209-2 "Extended-stress testing of photovoltaic modules – Part 2: Component materials and packaging," has now finished phase 2 where the draft of the standard is completed. This standard, once published, further aims to bridge the gap between materials, coupons, and full PV modules.

Other standards evaluate in depth the reliability and safety aspects of individual components, such as PV module electrical connections, including connectors (IEC 62852), cables (IEC 62930), diodes (IEC 62916), and junction boxes (IEC 62790).

Most of the standards listed above are so-called type-approval which means a new PV panel or any alteration to an existing setup of a PV panel must comply with the testing specifications as defined in the standards. Beside this large field of type-approval regulations additional test sequences and standards were developed which are field-relevant. The list of this field-relevant tests and standards is long and only a short selection can be introduced in this chapter.

As an important example IEC 62804 defines the procedures to test and evaluate for PID-delamination (PID-d) with focus on the laminate of crystalline silicon PV modules. PID can cause significant power losses up to total failure on module and system level if the wrong components and respectively materials and/or system configuration are used. In general, the detrimental PID effect is caused by the stress of an electrical potential applied between the active cell circuit and the external surfaces or parts of the module. Another standard worth to mention is IEC 61701 dealing with salt mist corrosion which becomes even more important due to the large dissemination of PV, with parks also installed close to, by the sea or as floating PV at the sea. For such installations PV modules experience, beside sunlight, temperature, and humidity an intensified exposure of salt mist which may act corrosive on module materials. The test sequences of the standard aim to evaluate if a panel and its individual components can withstand a larger concentration of salt mist over its lifetime. For all who have not sought a ride along urban areas for example in parts of Austria, Switzerland, and Germany (Bavaria) will be surprised to see how many PV installations are found on animal sheds. Livestock produce various gases such as methane, ammonia, CO_2, etc. From these gases, mainly ammonia (NH_3) was found to be detrimental to materials and components of PV modules and IEC 62716 was specifically developed to cover for this topic. The test sequences of the standard aim to determine the resistance of PV modules to ammonia. Procedures are combined to evaluate possible faults when operating under wet atmospheres with high concentration of dissolved ammonia.

4.3.4   Overview on existing measurement standards

As mentioned in the last section, the number of existing standards related to PV products is quite high and can only be partly covered in this chapter. The same is valid for all other available standards describing the application of measurement technologies or the interpretation of data.

Let us begin with IEC 60904: characterization of PV devices which comes as a series of more than 10 parts. The first part deals with the required procedures for the measurement of current-voltage characteristics (also called the characteristic current-voltage or in short form the I–V curves) of PV devices if illuminated by natural or artificial (simulated) sunlight. All described procedures are applicable to a non-concentrating single solar cell, a sub-assembly of solar cells or a PV module consisting out of a larger number of in series or/and in parallel interconnected solar cells for use in terrestrial environments. The reference terrestrial solar spectral irradiance distribution is defined in IEC 60904-3. IEC 60904-9 applies to solar illumination simulators no matter if the application is in PV test, calibration laboratories or in manufacturing lines of solar cells and PV modules. IEC 60904-9 lies out the classifications of solar simulators, including calibration, traceability, linearity, spectral responsivity, mismatch correction, and sun simulator requirements.

As stated in a previous section, the EL is used industry-wide in nearly 100% of all module manufacturing lines, capable of detecting electrical and mechanical defects related to the cell material and the electrical interconnection of the module parts (i.e., individual strings of cells). Even though no detailed defect catalogue exists, and it is more than doubtable if it ever will be published inside a standard due to the huge variety and number of potential defects and most of all unknown severity of individual defect patterns to the module longevity, power output and safety during operation. For standardization, part IEC 60904-13 was developed which lays down the fundamental requirements of the electroluminescence measurement procedure and gives certain information on some defects and root causes on cell level. To extend IEC 60904-13, which is written for indoor measurement, to outdoor application a new proposal was recently published.

Important for PV park operators is the family of IEC 62446. In the first part the required information and documentation of grid connected PV systems to be handed over to a customer are defined. Beside this, the commissioning procedure, additional information, and documents to verify the safe installation and correct operation of the system as well as the inspection criteria and documentation are written down. The second part goes a step further and describes the basic preventive and corrective as well as performance-related maintenance requirements and recommendations for grid-connected PV systems. As a highly important characterization technique for installed PV panels thermographic IR inspection of PV modules and plants is further described in part 3 of the standard. The IR can detect faulty bypass diodes, short circuits inside the solar module, local shunts on solar cells, larger current mismatching between individual cells and any other kind of excessive heat generation or uneven heat distribution inside the panel. This means IR covers capabilities to ensure safe operation and stable power outputs. Furthermore, part 3 lays down the requirements for measurement equipment, ambient conditions, inspection procedures and reports and personnel qualification.

Finally, it is worth to mention the IEC 61853 series (module performance testing and energy rating) and the IEC 61724 series (module performance in systems). In IEC 61853, the fundamentals to assess and predict the energy yield are addressed whereas IEC 61724 outlines equipment and methods for performance monitoring and analysis of PV systems and defines a procedure for measuring and analyzing the power production with the goal of evaluating the quality of PV system performance.

4.3.5   Major milestone in standardization

PV industry had to wait for more than 10 years until finally in 2016 a long-awaited revision on photovoltaic module standards was published. IEC 61215:2016 cancelled and replaced the second edition of IEC 61215:2005. The new standard introduced deep changes in the certification procedure which were overdue to comply with the requirements of PV industry [49]. The structure of the general requirements and test sequences and methods were updated, and its structure aligned with actual requirements. The new standard included in total 19 module quality tests (MQT) with some of them even requiring additional tests. Beside this, IEC 61646 and IEC 61215 have merged to standardize all flat panel technologies A few months after IEC 61215 was published in 2016 another major update on IEC 61730 came online to improve the minimum design requirements which finally guarantees the safety of the product during its operation. Beside this, the new standard updates serve as a link between European and American guidelines which were for many years a weak point for international panel manufacturers considering selling their product on both markets. The update on IEC 61730 was even deeper since it had been completely revised, new testing sequences added and the electrical requirements in the panel design re-evaluated to apply for higher system voltages. All changes served to fulfill the latest market trends.

Just 5 years, later IEC 61215:2016 was again updated by the publication of IEC61215:2021. The latest edition includes major technical changes such as:

- In previous IEC 61215 editions, panels only had to undergo static mechanical load testing which, in many cases, does not reflect the reality. For this, the new edition contains a cyclic (dynamic) mechanical load test (MQT 20).
- PID affected and still affects many solar modules during operation. To initially investigate modules capability to withstand PID a test for detection of potential-induced degradation is added (MQT 21).

- The variety of PV products has increased extremely over the last couple of years with new applications such as for example BIPV (building integrated), FPV (Floating Photovoltaics), VIPV (vehicle integrated), Agrivoltaics and so on. Additional test methods required for flexible modules were added for this reason. This includes a bending test (MQT 22) for flexible modules.
- A procedure for stress-specific stabilization for boron oxygen LID (MQT 19.3) was added.
- Including procedures for bifacial panels: addition of definitions, references, and instructions on how to perform the IEC 61215 design qualification and type approval on bifacial PV modules.
- Lying down the requirements for retesting to be performed according to IEC 62915.
- The nominal module operating test (NMOT) and associated test of performance at NMOT were taken out.
- An addition of a final stabilization procedure for modules subject to PID testing.

4.3.6   What makes current standards problematic

The fast development of some PV-related standards with the goal to keep track to the incredible fast speed of PV module development world-wide can be explicitly seen for IEC61215. For 11 years, no significant update on the standard from 2005 to 2016 and now after 5 years a new standard revision published just in 2021. Nonetheless despite all joint efforts of various groups, the final statement can be made that standards are always behind technological developments. Maybe the reason why there was no urgent need to publish a new IEC standard edition between 2005 and 2016 was for: cell and specifically module developments were at a slower pace. Crystalline silicon solar cells were mainly improved by adopting material developments and less by new cell architectures. For modules, the development pace was even less: the standard module design used since decades has changed only slightly, mostly in terms of encapsulation and backsheet materials. But with recent development in cell size, no matter if it is half-cut cells or varying sizes beyond 6-inch, cell architectures and various module designs with different interconnection technologies, certification standards cannot be updated in time. The steadily increasing world-wide production capacity with a strong rivalry between producers, quenching each quantum in terms of efficiency out of solar cells and modules cell to module losses, has introduced a development speed which was unimaginable for a long time in PV industry. The outcome of this is outdated test procedures, too general standard test definitions and in some cases even missing test sequences. The same is valid for standards related to measurement procedures as can be seen by bifacial solar cell and module products. In this case, both sides are used for harvesting sun light and converting it to electrical energy. In test labs, the question automatically arises: how to define an efficiency of such product, which measurement protocol best to apply for measuring the cell and module power output of the front and rear side: illuminating both sides at the same time or in a sequential way? The answers are not immediately as straight forward to obtain as they might appear. Many test labs which also act as certification bodies nowadays try to oblige by offering additional, customer-tailored services. They support the customer with adapting and developing suitable testing conditions for new product technologies. By studying the product changes and BOM such labs help identifying required modifications in the test sequence and procedure defini-tions which strongly increases the chances to fulfill the requirements. For this, it is mandatory that with any modification to a test protocol the original objective of the underlying standard is preserved.

Another important point to further discuss is the validity of standards in terms of climatic conditions. Specifically in the past with previous versions of IEC 61215 and IEC 61730 the following statement was correct: PV modules and its materials are developed for field application of moderate climatic conditions and derived test sequences and procedures of standards are defined to prove modules capability to withstand such climatic conditions in terms of product longevity and safety. But today, large PV systems reaching 1 GW, and more are often not installed in loca-tions with moderate climatic conditions. The systems are built in areas where the number of annual sun hours is highest and second where large land areas are available and cheap: arid areas with high ambient temperatures, large temperature gradients between day and night, very large UV exposure, little or no rainfall, heavy wind conditions to mention a few. Test sequences such as damp heat, UV exposure, light soaking, temperature cycling, humidity freeze, and load testing would have to be extremely extended (and combined) in terms of duration to cover for such climatic conditions. Beside this, there is no or only little scientific base for beforementioned tests and testing parameters to the life span of modules [50]. A very good overview for module failures and test sequences suited to validate module quality is given in [51].

Often extensions to the beforementioned tests are asked for by the customer and executed at test labs which help to determine failures regarding one stressor and help benchmarking different module and material designs. Specifically, for hot climates, the IEC 63126 standard was published in 2020 (guidelines for qualifying PV modules, components, and materials for operation at high temperatures). IEC 63126:2020 specifically defines further testing requirements for PV modules which are installed and operated under conditions, automatically leading to larger module temperatures which are beyond the scope of IEC 61215-1 and IEC 61730-1 and the relevant component standards, IEC 62790, and IEC 62852.

Finally, the author wants to stress that passing the certification requirements of IEC 61215 and IEC 61730 only proofs a specific technological level of module design, material, and production which is, to a certain degree, only applicable for certain climatic conditions. Passing the certification does not automatically define a specific lifetime of the module nor is it a guarantee that product power warranties can be achieved. A certain module development may last for 30 or even 40 years in countries such as Sweden or The Netherland but may fail after 15 or 20 years if installed in the Atacama Desert in Chile. All tests inside a standard focus on the identification of specific known weaknesses or failure modes as given for a certain set of constraints in terms of the module materials, its interaction, and the applied climatic conditions. Not to mention that the test sequences are only applied to a small number of chosen modules which may not always represent the module production on a typical nightshift. The certification process can currently only guarantee for a minimum in quality and reliability and sort out problematic products with faulty or wrong material combinations.

4.3.7   Standard requirements for advanced module concepts including bifacial solar modules

In the last chapters, the strength and weak points of current certification standards were discussed in detail. The conclusion is drawn that standards are always one step behind current technological developments. This is specifically valid for system applications such as FPV, VIPV, or Agrivoltaics (only for BIPV IEC 63092-1:2020 was published in 2020) or various PV module technologies with advanced module concepts such as half-cut, high power modules with extreme large size, shingling, novel anti-soiling deposition, a tiled module configuration—sometimes referred to as paved, conductive backsheet and multi-busbar setups with far more than 10 busbars per cell or last but not least for extreme climatic conditions such as desert or Arctic/Antarctic applications [52]. Producers currently can conduct comprehensive testing and certification services beyond the testing sequences required in standards at testing labs which somehow act as proof for modules to keep its power and lifetime promise.

In the following for three advanced module concepts which are specifically important for n-type solar cells challenges in terms of specific test sequences inside the IEC 61215 standards are discussed. The further discussed technologies are:

- Conductive backsheet.
- Tiling/paving ribbon technology (similar to the seamless soldering technology but originating from two different companies: Jinko and LONGi).
- Shingling.

With focus on six material quality tests inside IEC 61215:2021 (MQL 10, 11, 12, 13, 16 and 20), the beforementioned technologies are further discussed.

1. UV preconditioning test (MQT 10) and elevated ambient temperatures
    The UV preconditioning test is specifically important to be extended for all module technologies and all system applications if the location of the installation is set up under hotter climatic conditions such as in deserts and countries with significantly more than 1,500 sun hours per year. Mainly larger PV systems with capacities beyond 500 MW are currently installed at those locations.
    UV in combination with heat speeds up chemical reactions in the encapsulation material, the backsheet material and the interfaces between both [53]. Beside this UV in combination with heat increases material degradation. Neither for tiling/paving nor for shingling any specific weaknesses are directly accessibly. For conductive backsheets potential risks of material delamination (e.g., conductive copper layers from backsheet material itself) also appear to be

very small since most of the backsheet area is anyhow covered by solar cells thereby reducing the incoming UV exposure rate to a minimum. A different statement can be given for the electrical bonding method: if electrically conductive adhesives (ECA)'s instead of standard soldering alloys for busbar to cell interconnection are used, which is mostly the case for cell interconnection on conductive backsheets or via shingling, increased ambient temperatures can reduce the adhesion and hence the peel force between the interconnected materials (e.g. cell to copper layer of the backsheet material or cell front to cell backside when shingling is used) [54]. Depending on the layout and specific design UV can also additionally act detrimental. The main problem is that in terms of mechanical bondability weakened electrical contacts in general are difficult to detect: often the individual materials are mechanically pressed together inside the module's material sandwich. Even if a large adhesion loss in the solder or ECA takes place the pressure keeps the individual materials electrically together and results in no significant increase in series resistance. The situation becomes worse once mechanical load, e.g., by wind or snow load or temperature cycling, e.g., in wintertime or in deserts, takes place. Such forces act detrimental and the weakened electrical contact results in significant fill factor and hence power losses. For module makers with products based on conductive backsheet or shingling technology, the advice here is to voluntary take part in extended test sequences beyond the norm.

2. Thermal cycling test (MQT 11)

Thermal cycling intends to investigate if a solar module in its electrical and mechanical integrity is capable to withstand larger temperature gradients. The introduced thermo-mechanical forces, for a larger part caused by the varying coefficients of thermal expansion (CTE) of the individual module materials, degrade adhered surfaces. Backsheet compounds or the interface between encapsulant and glass, cell or backsheet may lose its integrity and delamination may result. The same appears to be valid for the electrical contact interfaces: cell busbar to solder (resp. ECA), solder (resp. ECA) to ribbon, and so on. The ECA interconnection typically comes with a far smaller adhesion hence peel force between the individual materials if compared to the standard soldering interconnection by a lead-tin based alloy [55]. Induced thermo-mechanical forces on material interfaces with varying CTE's on each interface side or in between layers/compounds play a significant role during thermo-cycling. Beside this electrical pressure contacts often loose electrical contact during cycling as well. In this matter, all three technologies are specifically prone to experience additional losses in terms of the electrical bonding behavior and the electrical qualities. For tiling/paving even slightly pre-damaged cell edges may result in crack growth over time initially caused by the thermal-mechanical stress.

The same statement as for UV and heat exposure can be made: Extended test sequences help validating the risk. The experienced reader can immediately argue here that products are typically sold with a valid certification in IEC 61215 and IEC 61730. But as mentioned before, the test sequences of the certification standards do only partly reflect the specific climatic conditions and the boundary conditions of any installation. A passed certification does not automatically result in a scientific statement that the tested product is capable to operate at any location in any system world-wide for a given lifetime.

3. Damp heat test (MQT 13)

The damp heat test with its static temperature application for 1,000 h does typically not introduce any larger thermo-mechanical forces to the module and its materials/parts. The different material CTEs play a less of a role. The strongly increased temperature level speeds up chemical reaction inside the materials and at its interfaces between materials. Beside this, the natural ageing process is strongly increased. This means the same conclusions and recommendations as made for UV preconditioning test (MQT 10) and elevated ambient temperatures can be drawn and given. Standard backsheet materials have a significant moisture vapor transmission rate (MVTR) of the air-side layer which plays a significant role on the performance and durability of the backsheet and consequently on the reliability of the PV module [56]. This means standard backsheet-glass PV modules typically experience a certain moisture ingress due to its water permeability inside the backsheet material compound and later into the module which acts further mechanically and electrically detrimental on the individual materials and its interfaces over time. Ambient conditions such as temperature and humidity play a major role for modules installed in the field and backsheet-glass solar modules are prone to accelerated ageing in countries with high levels of humidity and/or temperature. On the positive site, the almost perfect water barrier of the conductive backsheet module protects the interior of this specific module type to a far higher degree against accelerated ageing due to humidity and temperature. Modules with tiling/paving ribbon or shingling technology can be easily further protected if glass is used instead of a backsheet material, resulting in a so-called glass–glass module with water barrier qualities close to 100%. On the other hand, may glass–glass experience other reliability issues over time [57].

4. Humidity-freeze test (MQT 12)

The humidity-freeze test is a combination of damp heat and thermal cycling, even though the number of cycles is rather small. This means a temperature cycling at a high humidity rate is applied (at a certain threshold level of temperature). The failure codes introduced to modules by applying HF testing can be like the ones seen during/after damp heat and temperature cycling. But there is an important difference by applying high rates of moisture during testing: if moisture diffuses into the module, which is almost inevitable for backsheet-glass modules, water gets close to the boiling point and far beyond the freezing point during temperature cycles. Since the density changes with the temperature of water which is trapped in cavities inside module materials microscopic forces result. These forces, over time, can act destructive and degrade the module materials and its interfaces. If one of the above discussed technologies: conductive backsheet, shingling, or tiling/paving ribbon is more prone to potential degradation during HF testing if compared to standard crystalline silicon solar cell modules is unclear. From a technological point of view, the statement should be right that as long as no larger cavities in novel module technologies exist, e.g., gaps which are not filled by encapsulation material during lamination no effects are anticipated which are not also seen for standard module technology. This statement is valid for the impact of diffusing water into the module and not valid for the temperature cycling in general. Here, the same statement as before can be made.

The last two tests are discussed as one unit:

5. Static mechanical load test (MQT 16)
6. Cyclic (dynamic) mechanical load test (MQT 20)

Mechanical loads, if applied to the solar module, result in forces to the module materials which may act detrimental. A difference exists for solar cells embedded in glass–glass modules lying in the neutral plane and are especially protected against tensile and pressure loads [58]. Static loads result from snow loads or the mounting system itself. Often modules are mounted wrongly resulting in permanent static loads. Cyclic loads are mainly caused by wind loads. Areas with higher wind rates cause permanent cyclic loads to solar modules. The oscillation amplitude usually is small but strongly depends on system design, mounting structure, and system location. And of course, on the size of the solar module and the installation direction: landscape or portrait itself. This way, mechanical loads applied to solar modules can be seen similar to the strain a module experiences during thermal cycling with the difference that the module temperature is a result of the ambient temperature and the temperature generated inside the module by light absorption and power consumption inside electrical resistances. Mechanical damages on cell level, such as surface or volume defects, are often prone to experience crack growth with severe damages on cell level hence a strong degradation in power [59]. If ECA is used for bonding cells to conductive backsheets or for shingling cells less adhesion between the joints may result in bonding degradation over time. Typically, series resistance increases, and a power loss is the result. For tiling/paving the specific conditions of ribbon shape, ribbon thickness and cell arrangement require to put special attention on crack growth (surface and volume cracks) with potentially pre-damaged cell edge areas.

The emphasis is again placed on the fact that passing the certification process not automatically results in product lifetimes and power losses as given in the specification of the producers. Too many unknown environmental and material boundary effects exist no norm can test against to give 100% certainty.

4.3.8 *A step in the right direction*

Many companies world-wide started providing additional trust to customers by applying extended testing and certification on their module products beyond IEC 61215. An appropriate tool given to these companies was by publishing the IEC TS 63209-1:2021 standard in 2021 [60].

IEC 63209-1:2021 intends to provide additional information to supplement the baseline testing as defined in the pass/fail standard IEC 61215. Standardized methods are provided for evaluating longer term reliability of PV modules and for BOM's which may be used for the manufacturing process. The test sequences of the standard are intended providing additional information for comparative qualitative analysis by specifically applying stresses which are relevant to application exposure. The goal is to assess the long-term product reliability in three phases:

(a) Part 1—Extended Stress Testing
  • 2,000 h of damp heat testing, compared to 1,000 h in IEC61215
  • 600 thermal cycles instead of 200 cycles in IEC 61215
  • Extended stress test for mechanical load (static and dynamic)
  • Long-term PID testing
  • Long-term LeTID testing
  • UV light resistance testing of the backsheet material

(b) Part 2—Regular Surveillance Monitoring
  • Monthly Quality Testing Program
  • Random selection of required test sample lots by testing labs
  • Continuously monitoring of PV module products during mass production
  • Monthly monitoring and review of test protocols

(c) Part 3—Supplier and Materials Monitoring
  • Master-list of critical core suppliers and materials (material risk assessment)
  • Randomly sample picking for polymeric footprint (one test sample from a minimum of two suppliers per month)
  • Regular Supplier Audit (minimum one supplier audit per year for core materials)

Another approach, yet far away from being a certification standard is the NREL approach: a solar module stress test for all seasons [61]. NREL has developed a new stress testing protocol for PV modules, designed to simultaneously expose modules to multiple stresses, as they likely appear in the field. Putting modules through this test reveal new information regarding backsheet degradation.

4.4 Advanced n-type module technologies

4.4.1 Introduction

On the forefront of high-power modules are n-type modules, which combine highest performance cells with advanced interconnection technologies, optimized layout, and materials. Historically, the increase in module power has largely been driven by improving cell efficiencies. In the last 3–5 years, novel interconnection technologies and module materials changed the game. In addition, the disruptive change in wafer size further accelerates the scaling in module power, and further reduction of the LCOE. As the deployment of PV is about to reach, the TW scale

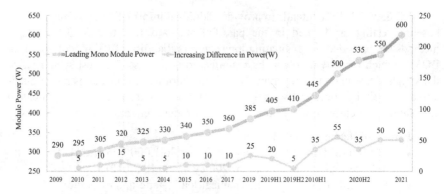

Figure 4.4   Evolution of module power in the last decade [63]

sustainability considerations are recently emerging as complementary key metric for PV modules [62].

Diversification in module technologies and application of tailored modules represents a paradigm shift that is set to continue. Flexibility to combine advanced interconnection technologies: multi-busbar with light reflecting ribbons, multi-wire, paving or gapless, shingling interconnections with cut cells, novel module layouts and advanced module materials, such as white backsheet grid, bring multiple options to design high-module efficiency, and/or power. Even more, module design is driven by not only peak performance but also considerations to reach the highest energy yield in a specific climate and/or application. In summary, module manufacturers leaped into the area of kWh and diversification of their product portfolio.

A PV module is composed of electrically connected solar cells that are laminated between an encapsulant on each side of the cell and glass as front and plastic sheet or glass as a rear cover to ensure its long-term reliability and durability. Cell interconnection, especially for advanced cell technologies using n-type wafers, is a key enabler for high module performance and demands co-design of the cell and module metallization. Selection of the materials and processes for packaging must balance protective functions for reliability and durability, while maintaining optical and electrical performance of the module.

The performance losses that arise from cell integration and packaging in modules are expressed by the term cell-to-module ratio (CTM). Following the definition of ITRPV, CTM is defined as module power divided by cell power multiplied by the number of cells:

$$CTM_{MPP} = \frac{P_{MPP\ module}}{nb\ cells\ x\ P_{MPP\ cell}} \qquad (4.10)$$

The CTM can also be defined for all the other IV parameters: CTMIsc, CTMVoc, CTMImpp, CTMVmpp, CTMFF, and CTMη. This is one of the key metrics to understand losses and to improve the initial performance of a module. The International Technological Roadmap for Photovoltaic (ITRPV) in its 2021

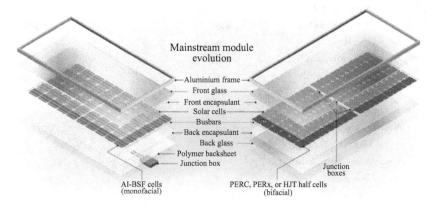

Figure 4.5   Schematics of module layup with traditional modules dominated the market (left) in the past and layup of current high-performance modules (right) [64]

edition expects the increase of the CTMPmpp from 98% to 100 % for full cells and 100% to 102% for half-cut cells in the next decade [65].

Finding optimum in terms of performance and cost is becoming more and more challenging owing to the increasing number of technologies, parameters, and their interdependence.

To assist the design of high-performance modules, several proprietary module power and/or CTM simulation tools have been developed. Some of the most known software packages available for the simulation of photovoltaic module performance are SmartCalc from FhG-ISE [66,67], Griddler Solar and Module from SERIS [68], SunSolve from PVLightHouse [69], and CTMOD from CEA-INES [70,71].

SmartCalc is a proprietary software developed by Fraunhofer ISE considering 15 factors based on analytical equations calibrated by experimental measurements. The losses and gains cover the module components, inactive area, optical and electrical effects and important interfaces as shown in Figure 4.6. Furthermore, balancing performance and module cost in module optimization adds an additional layer of complexity; hence automated optimization algorithms are also explored in SmartCalc [66].

CTMOD is a proprietary software developed by CEA based on analytical equations, and specifically calibrated for heterojunction cells and novel generation of module materials. This extended simulation tool enables a wide combination of inputs that provides complete optimized solutions (with different criteria such as performance, €/W, etc.) [71].

An example simulation with CTMOD is presented in Figure 4.7 and as expected, the main losses arise from both, the absorption in the glass, encapsulant and light falling in inactive areas. These losses are compensated and hence mini-mized by the reduction of the resistive losses which are lower at module level in the case for cut cells.

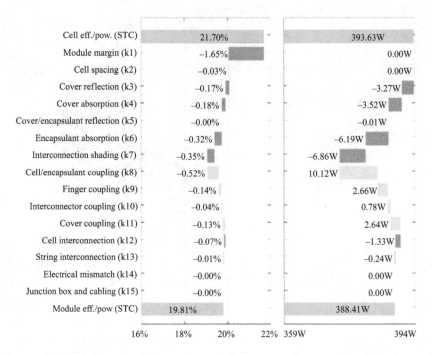

Cell eff./pow. (STC)	21.70%	393.63W
Module margin (k1)	−1.65%	0.00W
Cell spacing (k2)	−0.03%	0.00W
Cover reflection (k3)	−0.17%	−3.27W
Cover absorption (k4)	−0.18%	−3.52W
Cover/encapsulant reflection (k5)	−0.00%	−0.01W
Encapsulant absorption (k6)	−0.32%	−6.19W
Interconnection shading (k7)	−0.35%	−6.86W
Cell/encapsulant coupling (k8)	−0.52%	10.12W
Finger coupling (k9)	−0.14%	2.66W
Interconnector coupling (k10)	−0.04%	0.78W
Cover coupling (k11)	−0.13%	2.64W
Cell interconnection (k12)	−0.07%	−1.33W
String interconnection (k13)	−0.01%	−0.24W
Electrical mismatch (k14)	−0.00%	0.00W
Junction box and cabling (k15)	−0.00%	0.00W
Module eff./pow (STC)	19.81%	388.41W

Figure 4.6   Efficiency and power water fall diagram of conventional overlap cell modules [66]

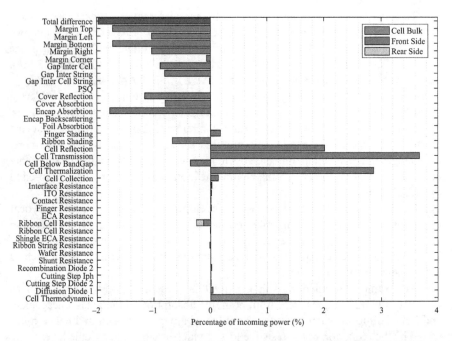

Figure 4.7   CTM gains/losses for a 60 ½ M12 cells module (620 W – 21.4%) with reflective backsheet

4.4.2 Advanced interconnection technologies

The most widely used interconnection technology today is *ribbon tabbing* where solder coated copper ribbons connect the front side of a cell to the rear side of its neighbor with a soldering process. Nevertheless, numerous other techniques have been developed, and their deployment has accelerated in the last years in the race to increase module power. According to ITRPV multi-wire based interconnection will be deployed in more than 50% of all modules by 2023, while overlap/shingling interconnection, as well as structured sheets or ribbon gluing, are projected to equally gain market share [65].

The main objective of the interconnection is to minimize resistive losses while maintaining minimal optical shading. Interconnection also plays a major role in the reliability of photovoltaic modules in the field. Typical degradation observed are partial or complete disconnection between ribbons and cells, corrosion of copper, and cell's cracks. All these degradation mechanisms could lead to the creation of hotspots and fatal failure of the modules [72–74].

To increase production capacity, cell producers are increasingly using n-type wafers and larger wafers up to G12 (210 × 210 mm^2). These new advanced cells may require specific interconnection techniques, while the increase in wafer size implies an increase in photogenerated current that must be managed at the interconnection level. On the one hand, increasing photocurrent demands increasing metal cross-section for interconnecting materials on module level. On the other hand, sustainability is becoming an increasing concern for solar industry which could lead to a shortage of certain critical materials (silver, indium) [62,75].

Today, looking at the latest modules developed in the market, it is possible to note a wide variety of cell technologies, interconnection strategies, wafer sizes, and cell cuts to improve module performance.

Table 4.3 Non-exhaustive list of the most efficient modules in 2021, demonstrating the variety of technologies

Company	Product	Cell	Interconnection	Wafer size	Full/half/third-cell	Power (W)	Efficiency (%)
Sunpower Maxeon	Maxeon 3	n-type IBC	Dogbone IBC	M0	Full	400	22.7
Sunpower Maxeon	Performance	Mono PERC	Shingle	M3	/	330	22.5
Suntech	HJT	HJT	MBB	M3	½	450	22.1
Jolywood	JW-HD156N	TOPCon	MBB	M10	1/2	615	22.0
REC	Alpha	HJT	SWCT	M4+	1/2	380	21.7
RISEN	TITAN	mono	MBB	M12	1/2	605	21.4
FuturaSun	Zebra	IBC	IBC ribbons	M3	Full	360	21.3

4.4.2.1   Soldering (high and low temperature) versus bonding with conductive adhesives

Soldering is the most commonly used interconnection technology in photovoltaic industry. The connection between the solder-coated ribbon or copper wire and the metallization of the cell is formed by heat treatment. The solder coating are mostly lead-based alloys such as SnPb or SnPbAg. Recently, to remove the lead from PV modules and to comply with RoHS directives, new alloys are emerging such as SnAg or SnAgCu [76,77]. Soldering can also be compatible with low process temperatures thanks to the use of specific low melting point alloys, such as bismuth or indium. Although the use of the low temperature process has been introduced for temperature sensitive cell architectures such as HJT, this technology might be interesting for all types of cells to limit the internal thermomechanical stress of the soldering process [73]. However, In and Bi remain expensive and are considered rare and the reliability of the interconnection is still under investigation. This latter concern arises from the fact that these alloys can develop a brittle inert metallic alloy, varying microstructure that can negatively affect their mechanical strength [78].

Wires can be used instead of ribbons which are thinner than ribbons, usually ranging between 250 and 350 μm compared to 0.8 mm for ribbons. This allows the number of interconnectors to be multiplied without impacting the photogenerated current. Typically, going from BB6 on M2 cells to MBB9 or even SWCT (14 wires) allows a gain in FF of nearly 1%. In addition, thanks to the round shape, the effective shading is reduced compared to flat ribbons (50–60% compared to 90%), which even allows a gain in photogenerated current by decreasing the shading by almost 2%.

The major challenge in the deployment of this interconnection technology is the soldering process to form reliable bonds between the cell metallization and the wire. To increase the reliability of soldering, the use of solder pads has been proposed by Schmid among other players.

ECAs are the main alternatives to soldering [79,80]. They have various advantages inherited from the family of adhesives, such as lowN toxicity, low processing temperature (of the order of 150°C compared with 220°C for lead-free solder), the ability to join different materials, to tolerate mechanical cycling, to make small-scale assemblies and to have a probably lower environmental impact than lead-free solder [81,82]. However, they also come with important trade-offs. Current ECAs are generally less conductive than solder alloys (by a factor of approximately 100), the impact resistance is lower, the mechanical strength is sensitive to various environmental parameters and the electrical conductivity degrades rapidly if exposed to ambient conditions (compared to solder) requiring low-temperature storage of the materials. This distinguishes ECAs from brazing alloys where mechanical and electrical failures are generally linked. The use of ECAs is particularly adapted for shingling applications in photovoltaic industry [83].

Thanks to latest developments of ECAs, a reduction of ECA consumption higher than 50% has been demonstrated by CEA and qualified without impacting the performance going from 400 to 420 W with GG or GBS 72-cell modules.

4.4.2.2   High-density module interconnection: paving, tiling and shingling

One of the main losses from cell to module is from optical losses in inter-cell and inter-string spaces. To counter these losses, the reduction or even removal of the intercell spaces is the logical evolution though bringing numerous technical challenges. Gapless strings give a homogenous appearance to modules, specifically for applications requiring aesthetics such as building integrated PV (BIPV) or vehicle integrated PV (VIPV).

Shingling has been the first gapless interconnection in the early 1960 [84]. New usages with this interconnection technologies emerged at the beginning of the 1990 for solar cars which lead later to the commercialization by larger solar companies such as SunPower [85] or Solaria [86]. Shingling requires cutting cells in usually 5–6 stripes for an M2-sized wafer, and then connecting cells through stacking using an ECA [83]. Shingling is an interconnection option that is well suited to homojunction cells. It could also be a good solution for heterojunction cells if the problems of laser loss and metallization conductivity were addressed.

Although the technology uses a low amount of ECA (around 20 mg per M2 cells), it requires a highly conductive cell metallization and hence an increasing silver consumption. Industrial uptake of shingling interconnection solution is

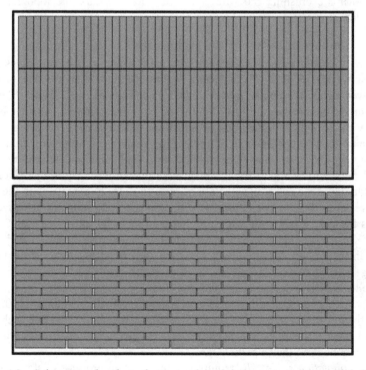

Figure 4.8   Schematic sketches of examined module layouts; top: parallel shingled string layout, bottom: matrix layout [88]

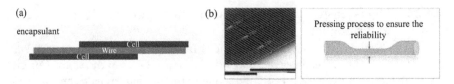

Figure 4.9   (a) Schematics showing the overlap; (b) picture of tiled cells and scheme of the wire shape after pressing [90]

therefore very dependent on finding alternative material for silver. Innovative shingling interconnection solutions have also been developed by FhG-ISE using an aluminum foil with a specific process [87].

Paving or tiling interconnection are more recent and offer a hybrid solution between shingling and ribbon/wire interconnection. As the overlapping cut cell pieces are connected with wire or ribbon, at the overlap area the two cells are therefore superimposed with an additional ribbon or wire. This stack can have a thickness of more than 600 μm, which implies an important increase in the local topography, bringing additional mechanical stress, and requiring adaptation in module packaging [89]. To partly overcome this issue, some manufacturers such as Jinko Solar are crushing the wire at the overlap to reduce the topography and mechanical stress.

4.4.2.3 Impact of different cell technologies

PERC and its evolution TOPCON are the main n-type cell technologies manufactured worldwide in 2021. Standard high-temperature metallization processes and silver pastes are compatible with these cell technologies. Widely deployed soldering techniques using SnPbAg ($\sim$180°C) coated copper ribbons became industry standards. Continued cell optimization and advances in printing technologies enabled the deployment of increasing number and thinner cell busbars, allowing for increased power at module level and improved reliability due to redundant interconnections. Environmental considerations lately triggered adapting the soldering processes to Pb-free alternatives in order to comply with the RoHS directive, even if the solar industry is still exempt for the moment. However, this implies the use of materials with a higher melting point alloys such as SnAgCu (217°C) or SnAg (221°C), which leads to higher internal stress [76,77]. Major disadvantage of this technology is the high thermal stress induced on the cell leading to bowing and micro-cracks. Importantly, advanced cell architectures require low stress interconnection technologies.

Low temperature coatings based on bismuth (140°C) or indium (120°C) are alternatives that can reduce the induced process temperature and hence the thermomechanical stress as explained earlier, though reliability, cost, and sustainability considerations limit their uptake.

Heterojunction solar cells require low-temperature processes below 200°C to avoid the alteration of the silicon amorphous layer. Since standard silver

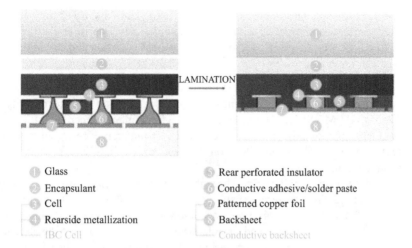

LAMINATION

① Glass	⑤ Rear perforated insulator
② Encapsulant	⑥ Conductive adhesive/solder paste
③ Cell	⑦ Patterned copper foil
④ Rearside metallization	⑧ Backsheet
IBC Cell	Conductive backsheet

*Figure 4.10   Schematic of the cross-section of a module layup with conductive
backsheet before and after lamination [97]*

metallization pastes with a firing step at 800°C are incompatible with HJT cells
low-temperature silver pastes became standard. These pastes are less conductive
and less suitable for soldering because of their higher wettability. Currently the use
of ECA to attach silver-coated ribbons on the cells is the most conventional method
[79–82]. Among other advantages, it reduces the internal thermo-mechanical stress
and is a lead-free technique. An alternative interconnection technology compatible
with HJT is the embedded electrode sheet such as smart wire connection technology (SWCT) patented by Meyer Burger. The multi-wire interconnection with
embedded electrode sheet where the wire is embedded in the ethylvinyl acetate
(EVA) sheet of the module is a variation of the multi-wire interconnection with
low-temperature solder alloys where the wires are soldered on cell fingers during
lamination [91]. Sticky Solar has developed a ribbon tape solution which also
allows for a higher throughput and to minimize the machine footprint. It also can be
applied with the use of 14–18 wires, allowing for light recycling and this way
improving the module performance, and giving a redundancy in the interconnection
which enhances the reliability. However, soldering is only effective on the combined surface of wires ($\emptyset = 250$–$350\ \mu$m) and fingers (30–$40\ \mu$m). The reduction of
finger width could then have a huge impact on reliability. Finally, new developments in silver pastes which are now emerging allow for standard soldering with
heterojunction cells. However, silver consumption for this kind of metallization
remains high. Shingling is also possible for HJT but currently less interesting
compared to homojunction cells due to the higher cell cutting losses and less
conductive finger. To overcome these limitations, novel HJT cell architectures with
reduced thermal sensitivity, and/or the use of copper plating are explored [92,93].

Back-contact cells with both polarizations on the same metallization plane
require a specifically tailored interconnection technology. Unlike for two-side

contacted cells, a standard interconnection technology and layout has not yet been deployed and hence a largely variety of technologies are under investigation. We can distinguish three types of interconnection technologies adapted to both MWT and IBC cell architectures [94]:

1. One layer metallization where the positive and the negative polarity are separated by physical distance. The most well-known approach is the technology of Sunpower's MAXEON technology with dogbone interconnection soldered at the edges of the cells. The major drawback of this approach is the stringent requirements on cell metallization design, and a highly conductive cell metallization.

2. Two/multi-layer metallization where the opposing polarities are separated by an insulating layer/material enabling a large flexibility in cell and module design. Its most common implementation is with ribbon soldering/gluing between neighboring cells and a printed insulating paste to ensure the electrical insolation between opposing polarities. This interconnection has been integrated by many players, for example by the Valoe/Solitek joint venture which is using ZEBRA cells, and even the multi-wire interconnection by Meyer Burger [95]. An interesting alternative has been presented by imec using glass fiber as insulation layer and solder paste for interconnection [96].

3. Module-level interconnection technologies brought a disruptive change as both stringing and bussing are implemented in one process step while keeping the advantages of multi-level metallization. Interestingly, most module-level interconnection replaces soldering as interconnection by low-stress approaches such as ECA-based gluing or welding. As in back-contact cells the high-temperature solder process further increases the thermal stress and bowing of the cell.

The conductive backsheet technology uses a patterned conductive backsheet as an integrated electrical circuit for interconnection, ECA-based gluing for interconnection and a polymer layer for electrical insulation. This technology has been recently commercialized by Endurans Solar/Silfab [97]. Alternatively, the aluminum-based mechanical and electrical laser interconnection (AMELI) technology developed by ISFH proposes busbar–free back-contact cells with multi-level metallization by contacting fingers to an Al foil. Similar to the backsheet technology, a polymer layer insulates the base from the emitter metallization [98]. Remarkably, this technology introduces the use of Al and welding technology, both are novel in the PV sector.

4.4.2.4   Impact of thinner wafers

Thin wafers are a promising path to reduce both, the cost and environmental impact of cells. Silicon heterojunction cells (a-Si:H) are particularly suitable for realizing very thin c-Si cells, because of its excellent surface passivation capacity. It is possible to reduce the thickness from the current standard at 160–180 μm down to 80 μm with almost no loss in efficiency, with an optimum of around 120 μm determined by the mechanical resistance of the wafer. Several studies have already demonstrated that decreasing the wafer thickness results in an increase of the bow and residual stress induced in the solar cell, while other studies consider that the solder process solely induces thermomechanical stress [99].

4.4.2.5 Impact of larger wafer

Recently, the wafer size in PV industry has increased from M2 (156.75 mm × 156.75 mm) to M10 (181 mm × 181 mm) or even M12 (210 mm × 210 mm). This increase has double impact on interconnection:

- Increase of the inter cell-busbar distance at constant BB number. For M2 wafers, the standard had evolved to half-cell with about five ribbons or nine wires. To avoid additional resistive losses in the fingers, the inter cell-busbar should remain constant, hence the standard will have to evolve to 6 ribbons or 10 wires for M10, and to 8 ribbons/12 wires for M12.
- Increase of the photogenerated current. This increases from about 9 A to 12 A for M10 and 16 A for M12. The addition of a cutting step becomes necessary to limit the resistive losses. However, cutting losses need to be carefully measured and included in the optimization as they may change depending on the cell technology. High efficiency passivated cell technologies may be more sensitive to cutting steps and/or might require adapted cutting and edge passivation.

Figure 4.11 shows the evolution of performance depending on the number of cutting steps. If the cutting step does not induce any loss, more cuts are better in terms of performance. However, the efficiency of the module must be evaluated carefully because increasing the number of cut cells leads to an increase in

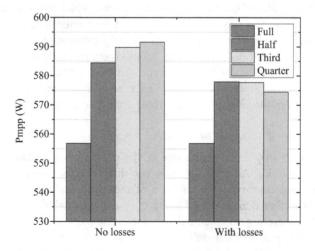

Figure 4.11   Simulated module performances of a GG module (eq 60 cells with HJT and M12 wafers) depending on the cutting losses (using CTMOD)

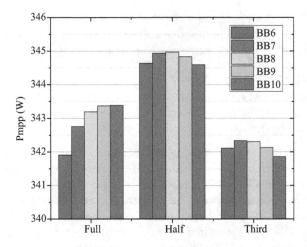

Figure 4.12   Simulated module performance for GG module (eq 60 cells with HJT cell and M4 wafers) depending on the number of BB and the number of cutting step using CTMOD (considering cutting losses)

space losses in the module. Additionally, more stringing steps can reduce the throughput of a module line and increase the scrap rate. Finally, it can have an impact on the output voltage of the module. If losses linked to cell cutting are considered, there is an optimum between ½ and ⅓ cell as estimated by CTMOD for HJT cells.

On the one hand, to minimize silver consumption at metallization level an increase in the number of interconnections is required, which leads to an increase of the copper consumption. Figure 4.12 shows that for performance considerations, the optimal BB number on a M2 cell is strongly influenced by the number of cutting steps. For a full cell of M2 size, an optimum performance at BB9 can be found, while it decreases to BB8 for ½ cells and BB7 for ⅓ cell. The losses associated with the cutting process must carefully be assessed, and, if possible, mitigated thanks to repassivation, to enable more cutting to take place [100].

4.4.3   Advanced module materials

Module materials play an essential role in the performance, reliability, and durability of PV modules. Additionally, they can directly affect the cost through raw material cost and determine the cadence of the layup and lamination. Material innovation is further complicated by the complex interaction, which in case of novel cell types are not fully elucidated; that validation of all new materials in modules is preferred, and not only through qualification of independent components [101].

Frontsheets provide transparency and structural protection as well as electrical insulation and barrier for moisture and oxygen ingress. Low iron tempered flat glass with an antireflective coating became the industry standard. Scaling the glass size to large modules means modules are getting heavier, which triggered the innovation of thinner glass and plastic front sheets. UV resistance of polymer front sheets remains the most challenging step in qualification [101] though still numerous products are emerging for roof application. Standard glass modules weigh 12–16 kg/m^2 which this way can be scaled below 8 kg/ m^2 [102].

Coupling anti-reflective and anti-soiling coating are becoming particularly critical for desert areas where performance loss peaks up to 39% due to soiling. The abrasion of this layer has also been extensively investigated if installed at harsh environmental conditions and a 25-year lifetime cannot be guaranteed [103].

Encapsulants protect the interconnected cells from oxygen, moisture, mechanical shocks, and must provide adhesion between the various layers. Ethylene vinyl acetate (EVA) is the most commonly used material with a large variety of chemical formulations. However, the basic compound triggers the formation of acetic acid in the presence of humidity, leading to corrosion of the metallization. This effect worsens in glass–glass modules with low diffusion rates. To ensure durability, polyolefin (PO) and thermoplastic elastomers (POE, TPO) were more often used in glass–glass modules and even glass-backsheet modules with premium quality.

The integration of white grid patterned encapsulants in rear side record efficiency bifacial modules is an emerging innovation, which aimed at allowing bifacial light collection and light recycling in between cells for an ultimate performance [104].

Backsheets complete the electrical insulation of the module and provide the rear side protection from moisture and environmental pollutants such as dirt, sand, etc. Multi-layered sheets are the most common including polyvinyl fluoride (PVF), polyethylene terephthalate (PET), polyamide (PA), polypropylene (PP), among others. Numerous innovations have been launched in this field. On the one hand, the durability of the backsheets has been increased through co-extrusion of the various layers and more stringent qualification tests. On the other hand, interface engineering between the various layers enables backsheet manufacturer to improve the barrier properties and adhesion of their material [104].

4.4.4   Integration of module level electronics

As photovoltaic modules are increasingly integrated in buildings, shade-tolerance, integrated functionality, sensors for measuring the power and temperature of the module and intelligence are becoming part of the module itself. Some module makers already combine PV with digital technologies, named as photovoltratronics, which is aimed at maximizing energy generation and its local utilization [105].

In PV modules, the series connection of the cells means that shading over just one cell or part of the module can limit the current of the entire device. To overcome this problem, standard 60-cell PV module architecture is composed of three substrings protected by bypass diodes placed in a common junction box. Shading over part of this type of module triggers the bypass diode and the shaded substring is bypassed reducing the module output by at least 30%.

Emergence of bifacial and half-cut cells paved the way toward a novel module architecture, separating the top and bottom part of the module with six substrings and three bypass diodes in three separate "split" junction boxes placed at the middle of the module [106]. This layout is particularly suited for static shading, while for dynamic shading, numerous innovations are explored in form of the I-module [107], TESSERA module [108], and reconfigurable modules [109]. Furthermore, integrating smaller cells and substrings coupled with integrating control circuit, power optimizers, and converters in the module. These latter technologies currently remain prototypes as the energy gain does not compensate the additional cost, and importantly the unknown reliability creates a barrier for its large-scale uptake.

Additionally, exploration of integrated energy storage in PV modules to create truly autonomous devices has long-standing history with trial of battery and supercapacitors. However, they are still facing major challenges related mainly to thermal management and its impact on both, PV and battery lifetime, and safety to name a few.

4.5 The end of the operational life of photovoltaic panels

Photovoltaic technology is certainly an excellent option to meet the growing demand for clean electricity in the world. In particular, crystalline silicon panels had global market penetration rates of around 92% of the total number of different PV panel technologies available in the world in 2014, and their PV market presence is expected to be at least 73.3% by 2030 [110]. This is confirmed by the fact that during 2017 crystalline photovoltaic panels accounted for over 95% of the total world production of the global photovoltaic market [111].

However, it is unclear what will happen to those millions of crystalline PV panels, already installed and to be installed, once their useful life cycle comes to an end (i.e. end of generation, after 25 and 35 years of operation), which despite the environmental benefits can negatively impact the environment if they are discarded as waste generating high toxic chemical pollution and substantial waste at the end of a panel's operational life cycle. It is expected that the amount of waste from the PV sector (see Figure 4.13) increases from 1.7 (8) Mt in 2030 to 60 (78) Mt in 2050 [112].

On the other hand, the current recycling of photovoltaic waste at the end of its operational life is limited and toxic [114], it does not exceed 10% of the total PV materials in the world [115]. This is due to the cradle-to-grave (C2G) paradigm that currently prevails in the industrial PV sector, assuming a "linear" life cycle for crystalline silicon photovoltaic panel technology (PV-SiC). (Figure 4.14), from the extraction of raw materials used in their manufacture (cradle) to the end of their useful life and must be disposed of in a landfill (grave).

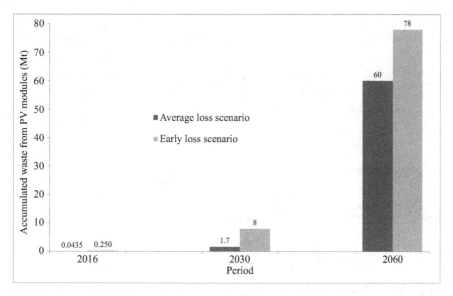

Figure 4.13   Estimation of the volume of global waste accumulated at the end of the lifetime of a photovoltaic panel approach [113]

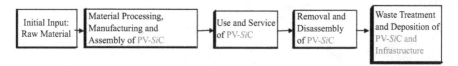

Figure 4.14   Simplified linear material flow in a C2G system approach [113]

However, the concept of the Cradle-to-Cradle (C2C) philosophy was coined in the book C2C: *Remaking the Way We Make Things* [116], where a C2C product was defined as one that is designed and manufactured to avoid environmental pollution, not only during its creation phase, but throughout its entire life cycle and also at the end of its useful life (EoL), where all its materials are recovered and re-used in a closed cycle of materials (CCM). This new paradigm emphasizes re-thinking the way we currently do things; from design, engineering, manufacturing, use, collection and then re-entry of materials into the industrial production cycle after the end of life of a solar panel (see Figure 4.15), i.e. "up-cycling" or supra-recycling.

A truly sustainable closed material cycle model must consider the importance of prioritizing waste management options. Major efforts should focus on redesigning PV panels, replacing toxic materials with environmentally sound alternatives, and incorporating social, economic, and environmental benefits [114]. Figure 4.16 presents this new approach, where the basis is that redesign and supra-recycling have the highest priority [113].

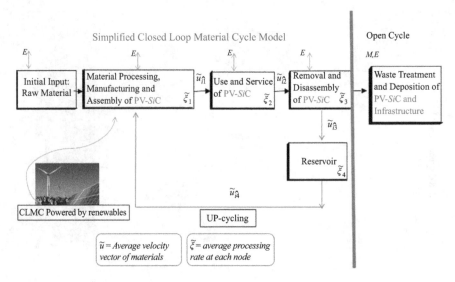

Figure 4.15 Simplified flow in a C2C system approach in a closed-loop system [113]

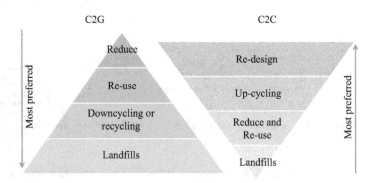

Figure 4.16 Current process prioritization in a C2G system versus C2C approach [113]

Thermodynamic implications of a closed material flow system

Ideally, a PV panel based on C2C principles should be designed taking into account the supra-recycling phase (i.e. recovery and re-use) for all materials used in the product. The implementation of such a system requires a substantial improvement in the efficiency of material selection, recovery, and re-use.[1] However, from a thermodynamic

---

[1] Of the five C2C principles, this work focuses mainly on the reuse of materials.

point of view, a PFV-SiC designed under C2C principles can be considered as a non-random mixture. If these building blocks are systematically distributed and ideally maintain the same volume, we would significantly reduce any increase in the entropy of the mixture of components. But to keep the work during the material separation phase as low as possible, the materials used in the product must be highly pure, something that happens not to be a major constraint for PV-*SiC*.

Currently, industry is trying to recover materials, but these thermochemical processes are still inefficient and toxic. We therefore initially suggest that, in order to make the PFV-SiC technology more sustainable, industry would have to comply with three fundamental principles:

1. Apply the C2C philosophy during the design phase of the panels, complying with the requirements of C2C certification and using the LCA-sustainable tool.
2. The PFV-SiC must be manufactured with high-purity materials.
3. A supra-recycling plant should be powered by low entropy sources (i.e. solar energy).

It is possible to assert that the work required to separate the components of a PFV-*SiC* designed under C2C principles can be considered minor compared to the amount of work required to separate the materials of a PFV-SiC designed without considering C2C principles. We now proceed to verify our assumption. For the purpose of demonstration, consider two pure components, in this case two ideal gases. We assume that each of these components is restricted to an equivalent volume that is regular. If we then consider the reversible equation for the total isothermal work for a non-reactive system, it can be written as follows for two components:

$$\delta W_{rev} = \frac{\partial A}{\partial P} dP + \sum \mu_{pi} dN_i + \sum A_i^o \, dN_i \qquad (4.11)$$

It is possible to prove that,

$$W_s = 0 \qquad (4.12)$$

The above demonstration offers a formal solution (4.11 and 4.12), which is fully available in [117], providing convincing support for the claim that it is possible to reduce the energy required to separate the embedded materials in a C2C product. It is found that the amount of work required to separate the materials in a PFV-*SiC* based on C2C principles is lower compared to the amount of work required to separate the materials in a non-C2C panel, due to the random mixing of its materials, where the minimum work for an eventual separation of its materials is given by $W_m = RT \left[ln \frac{1}{x} \right]$ [118].

Our solution, however, reveals a fundamental limitation for a closed-loop material flow or circular economy system. In order to get the work during the material separation phase as low as possible, the materials used in the product must be highly pure, something that coincidentally is not a major constraint for PFV-*SiC* [117].

Finally, in a closed material flow system in the photovoltaic industry, the materials must flow in circles forming a closed flow. Therefore, the physical constraint for a closed material flow can be expressed in terms of the equation for the conservation of mass. This equation determines the temporal rate of change in the density of any given volume of any given raw material matching in PFV-SiC, which we will call non-renewable abiotic primary resources (NRAPR) and flowing through each industrial phase (see Figure 4.15).

Let us consider an imaginary volume of NRAPR[2] finite with dimensions δx, δy, δz, containing a fixed collection of matter shifted by means of a field property ε (x,y,z,t) through any of the stages or phases in the arbitrary manufacture of something. Since the system is composed of a fixed amount of material, the form of the finite volume of material—NRAPR—will be distorted by the various forces applied during the manufacturing processes. However, as the density does not change with time, we can write the conservation of mass equation as:

$$\frac{\partial \rho}{\partial t} + \nabla \cdot (\rho) = 0 \tag{4.13}$$

Equation (4.13) is the Euler form of a first-order hyperbolic equation describing the mass conservation equation for certain NRAPR bodies, where ρ represents the NRAPR or its material density in terms of ρ per unit length and Υ represents the three-dimensional velocity vector. However, a further step is necessary to fulfill the first assumption that at any given time and spatial scale, the flux of a mass of NRAPR observed on a global scale can be considered as continuous [117].

Hence, it is possible to solve Eq. (4.13) numerically, using an explicit finite-difference approximation technique of donor cells to derive its discrete representation and a regressive scheme has been chosen to solve it numerically for the spatial derivative $\frac{\partial}{\partial x}$, because it is considered more stable for $\tilde{u} > 0$, whose results can help to understand the impact of space and time scales in a closed flow system of photovoltaic materials. Details and the numerical solution can be seen in [117].

4.6   Summary

Chapter 4 worked out the general requirements for modules based on n-type silicon solar cells, introduced module concepts and characterization techniques for this cell type and gave a general consideration on standards and recycling schemes. It was concluded that on the one hand n-type cells offer a very high efficiency potential but come on the other hand with the difficulty for cell interconnection during module processing to overcome by applying more advanced processing and interconnecting techniques. This statement is valid for some of the high efficiency cell concepts such as IBC or HJT cell design which require major modification to the module concept rather than the PERT and TOPCon cell concept, which only come with minor conceptual change requirements. It must be stressed out that for no

---

[2]For an imaginary volume of NRAPR, no precise numerical definition can be given, however, a cubic centimetre of NRAPR would fit well in most contexts where a volume of NRAPR is discussed, particularly those related to the photovoltaic industry.

n-type cell concept major restrictions can be named which would prevent the use of this high potential cell design. This is even more important when looking at the named advantages like higher conversion efficiency, reduced temperature coefficients and smaller field power degradation if compared to module concepts based on p-type cells. n-Type cells are also the best choice to be used inside bifacial module concepts which market penetration share steadily increases. No disadvantages were found to be named for product reliability and lifetime. On the contrary, lower temperature coefficients and reduced power degradation would lead to the assumption that energy production and power warranty might be significantly higher if compared to p-type cell modules.

Some work has already been performed on implementing certain n-type cell and module specific properties and requirements to standards. On the other hand, did thorough consideration of existing standards proof that only few module concepts would really require major updates on standard test protocols. Since almost all module concepts anyhow were designed and are currently applied to both cell types, n-type as well as p-type, any restriction or requirement named here is generally valid for high efficiency cell concepts. The same statement can be made for characterization and monitoring techniques. If a specific module type requires a specific characterization, measurement plan or application technique strongly depends on the underlying module concepts. As for example, a longer time for the I–V sweep is required for the highly efficient n-type cells and modules. The general requirements for bifacial module measurement were given and the impact of the junction box, frame, and mounting position to the measurement results further discussed. As a best practice, a close to zero rear side shading effect for bifacial module concepts can be named. In terms of outdoor characterization and plant inspection the same tools, inspection routines and intervals can be used as for crystalline silicon solar cell modules based on p-type cells.

The discussion in the recycling chapter showed that in a best situation, the C2C philosophy is applied to design phase of the PV panel, no matter if it is based on n-type or p-type solar cells or how the detailed module design looks like. Very important to note is the manufacturing process which shall be based on high purity materials for which the implementation of such a system requires a substantial improvement in the efficiency of material selection, recovery, and re-use.

We would like to conclude this chapter with the following statement: the likelihood of detecting the unexpected on the negative side for n-type silicon solar cell modules is as high as it is for current p-type technologies BUT the chance to be surprised on the positive side with an increase in longevity and most of all power output is far higher for module concepts based on n-type cells.

References

[1] Fischer M, Woodhouse M, Herritsch S, and Trube J. Internacional Technology Roadmap for Photovoltaic (ITPR), 2020 Results. VDMA e.V, 2021.

[2]   (IEC) International Electrotechnical Commission. IEC TS 60904-1-2:2019. Photovoltaic devices – Parts 1 and 2: Measurement of Current–Voltage Characteristics of Bifacial Photovoltaic (PV) Devices, 2019.

[3]   Muehleisen W, Loeschnig J, Feichtner M, *et al.* Energy yield measurement of an elevated PV system on a white flat roof and a performance comparison of monofacial and bifacial modules. *Renew. Energy* 2021;170:613–619, DOI:10.1016/j.renene.2021.02.015.

[4]   Shan W. p-type PERC with Larger Wafers vs. n-type Cells with Higher Efficiency. PV CELLTECH Mapp. Transit. to N-type, 2021.

[5]   Jia C. The benefits and applications of Jolywood N-type TOPCon Technology. PV CELLTECH Mapp. Transit. to N-type, 2021.

[6]   Murgioni R. Can N-type modules conquer the project market? *PV Magazine Webinar,* 2021.

[7]   Šlamberger J, Schwark M, Van Aken BB, and Virtič P. Comparison of potential-induced degradation (PID) of n-type and p-type silicon solar cells. *Energy* 2018;161:266–276, DOI:10.1016/j.energy.2018.07.118.

[8]   Kopecek R and Libal J. Bifacial photovoltaics 2021: status, opportunities and challenges. *Energies* 2021;14:2076, DOI:10.3390/en14082076.

[9]   Benda V and Černá L. PV cells and modules – state of the art, limits and trends. *Heliyon* 2020;6:e05666, DOI:10.1016/J.HELIYON.2020.E05666.

[10] Green MA, Dunlop ED, Hohl-Ebinger J, Yoshita M, Kopidakis N, and Hao X. Solar cell efficiency tables (Version 58). *Prog. Photovolt. Res. Appl.* 2021;29:657–67, DOI:10.1002/pip.3444.

[11] Clean Energy Reviews. Best Solar Panels, 2021.

[12] Ken-ichi I, Yasuo K, and Niwano M. An extensively valid and stable method for derivation of all parameters of a solar cell from a single current-voltage characteristic. *J. Appl. Phys.*, 2008;103(9):094507. 1–6, DOI:10.1063/1.2895396

[13] Zhang C, Zhang J, Hao Y, Lin Z, and Zhu C. A simple and efficient solar cell parameter extraction method from a single current-voltage curve. *J. Appl. Phys.* 2011;110(6):64504:1–7, DOI:10.1063/1.3632971

[14] Emery K. Measurement and characterization of solar cells and modules. In: Antonio L and Hegedus S (eds.), *Handbook of Photovoltaic Science and Engineering*, 2nd ed., New York, NY: John Wiley & Sons, Ltd, 2011, pp. 797–840 (Chapter 18). IBN: 978-0-470-97466-7, DOI: 10.1002/9780470974704.ch18

[15] Gueymard C. The Sun's total and spectral irradiance for solar energy applications and solar radiation models. *Solar Energy*, 2004;76(4):423–453, DOI: 10.1016/j.solener.2003.08.039

[16] IEC 60904-1: Measurement of Photovoltaic Current–Voltage Characteristics. Standard. Geneva, CH: International Electrotechnical Commission, 2006.

[17] Nigel T (ed.). Guidelines for PV Power Measurements in Industry. *EUR 24359 EN.* Luxembourg: Publications Office of the European Union; JRC57794, 2010, DOI: 10.2788/90247, http://publications.jrc.ec.europa.eu/repository/handle/JRC57794

[18] Edler A. Development of bifacial n-type solar cells for industrial application. PhD thesis. Universität Konstanz, 2014, http://nbn-resolving.de/urn: nbn:de:bsz:352-275017

[19] Joshua S, Christian R, Johanna Bonilla C, *et al.* Bifacial photovoltaic modules and systems: experience and results from international research and pilot applications, Report IEA-PVPS T13-14:2021, IEA-PVPS, 2021, ISBN: 978-3-907281-03-1.

[20] Libal J and Kopecek R (eds.). Bifacial Photovoltaics: Technology, Applications and Economics. Institution of Engineering and Technology, 2018, ISBN: 9781785612749.

[21] del Cueto J and McMahon TJ. Analysis of leakage currents in photovoltaic modules under high-voltage bias in the field. *Prog. Photovolt: Res. Appl.* 2002;10:15–28, DOI:10.1002/pip.401

[22] Dhere NG, Shiradkar NS, and Schneller E. Device for detailed analysis of leakage current paths in photovoltaic modules under high voltage bias. *Appl. Phys. Lett.* 2014;104(11):112103, DOI:10.1063/1.4869028

[23] Luo W, Yong Sheng K, Peter H, *et al.* Potential-induced degradation in photovoltaic modules: a critical review. *Energy Environ. Sci.* 2017;10:43– 68, DOI:10.1039/C6EE02271E

[24] Halm A, Schneider A, Mihailetchi VD, *et al.* Potential-induced degradation for encapsulated n-type IBC solar cells with front floating emitter. *Energy Proc.* 2015; 77:356–363, DOI: 10.1016/j.egypro.2015.07.050

[25] Yamaguchi S, Van Aken B, Masuda A, and Ohdaira K. Potential-induced degradation in high-efficiency n-type crystalline-silicon photovoltaic modules: a literature review. *Solar RRL* 2021;5:2100708, DOI: 10.1002/ solr.202100708

[26] Friesen G, Herrmann W, Belluardo G, and Herteleer B. Photovoltaic module energy yield measurements: existing approaches and best practice. Report IEA-PVPS T13-11:2018, *IEA-PVPS*, 2018, ISBN: 978-3-906042-52-7

[27] Mishima T, Taguchi M, Sakata H, and Maruyama E. Development status of high-efficiency HIT solar cells. *Solar Energy Mater. Solar Cells* 2011;95:18–21, DOI:10.1016/j.solmat.2010.04.030

[28] Mitterhofer S, Glažar B, Jankovec M, and Topič M. The development of thermal coefficients of photovoltaic devices. *J. Microelectron. Electron. Compon. Mater.* 2019;49(4):219–227, DOI: 10.33180/InfMIDEM2019.404

[29] Lopez-Garcia J, Pavanello D, and Sample T. Analysis of temperature coefficients of bifacial crystalline silicon PV modules. *IEEE J. Photovolt.* 2018:8(4):960–968, DOI:10.1109/JPHOTOV.2018.2834625

[30] Nguyen T-P, Molinie P, and Destruel P. Organic and polymer-based light-emitting diodes. In: Nalwa HS (ed.) *Handbook of Advanced Electronic and Photonic Materials and Devices*, London: Academic Press, 2001, pp. 1–51 (Chapter 1), ISBN: 9780125137454, DOI: 10.1016/B978-012513745-4/ 50081-0

[31] Piper WW and Williams FE. Electroluminescence. In: Seitz F and Turnbull D (eds.), *Solid State Physics*, London: Academic Press, Vol. 6,

1958, pp. 95–173, ISSN 0081-1947, ISBN: 9780126077063, DOI:10.1016/S0081-1947(08)60726-2

[32] Breitenstein O, Bauer J, Bothe K, *et al.* Luminescence imaging versus lock-in thermography on solar cells and wafers. In: *26th EU PVSEC.* Hamburg, Germany, 2011, pp. 1031–1038. ISBN: 3-936338-27-2, DOI:10.4229/ 26thEUPVSEC2011-2CO.13.5

[33] Bothe K and Hinken D. Quantitative luminescence characterization of crystalline silicon solar cells. In: Willeke GP and Weber ER (eds.), *Advances in Photovoltaics: Part. 2. Semiconductors and Semimetals,* New York, NY: Elsevier, Vol. 89, 2013, pp. 259–339 (Chapter 5). ISBN: 9780123847027, DOI: 10.1016/B978-0-12-381343-5.00005-7

[34] Mansouri A, Zettl M, Mayer O, *et al.* Defect detection in photovoltaic modules using electroluminescence imaging. In: *27th EU PVSEC,* Frankfurt, Germany, 2012, pp. 3374–3378. ISBN: 3-936338-28-0, DOI: 10.4229/27thEUPVSEC2012-4BV.2.45

[35] Köntges M, Kurtz S, Packard C, *et al.* Review of failures of photovoltaic modules. In: *IEA-PVPS,* 2014.

[36] Eder GC, Voronko Y, Dimitriadis S, *et al.* Climate specific accelerated ageing tests and evaluation of ageing induced electrical, physical, and chemical changes. *Prog. Photovolt.: Res. Appl.* 2019:27(11):934–949, DOI:10.1002/pip.3090.

[37] Köntges M, Morlier A, Eder G, *et al.* Review: Ultraviolet fluorescence as assessment tool for photovoltaic modules. *IEEE J. Photovolt.* 2020:10 (2):616–633, DOI: 10.1109/JPHOTOV.2019.2961781

[38] Ravi P, Muller M, Simpson LJ, *et al.* Indoor soil deposition chamber: evaluating effectiveness of antisoiling coatings. *IEEE J. Photovolt.* 2019:9:227–232, DOI: 10.1109/JPHOTOV.2018.2877021

[39] Reza M, Hizama H, Gomes C, *et al.* Power loss due to soiling on solar panel: a review. *Renew. Sustain. Energy Rev.* 2016: 59:1307–1316, DOI: 10.1016/j.rser.2016.01.044

[40] Kimber A, Mitchell L, Nogradi S, and Wenger H. The effect of soiling on large grid-connected photovoltaic system in California and the southwest region of the United State. In: *2006 IEEE 4th World Conf. Photovolt. Energy Conf.,* Vol. 2, 2006, pp. 2391–2395, DOI: 10.1109/WCPEC. 2006.279690

[41] Massi Pavan A, Mellit A, and De Pieri D. The effect of soiling on energy production for large-scale photovoltaic plants. *Solar Energy* 2011;85 (5):1128–1136, ISSN 0038-092X, DOI: 10.1016/j.solener.2011.03.006

[42] Rabanal-Arabach J and Schneider A. Anti-reflective coated glass and its impact on bifacial modules' temperature in desert locations. *Energy Proc.* 2016;92:590–599, DOI: 10.1016/j.egypro.2016.07.024

[43] Ilse K, Jorge R, Lukas S, *et al.* Comparing indoor and outdoor soiling experiments for different glass coatings and microstructural analysis of particle caking processes. *IEEE J. Photovolt.* 2018;8(1):203–209, DOI: 10.1109/JPHOTOV.2017.2775439

[44] Hagendorf C and Bagdahn J. Comparing indoor and outdoor soiling experiments for different glass coatings and microstructural analysis of particle caking processes. *IEEE J. Photovolt.* 2018;8(1):203–209, DOI: 10.1109/JPHOTOV.2017.2775439

[45] Conceição R, Silva H, Mirão J, and Collares-Pereira M. Organic soiling: the role of pollen in PV module performance degradation. *Energies* 2018;11(2):294, DOI: 10.3390/en11020294

[46] Olivares D, Ferrada P, Bijman J, *et al.*, Determination of the soiling impact on photovoltaic modules at the coastal area of the Atacama Desert. *Energies* 2020;13(15):3819, DOI: 10.3390/en13153819.

[47] International Electrotechnical Commission, IEC.

[48] Hauch JA, Brabec CJ, Fabricius N, and Bergholz W. Standardization as an Instrument to Accelerate the Development of Stable Emerging Photovoltaic Technologies—The IEC TS 62876-2-1:2018—Technical Specification for the Stability Testing of Photovoltaic Devices Enabled by Nanomaterials, Progress Report in Energy Technology, 2020.

[49] https://insights.tuv.com/blog/how-to-achieve-revised-iec61215-iec-61730.

[50] Kim J, Rabelo M, Padi SP, Yousuf H, Cho E, and Yi J. A review of the degradation of photovoltaic modules for life expectancy. *Energies*, 2021;4:042009.

[51] International Energy Agency Photovoltaic Power Systems Programme Review of Failures of Photovoltaic Modules, Report IEA-PVPS T13-01, 2014.

[52] Advanced Module Technologies 2019: Taiyang News Advanced Modules 2019 Report Shows Surprising Developments Leading to Significantly Higher Power Ratings of Solar Panels..

[53] Omazica A, Oreski G, Halwachs M, *et al*, Relation between degradation of polymeric components in crystalline silicon PV module and climatic conditions: a literature review. *Solar Energy Mater. Solar Cells* 2019; 192:123–133.

[54] Hoffmann S, Geipel T, Meinert M, and Kraft A. Analysis of peel and shear forces after temperature cycle tests for electrical conductive adhesives. In: *Proceedings of the 33rd European PV Solar Energy Conference and Exhibition*, 2017.

[55] Eitner U, Geipel T, Holtschke S-N, and Tranitz M. Characterization of electrically conductive adhesives. *Energy Proc.* 2012;27:676–679.

[56] Singh R and Singh AK. Influence of MVTR of air-side layer of backsheet on the reliability of backsheet itself and consequently on the reliability of solar PV modules. In: *Proceedings of the 31st European Photovoltaic Solar Energy Conference and Exhibition (EU PVSEC 2015).*

[57] Sinha A, Sulas-Kern D, Owen-Bellini M, *et al*. Glass/glass photovoltaic module reliability and degradation: a review. *J. Phys. D: Appl. Phys.* 2021;54:413002, https://doi.org/10.1088/1361-6463/ac1462.

[58] Rabanal-Arabach J. Development of a c-Si photovoltaic module for desert climates, PhD Thesis, University of Konstanz, 2019.

[59]   Sander M, Dietrich S, Pander M, *et al.* Investigations on cracks in embedded solar cells after thermal and mechanical loading. In: *Proceedings of the 27th European Photovoltaic Solar Energy Conference and Exhibition*, 2012.

[60]   Jakisch L. Quality Controlled PV. TÜV Rheinland.

[61]   Zhang Y, Kim M, Wang L, Verlinden P, and Hallam B. Design considerations for multi-terawatt scale manufacturing of existing and future photovoltaic technologies: challenges and opportunities related to silver, indium and bismuth consumption. *Energy Environ. Sci.* 2021;14:5587–5610. https://pubs.rsc.org/en/content/articlelanding/2021/ee/d1ee01814k.

[62]   Zhang Y, Kim M, Wang L, Verlinden Pand Hallam B. Design considerations for multi- terawatt scale manufacturing of existing and future photovoltaic technologies: challenges and opportunities related to silver, indium and bismuth consumption. *Energy Environ. Sci.* 2021, DOI:10.1039/D1EE01814K.

[63]   Solar T. Vertex 550+ and Beyond Webinar, September 2020.

[64]   Ovaitt S, Mirletz H, Seetharaman S, and Barnes T. PV in the circular economy, a dynamic framework analyzing technology evolution and reliability impacts. *iScience* 2022;25(1):103488, ISSN 2589-0042, https://doi.org/10.1016/j.isci.2021.103488.

[65]   International Technology Roadmap Photovoltaics, edition 2021.

[66]   Shahid J, Karabacak AÖ and Mittag M. Cell-to-module (CTM) analysis for photovoltaic modules with cell overlap *30th International Photovoltaic Science and Engineering Conference*; 2020.

[67]   Shahid J, Mittag M, and Heinrich M. A multidimensional optimization approach to improve module efficiency, power, costs. In: *35th European Photovoltaic Solar Energy Conference and Exhibition*.

[68]   http://griddlersolar.com/index.php/design/

[69]   https://www.pvlighthouse.com.au/

[70]   Eymard J. Contribution to the investigation of cell-to-module performance ratios in advanced heterojunction silicon photovoltaic technologies. Thesis, *Lyon*, 2018.

[71]   Eymard J, Barth V, Sicot L, *et al.* CTMOD: a cell-to-module modelling tool applied to optimization of metallization and interconnection of high-efficiency bifacial silicon heterojunction solar module. *AIP Conf. Proc.* 2021;2367(1):020012, DOI:10.1063/5.0057000.

[72]   Xia L, Wen Z, Liao K, Chen J, Li Q, and Luo X. Unveiling the failure mechanism of electrical interconnection in thermal-aged PV modules. *IEEE Trans. Device Mater. Reliab.* 2020;20(1):24–32, DOI:10.1109/TDMR.2019.2956506.

[73]   Eslami Majd A and Ekere NN. Crack initiation and growth in PV module interconnection. *Solar Energy* 2020;206:499–507, DOI:10.1016/j.solener.2020.06.036.

[74]   Colvin DJ, Schneller EJ, and Davis KO. Impact of interconnection failure on photovoltaic module performance. *Prog. Photovolt.: Res. Appl.* 2021;29(5):524–532, DOI:10.1002/pip.3401.

[75]   Grandell L and Thorenz A. Silver supply risk analysis for the solar sector. *Renew. Energy* 2014;69:157–165, DOI:10.1016/j.renene.2014.03.032.

[76]   Anderson IE, Cook BA, Harringa JL and Terpstra RL. Sn-Ag-Cu solders and solder joints: alloy development, microstructure, and properties. *JOM* 2002;54:26–29.

[77]   Anderson IE. Development of Sn–Ag–Cu and Sn–Ag–Cu–X alloys for Pb-free electronic solder applications. *J. Mater. Sci.: Mater. Electron.* 2007, 1855–76, https://doi.org/10.1007/s10854-006-9011-9.

[78]   Kariya Y and Otsuka M. Effect of bismuth on the isothermal fatigue properties of Sn-3.5mass%Ag solder alloy. *J. Electron. Mater.* 1998;27(7):866–870, DOI: 10.1007/s11664-998-0111-6.

[79]   Morris JE and Liu J. Electrically conductive adhesives: a research status review. In: *Micro- and Opto Electronic Materials and Structures: Physics, Mechanics, Design, Reliability, Packaging.* New York, NY: Springer, 2007, pp. 527–570.

[80]   Geipel T, Meinert M, Kraft A, and Eitner U. Optimization of electrically conductive adhesive bonds in photovoltaic modules. *IEEE J. Photovolt.* 2018;8(4):1074–1081, DOI: 10.1109/JPHOTOV.2018.2828829.

[81]   Kaiser C. Reduction of ECA amount for the ribbon interconnection of heterojunction. In: *Proceedings of the 37th EU PVSEC (Virtual Event),* 2020, p. 5.

[82]   Geipel T, Rendler LC, Stompe M, Eitner U, and Rissing L. Reduction of thermomechanical stress using electrically conductive adhesives. *Energy Proc.* 2015;77:346–355, DOI: 10.1016/j.egypro.2015.07.049.

[83]   Theunissen L, Willems B, Burke J, *et al.*, Electrically conductive adhesives as cell interconnection material in shingled module technology. *AIP Conf. Proc.* 2018;1999:080003.

[84]   Myer JH. Patent US3369939A, 1962.

[85]   Webpage 'SunPower Introduces New Solar Panel: The Performance Series', http://us.sunpower.com/blog/2015/11/12/sunpower-introduces-performance-series-solar-panel/.

[86]   Press release 'SunEdison inks deal with Solaria to license technology', 2015, http://static1.squarespace.com/static/568f7df70e4c112f75e6c82b/t/56b9b619c6fc081fd55b5d9b/1455011354392/Solaria+SunEd+FINAL_10.26.15.pdf.

[87]   Paschen J, Baliozian P, John O, Lohmüller E, Rößler T, and Nekarda J. FoilMet®-Interconnect: busbarless, electrically conductive adhesive-free, and solder-free aluminum interconnection for modules with shingled solar cells. *Prog. Photovolt.: Res. Appl.* 2022;30:889–898, DOI:10.1002/pip.3470.

[88]   Mondon A, Klasen N, Fokuhl E, Mittag M, Heinrich M, and Wirth H. Comparison of layouts for shingled bifacial PV modules in terms of power output, cell-to-module ratio and bifaciality. In: *Proceedings of the 35th European PV Solar Energy Conference and Exhibition,* 2018.

[89]   Schulte-Huxel H, Blankemeyer S, Morlier A, Brendel R, and Köntges M. Interconnect-shingling: maximizing the active module area with conventional module processes. *Solar Energy Mater. Solar Cells* 2019; 200:109991, DOI:10.1016/j.solmat.2019.109991.

[90]   Jinko Webinar

[91]   Park JE, Choi WS, and Lim DG. Multi-wire interconnection of busbarless solar cells with embedded electrode sheet. *Energies* 2021;14(13):1–19, DOI:10.3390/en14134035.

[92]   Lachowicz A, Andreatta G, Blondiaux N, *et al.* Patterning techniques for copper electoplated metallization of silicon heterojunction cells. In: *2021 IEEE 48th Photovoltaic Specialists Conference (PVSC)*, 2021, p. 1530–1533, DOI: 10.1109/PVSC43889.2021.9518493.

[93]   Hatt T, Bartsch J, Schellinger S, *et al. Stable Copper Plated Metallization on SHJ Solar Cells & Investigation of Selective Al/AlOx Laser Patterning*, 2021, DOI: 10.4229/EUPVSEC20212021-2DV.3.27.

[94]   Govaerts J, Baert K, Poortmans J, Borgers T, and Ruythooren W. An overview of module fabrication technologies for back-contact solar cells. In: *PVTech*, 2012.

[95]   Faes A, Paviet-Salomon1 B, Tomasi A, *et al.* Multi-wire interconnection of back-contacted silicon heterojunction solar cells. In: *7th Workshop on Metallization & Interconnection for Crystalline Silicon Solar Cells*, 2017.

[96]   Van Dyck R, Borgers T, Govaerts J, Voroshazi E, Van Vuure AW, and Poortmans J. A woven fabric for interconnecting back-contact solar cells. *Prog. Photovolt: Res. Appl.* 2017;25:569–582, DOI: 10.1002/pip. 2851.

[97]   https://www.dsm.com/content/dam/dsm/dsm-in-solar/en_us/documents/ brochure-conductive-backsheets.pdf

[98]   Schulte-Huxel H, Blankemeyer S, Merkle A, Steckenreiter V, Kajari-Schröder S, and Brendel R. Interconnection of busbar-free back contacted solar cells by laser welding. *Prog. Photovolt.: Res. Appl.* 2015;23:1057–1065, DOI: 10.1002/pip.2514.

[99]   Nasr Esfahani S, Asghari S, and Rashid-Nadimi S. A numerical model for soldering process in silicon solar cells. *Solar Energy* 2017;148:49–56, DOI: 10.1016/j.solener.2017.03.065.

[100] Gérenton F, Eymard J, Harrison S, Clerc R, and Muñoz D. Analysis of edge losses on silicon heterojunction half solar cells. *Solar Energy Mater. Solar Cells* 2020;204:110213, DOI:10.1016/j.solmat.2019.110213.

[101] Performance, Operation and Reliability of Photovoltaic Systems Designing New Materials for Photovoltaics: Opportunities for Lowering Cost and Increasing Performance through Advanced Material Innovations Report IEA-PVPS T13-13, April 2021.

[102] Commault B, Duigou T, Maneval V, Gaume J, Chabuel F, and Voroshazi E. Overview and perspectives for vehicle-integrated photovoltaics. *Appl. Sci.* 2021;11:11598, https://doi.org/10.3390/app112411598.

[103] Cordero RR, Damiani A, Laroze D, *et al.* Effects of soiling on photovoltaic (PV) modules in the Atacama Desert. *Sci. Rep.* 2018;8:13943, https://doi.org/10.1038/s41598-018-32291-8.

[104] Chunduri SK and Schmela M. Advanced module technologies, *TaiyangNews*, 2021.

[105] Ziar H, Manganiello P, Isabella O, and Zeeman M. Photovoltratronics: intelligent PV-based devices for energy and information applications. *Energy Environ. Sci.* 2021;14:106.

[106] Webpage https://www.recgroup.com/sites/default/files/documents/white-paper_twinpeak_technology.pdf?t=1635421299

[107] Govaerts J, *et al.* Developing an advanced module for back contact cells. *IEEE Trans. Compon. Packag. Manuf. Technol.* 2011;1:1319–1327

[108] Sloof LH, *et al.* Shade response of a full size TESSERA module. *Jpn. J. Appl. Phys.* 2017;56:08MD01.

[109] Baka M-I, Catthoor F, and Soudris D. Near-static shading exploration for smart photovoltaic module topologies based on snake-like configurations. *ACM Trans. Embedded Comput; Syst.* 2016;1:1–21.

[110] IRENA. *End-of-Life Management: Solar Photovoltaic Panels,* Abu Dhabi, UAE, 2016. http://www.irena.org/DocumentDownloads/Publications/IRENA_IEAPVPS_End-of-Life_Solar_PV_Panels_2016.pdf

[111] Fraunhofer ISE. Photovoltaics Report, 2019. https://www.ise.fraunhofer.de/en/publications/studies/photovoltaics-report.html. Accessed November 3, 2019.

[112] IRENA. *End-of-Life Management: Solar Photovoltaic Panels,* Abu Dhabi, UAE, 2016. www.irena.org/DocumentDownloads/Publications/IRENA_IEAPVPS_End-of-Life_Solar_PV_Panels_2016.pdf.

[113] Contreras-Lisperguer R, Muñoz-Cerón E, Aguilera J, and de la Casa J. Cradle-to-cradle approach in the life cycle of silicon solar photovoltaic panels. *J. Cleaner Product.* 2017;168:51–59.

[114] Contreras Lisperguer R, Muñoz Cerón E, de la Casa Higueras J, and Martín RD. Environmental impact assessment of crystalline solar photovoltaic panels' end-of-life phase: open and closed-loop material flow scenarios. *Sustain. Product. Consumpt.* 2020;23:157–173.

[115] Lunardi MM, Alvarez-Gaitan JP, Bilbao JI, and Corkish R. A review of recycling processes for photovoltaic modules. In: *Solar Panels and Photovoltaic Materials*. London: InTech, 2018.

[116] McDonough W and Braungart M. *Cradle to Cradle: Remaking the Way We Make Things*. New York City, NY: Farrar, Straus and Giroux, 2002.

[117] Contreras-Lisperguer R, Muñoz-Cerón E, Aguilera J, and de la Casa J. A set of principles for applying circular economy to the PV industry: modeling a closed-loop material cycle system for crystalline photovoltaic panels. *Sustain. Product. Consump.* 2021;28:164–179.

[118] Gutowski TG. Thermodynamics and recycling, a review. In: *2008 IEEE International Symposium on Electronics and the Environment*, 1–5 May 2008, IEEE. http://ieeexplore.ieee.org/document/4562912/. Accessed September 8, 2017.

Chapter 5

n-type silicon systems

Jérémie Aimé[1], Stéphane Guillerez[1] and Hervé Colin[2]

Even if usually n-type technologies are referred to cells & modules based on n-type silicon, the photovoltaic (PV) system adapted to n-type has evolved to be adapted to the new requirements. In this chapter, we will present an overview of PV systems considering all components to reach the connection and provide energy. We will also discuss about the efforts and innovation done to decrease levelized cost of energy (LCOE) considering that today, module and balance of system (BoS) are the main drivers of LCOE. Around 30% of the PV system cost is the module and over 60% concerns other components [1]. Thus, working in the BoS part and adapting components to the installation become more and more important for PV cost reduction and competitiveness. Finally, an improved design can also improve operation and maintenance (O&M) strategy and costs, making a double benefit in LCOE reduction.

5.1 General description of PV systems

A PV system is a set of components making it possible to generate electricity with PV modules, potentially convert it (DC power to AC power, low voltage to medium voltage, etc.) and deliver it to the consumer (from the local consumer to the electrical grid), with also the possibility to store it.

PV systems have a modular architecture providing the possibility for the power to range from few kilowatts to tens of megawatts.

PV systems can be classified according to different criteria:

- The connection: off-grid systems or grid-connected systems.
- Their size: residential (up to few kWp), industrial/commercial (up to few MWp), and utility scale (from few MWp to tens of MWp).
- The integration: free-standing or building integrated.
- The location: roof top, façade, ground, floating.
- The layout of modules: fixed-tilted, fixed-vertical, tracking (1D, 2D).

[1]Direction of Energy Programs (DPE) of the Commissariat à l'Energie Atomique et aux Energies Alternatives (CEA), France
[2]French Institute for Solar Energy (INES), Le Bourget-du-lac, France

- The electrical architecture: centralized systems (with inverters of power higher than 500 kVA) and decentralized systems (with "string" inverters of power ranging from few kVA to few hundreds of kVA).
- Others like: voltage level, with/without storage, direct-DC coupled or AC systems, hybrid systems (i.e. PV generator + backup generator), etc.

Let us focus on the description of a utility-scale PV plant, which is the type of system providing the largest amount of GWp currently installed over the world.

Such a system is composed of a PV array (i.e. PV modules + supporting structures), DC cables, converters (inverters + transformers), AC cables, and a set of protection devices (Figures 5.1 and 5.2).

Depending on the size of the PV system, the array of module may be divided in sub-arrays, each one feeding one inverter or one input of an inverter. A sub-array may be itself divided in smaller sub-arrays. In the end, the smallest group of modules is the PV string, which is made of modules connected in series. Systems having to comply with national electric codes (i.e. low-voltage limitation, 1,500 V_{dc}), the length

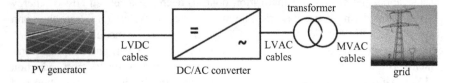

Figure 5.1   Typical schematic of the architecture of a PV system

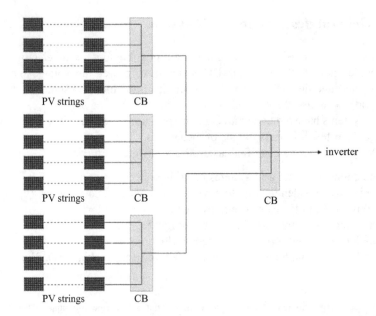

Figure 5.2   Schematic of the cabling of PV strings

of the strings is made accordingly in order to prevent to exceed the limitation at any time. Another constraint is the MPPT input voltage range of the inverter: any voltage lower than the low-voltage limit or higher than the upper voltage limit will lead to a disconnection of the inverter. With PV modules of V_{mpp} nearing 40 V, strings are usually composed of about 25 modules.

Then PV strings are connected in parallel in the array or sub-array. Obviously, all the strings connected to a same inverter input should be similar in terms of modules number. In order to limit the amount of DC cables and fit to the number of the inverter inputs, strings may be connected in parallel in combiner boxes, from which only two cables (plus/minus) come out. Such cables may also be connected in other combiner boxes as long as it is required to gather PV sub-arrays.

The supporting structures are mainly made of galvanized steel bar assemblies that are grounded through concrete foundations, but more often driven piles.

The cross sections of the DC cables are determined to ensure a maximum voltage drop from string output to inverter limited to a few percent of the rated voltage (1% is often observed in the field) according to their lengths. These cables may be aerial, at module level, and then buried. Sizes of AC cables are also optimized to limit the voltage drop.

In centralized utility-scale PV plants, inverters are often grouped by two in a technical housing with a dedicated transformer, the latter enabling the elevation of the voltage level (three-phase medium voltage of about 20 kV) for transport purposes. In decentralized architectures, inverters are close to the PV arrays and are connected to centralized transformers.

Protection devices are required at different locations in a PV plant:

- At string level, in the combiner box, fuses against overcurrent, DC surge arresters in case of lightning and disconnecting switches.
- At general combiner box level, circuit breakers and disconnecting switches.
- At inverter DC input, DC surge arresters and overall disconnecting switches.
- At inverter AC output, AC surge arresters and disconnecting switches.

5.2 PV inverters

The PV inverter is a major component of PV plant as this type of converters permits to convert DC power from PV to AC power in order to inject it into electrical networks. Regarding inverters, the market is shared between strings, central, microinverters, and power optimizers [2] (Figure 5.3).

A clear trend is the growth of string inverters market in all geographical regions [3]. String PV inverter market will be promoted through the high operational flexibility and ability to provide auxiliary line support and power backup.

230 V or less PV inverter (or micro inverter) market trend is a rapid growth with annual installations exceeding 15 GW by 2026. The growth is supported by the rising of small-scale residential PV installations [3].

The inverters are considered the brain of PV systems. It is now a common place to provide advanced features such as voltage and frequency control, reactive power at night. Advanced diagnostics such as IV curve are provided for string inverters to detect defaults for PV strings. The trend for the next decade is a strong growth of 1,500 V_{dc} strings (Figure 5.4). In the field of PV, the first demonstrators and technical and economic studies have shown the relevance of 3k V_{dc} power plants, particularly in linear installations. In the field of stationary storage, the need for centralized solutions on the grid places the cursor on several hundred KWh or even MWh. Finally, the advent of hydrogen is pushing developments toward solutions of several MW with announced objectives of several GW installed at European level (Greendeal call 2.2).

Centralized inverters (Figure 5.5) are devices operating from around 250 kVA. The power threshold making connection to the low voltage or 10–20 kV grid

Inverter /Converter	Power	Efficiency	Market Share (Estimated)
String inverters	Up to 150 kWp	Up to 98%	61.6%
Central inverters	More than 80 kWp	Up to 98.5%	36.7%
Micro-inverters	Module Power Range	90%-97%	1.7%
DC/DC Converters (Power optimizers)	Module Power Range	Up to 99.5%	5.1%

Figure 5.3   Market share of PV inverters

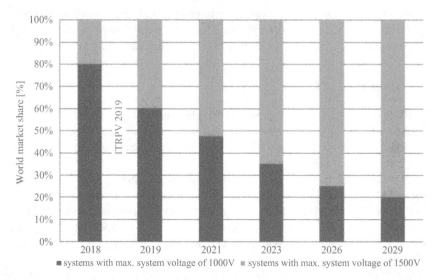

Figure 5.4   Evolution of market share of systems regarding system voltage (source [1])

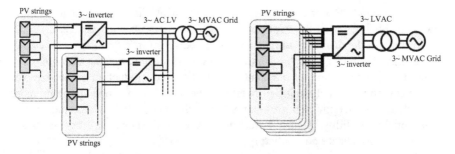

Figure 5.5   Single phase centralized inverter (left) and three-phase centralized inverter (right)

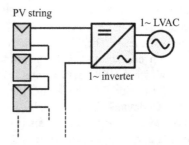

Figure 5.6   String inverter

mandatory is according to the grid codes applicable in each country. For instance, in France 3~ connection to low-voltage distribution grid is possible up to 250 kVA but connection to high-voltage distribution grid (up to 20 kV generally) is mandatory for higher power and up to 12 MVA. Central inverters combine thousands of PV strings in parallel, mostly operating under 1,500 V up to recently (output voltage varies from 300 V to 800 V RMS). They are dedicated to large ground based PV plant. The maximum efficiencies are closed to 99%.

String inverters (Figure 5.6) deal with power ratings up to a few tens of kW or even around 200 kW. The MPPT conversion function is multiplied (to extract the power on the scale of a PV string) and a maximum efficiency ratio is closed to 98.5% [4].

Inverters for residential are extremely popular in the market with several thousand references from various manufacturers. The observed efficiency ranges are 90–99% for this power class around 3–10 kW. The principle is to associate a string (relatively short, that is to say 600 V, and with a maximum up to 1 kV for the most powerful systems) with a single-phase inverter connected to the low-voltage domestic grid. Three-phase inverters are also available for low-voltage domestic grid.

Microinverters (Figure 5.7) cover from a few hundred watts to kW. Their role is to extract power as close as possible from the module (around 100 V maximum). Photovoltaic installations where the modules are in heterogeneous conditions of operation (shading, dirt, etc.) are preferred for these technologies. The observed yields are around 96% or even lower. For the moment, their market share is quite low with about 1–2% of the total power of PV inverters installed per year.

A trend in the market is to plan for a peak power of installed PV modules that is significantly higher than the nominal power of the inverter, as this approach allows for both smoothing of the plant's production and a reduction in the sizing of certain components, including the inverters. The inverters then operate much more regularly at their maximum power, but with less thermal cycling. The impact of this choice on the reliability of the inverters needs to be studied on a case-by-case basis, as these two effects are contradictory.

A challenge is the search for compactness with the increase in frequency while preserving particularly high yields and controlling or even reducing costs. The integration of the latest generation of semiconductor technologies provides a response to these technical and economic objectives. The electric mobility accompanies these developments with the adoption by the players in the semi-conductor market of solutions that are still not considered to be sufficiently expensive or mature in the short term. There is a consensus in favor of this advent with market shares for silicon carbide and gallium nitride growing significantly over the next few years. Mainstream SiC devices are available and PV inverters equipped with hybrid IGBT Si with SiC diodes or full Mosfet SiC sold on the market. The roadmap of SiC devices is well established [5] and GaN will be massively introduced on the market the coming years [6].

Main advantages of SiC and GaN power devices are high electric field and energy gap, high electron velocity, low-power losses, lower switching losses, high thermal conductivity, smaller size for equivalent breakdown voltage. The applications are high voltage, high temperature, and high switching frequency as illustrated by the Kiviat diagram in Figure 5.8.

The high-temperature operation is key to reduce cooling size. High switching frequency with a minimum of reverse recovery gives advantage to reduce the size of inductors. Coupled with low on-resistance, it is expected that SiC will replace Si for applications having rated voltages from 1.2 to 2.2 kV in a short-term perspective and from 3.3 kV and above in a medium-term perspective.

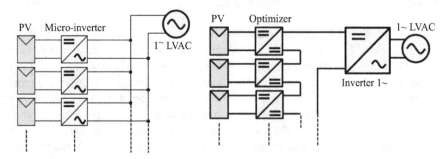

Figure 5.7   Microinverters (left) and DC optimizer with string inverter (right)

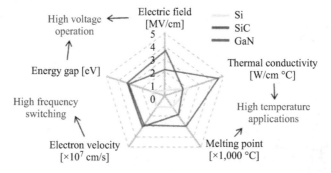

Figure 5.8 GaN vs. SiC vs. Si applications [19]

Table 5.1 Electrical and material properties of major semi-conductors [7]

	Si	GaAs	4H-SiC	GaN
Bandgap (eV)	1.12	1.43	3.26	3.39
Electron mobility (cm²/Vs)	1,350	8,000	1,000	1,200
Breakdown field (MV/cm)	0.3	0.4	2.8	3.0
Saturation velocity (cm/s)	1×10^7	1×10^7	2×10^7	2×10^7
Thermal conductivity (W/cm K)	1.5	0.5	4.9	2.5
Johnson's figure of merit	1	1.8	400	480
Baliga's figure of merit	1	15	610	730

Table 5.1 summarizes the electrical and material properties of major semi-conductors [7].

The 4H-SiC is used as a basis for comparison as it is the mostly used among SiC polytypes. Johnson's figure of merit is inherited from the definition of the basic limit on the device performance of a typical transistor set by the product of the critical breakdown field, E_c, and the saturated electron drift velocity, v_{sat}. The figure of merit is $(E_c.v_{sat}/\pi)^2$ [8]. Baliga's figure of merit is given by $\varepsilon.\mu.E_B^3$ where ε is the dielectric constant, μ the mobility, and E_B the breakdown field [7].

For GaN, the PV market should be addressed with microinverters and DC optimizers. Recent projects have shown the feasibility of 100 V HEMTs and 650 V FETs power devices in a 400 W current-fed full bridge DC/DC converter with H4 topology DC/AC [9] (Figure 5.9).

By the way, reliability with in-deep analysis of failure mechanisms is still a hot topic. Recent investigations have revealed current collapse mechanism for GaN [10]. The phenomenon has been recently characterized for GaN operating at 175°C [11].

The adoption of single-stage conversion topologies has been recently investigated [12] (Figure 5.10). Usual topologies are rather double-stage: Boost (DC/DC, achieving MPPT) + VSI (DC/AC, voltage inverter). These single-stage topologies can, among other things, allow to increase the conversion efficiency, to reduce the

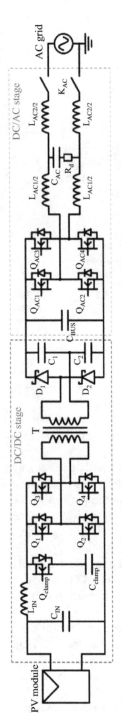

Figure 5.9 400 W current-fed full bridge DC/DC converter with H4 topology DC/AC

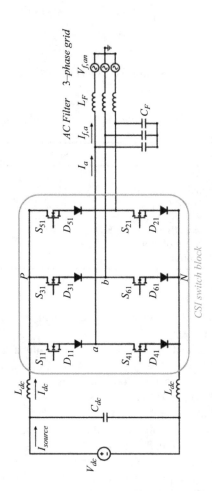

Figure 5.10 Three-phase current source inverter [4]

number of components, and to increase the compactness of the converter. This can eventually lead to an increase in the reliability of the converter, although this is not systematic.

Another strong topological variation between PV inverters is the number of phases in their grid connection. Indeed, for power levels above 10 kW, inverters are very generally connected to the three-phase grid, for power levels below 3 kW, they are very generally connected to the single-phase grid, and both solutions are possible for intermediate power levels. The major impact of using three-phase on component reliability is the greatly reduced need for capacitors in three-phase, since the need to store the fluctuating power of single-phase disappears. Thus, while in single-phase, it is often the voltage oscillations allowed that determine the size of the capacitors (functional limitation), in three-phase inverters, the limit is often the lifetime of the capacitors. As the need in terms of capacity is quite low, it would be possible to reduce the number of capacitors, thus concentrating the stress of the RMS current (due to network harmonics and switching pulses) on few capacitors, often with thermal consequences degrading their lifetime. As no certainty is allowed in advance, both approaches are usually carried out, the number and type of capacitors being chosen to satisfy both the functional criteria and the estimated failure rate criteria.

Electrical architectures require the use of 50/60 Hz transformers. The growth of offshore sites requires modules that are easier to integrate. The 50 or 60 Hz transformers are bulky, have a limited efficiency, and have a high installation cost. Historically, the railway industry has initiated work to integrate more efficient transformers with high efficiency (>99%) and less volume by increasing the operating frequency from 50 Hz to a few kHz. This is the concept of the so-called medium-frequency electronic transformers with the association of a magnetic component with power electronics. Today, this concept has been adapted for high voltage alternating current (HVAC) networks with solid-state transformers, which are experiencing a worldwide craze around Smart Grids. The high voltage direct current (HVDC) is of particular interest to the scientific community because of its native connectivity to RE sources, storage, limitation of losses on the lines, and increased resilience compared to HVAC networks. The MF transformer is the fundamental building block for power electronics interfaces requiring galvanic isolation. The expected compound annual growth rate (CAGR) is 24.8% over the period 2018–2024 [13]. The application of this brick is multiple with renewable energies such as wind, photovoltaic or hydroelectric, Smart Grids, or even rail traction.

5.3 Overview of PV systems' types and recent evolution

PV systems are specified for different consumers, environments and land situations. Renewables are distributed and match with decentralized supply of energy. PV power plants are connected to the grid or designed to address stand-alone systems and island grids for off-grid applications and consumers. PV systems are alone or coupled with battery storage in utility grids. For off-grid, two categories are addressed, PV battery systems and PV hybrid systems. The off-grid solar sector has seen a real growth over the past 10 years with a US$ 1.75 billion annual market. More than 400 million users are served. The sector will have a CAGR in sales of 6% over the next decade. More than 800 million users by 2030 should access to sustainable energy to satisfy the sustainable Development Goal 7 (SDG7) from United Nations aims to ensure access to affordable, reliable, sustainable and modern energy for all (source [14]). The present situation requires to step up efforts to reach the target by 2030, which implies continuous improvements of the deployed systems.

Adaptability of PV systems to the considered environment is a current trend (Figure 5.11). Modules are designed for specific markets such as desert, tropical climates or floating environment. In the next 10 years, 25% of market shares should be addressed by specific modules.

Trackers are expending as expected gains on production are of high interest (Figure 5.12). The overcost is more or less compensated depending the exposition.

Concerning trackers, the objective is to reduce BOS. Future developments consist of adapting the structures to new, more powerful and larger and sometimes two-sided modules, and reducing the number of foundations and the wiring and connections per installed power (BOS) (Figure 5.13). At the mechanical level, it is therefore necessary to take into account heavier and wider modules. At the electrical level, the number of modules per string must be increased. The trackers will also have to have greater flexibility to adapt to different type of module. The developments also concern the tracking optimization software to limit mutual shadows.

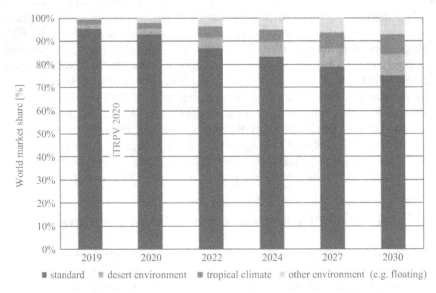

Figure 5.11   Evolution of share of systems for special regional applications (source [1])

New usages are emerging with the tension on lands occupation. Usual occupation is 0.5 m² of PV modules per square meter of land but a clear trend is surface densification with 0.75 up to 1 m² of PV for 1 m² of land.

In France, but it is more and more the case in other countries, when PV plants are planned to be installed, a review of environmental and economic impacts is

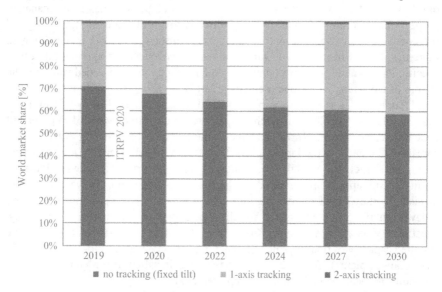

Figure 5.12   Evolution of market share of systems regarding tracker technology (source [1])

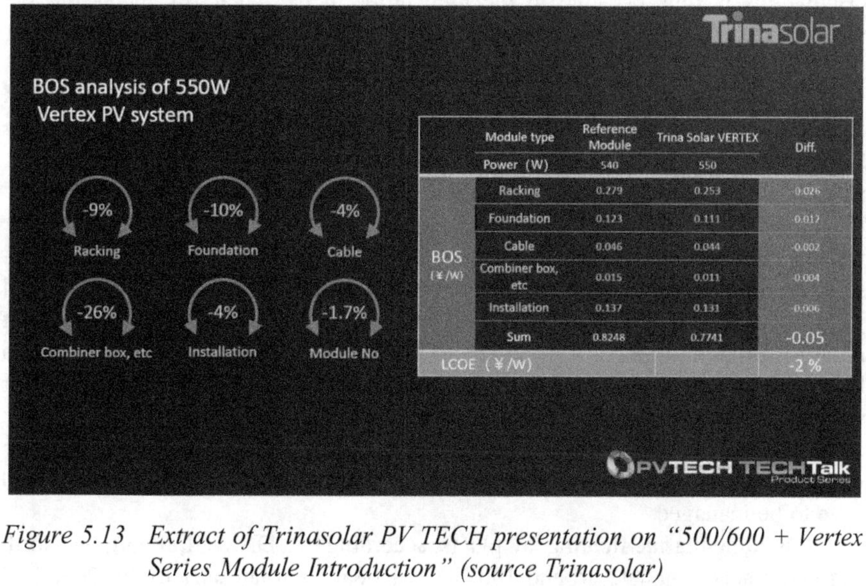

Figure 5.13   Extract of Trinasolar PV TECH presentation on "500/600 + Vertex Series Module Introduction" (source Trinasolar)

needed. Before the legislation put in place with CRE, the French Energy Regulatory Commission, created on 24 March 2000, the image of PV business in France was depreciated due to erratic installations. Nowadays, the previsions show that new territories should be investigated to match with planned PV capacity. Agrivoltaism, floating and *linear plants* are emerging solutions. Bifacial technology has advantage with vertical installations. The co-activity has to be managed properly. Regarding agrivoltaism the priority is given to the exploitation beyond the energy production. The dramatic recent events (April 2021) in France with more than half of fruits and vineyard productions frozen show the rising interest of specific PV installations aim to protect the plants along with an added value for the supply of electricity in Europe (sources [15,16]).

Floating PV is another way to consider MW scale plants. The first floating PV system was built in 2007 in Aichi, Japan. After an initial wave of deployment concentrated in Japan, Korea, and the United States, the floating solar market spread to China (now the largest player), Australia, Brazil, Canada, France, India, Indonesia, Israel, Italy, Malaysia, Maldives, the Netherlands, Norway, Panama, Portugal, Singapore, Spain, Sweden, Sri Lanka, Switzerland, Taiwan, Thailand, Tunisia, Turkey, the United Kingdom, and Vietnam. The expansion is a reality today even if ground plants remain the major part of the market share today. Floating PV is growing significantly: from 10 MWp in 2014 to a cumulative installed capacity of floating solar approaching 1.32 GWp as of mid-2018 (with 150 MWp power plants). About 2.5 GWp have been installed at the end of 2019. In addition, first salt water projects are reported [17].

Today, we are still a little above with the free float, with costs higher by 15–20%. However, cost reductions of around 10% have already been recorded in 2 years, with projects developed in the Netherlands. Closing the gap with the ground installations is thus possible in the long term. In particular, by increasing the volumes installed. Today, hundreds of MW of floating PV plant are working, far from the gigawatts of solar on the ground.

The difference in cost affects the costs of the metal structures, floats, and anchors, which can be quite variable from one body of water to another as well as the flotation system. All floating PV plants' components are on or under the water, and a cable is connected to the delivery station to inject the current.

Floating solar technology has advantages that largely offset these higher installation costs: reduced evaporation from water bodies, quick and easy construction and maintenance, no competition with agricultural land, or even improved yields thanks to the cooling effect of the water (theoretical references report improvements of 10–25% with respect to the ground). In general, OPEXs are also lower than for ground installations. In case of hybridization with hydropower, it also uses existing electrical infrastructures, increases the capacity factor at the connection point, and offers synergies with hydroelectric storage; nevertheless water level variation and snow charges (for hydro reservoirs in mountain areas) have to be managed.

Regarding marine floating PV plants, according to a DNV GL report, the North Sea could accommodate around 100 MW of floating solar capacity by 2030 and

500 MW by 2035. The LCOE of offshore photovoltaic systems is currently estimated at around € 354/MWh but in the future, it should be close to that of ground-
based solar parks (see next subpart).

Its development will articulate in two stages. First, assess bodies of water with
salted water inlets, such as estuarine areas, or ponds with marine interfaces.
Second, respond to the much more complex technological challenges integrating
swell conditions, tensions on anchoring, connection to the grid, problems that exist
in offshore wind power, but which, in solar with waterfront installations present
major challenges.

Acciona, part of H2020 DRES2Market consortium, as some experience on
floating PV plant with the first plant in Spain. The Sierra Brava reservoir, near
Zorita, Caceres, is Spain's third-largest wetland, preceded only by Doñana and the
Ebro Delta. It is an ornithological paradise and it is also an experimental energy
lab: it is here that the country's first floating grid-connected PV plant has been
launched. This Acciona innovation maximizes solar energy exploitation in the PV
niche. Mounted on the reservoir's surface, it's reminiscent of Chile's stilt houses
and occupies 12,000 square meters, or 0.07% of the reservoir's total area. The fact
that it floats on cold water, coupled with the effect of the breeze and evaporation,
improves cooling of the modules and increases their efficiency compared to land-
mounted panels. Site location is one of the problems with solar energy.
Competition from agricultural activities often reduces land availability, and floating plants also offer an alternative in areas with limited land space or underutilized
bodies of water. The fact that they can be installed in flat places like reservoirs
deals a blow to PV power's greatest enemy: shade. They reduce the evaporation of
water intended for consumption or irrigation and decrease algae growth by preventing sunlight from reaching the bed of the water body, thereby improving water
quality. According to Enrique Iriarte, Director of Technological Innovation
Projects for ACCIONA's Energy business, "further competitive advantages include
the fact that installing them doesn't require heavy machinery or earthworks, construction materials are recyclable and compatible with water usage, and they can
act as an economic stimulus for local governments and companies." As the plant
supplies power to the grid, ACCIONA monitors parameters like the energy production and efficiency of different technologies, the environmental impact on the
body of water and the local fauna, and the structural effects of mounting a solar
farm on the reservoir's surface. The plant takes measures to ensure peaceful
coexistence with the many birds in Sierra Brava reservoir, including signs that
provide information on natural resources, marker buoys that delimit designated
navigable areas, nest boxes, and floating islands to encourage nesting. Monitoring
is also performed to determine how the birds are interacting with the facility.

5.4   Integration of high-efficiency modules in systems

As discussed and illustrated in numerous examples above, the typology of PV
systems is vast and every PV plant is almost a unique case that must accommodate

technical criteria on the one hand, i.e. location, size, shape, PV components, plant architecture, grid requirements, and economic criteria, on the other hand, i.e. financing and operating costs, and expected revenues. It would be tedious to examine all the possibilities offered and the following paragraph will concentrate on the energy production gain brought by n-type module technologies on three different types of systems. The first one is fixed tilted south oriented ground mounted system, the second one is fixed vertical east-west oriented and the third one is an horizontal single axis tracking system oriented south-north.

The system studied in the simulation work is constituted by a single string of modules inter-connected in series to a string inverter as shown in Figure 5.14.

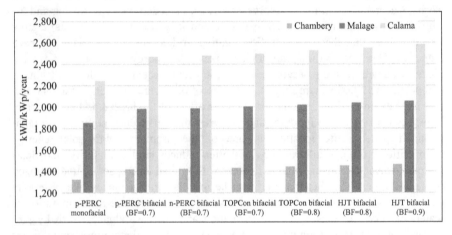

Figure 5.14   Annual producible in kWh/kWp calculated for different modules technologies

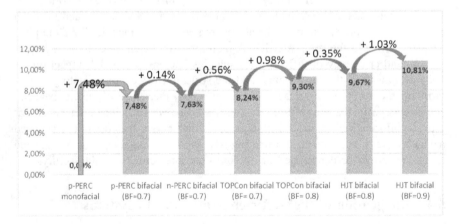

Figure 5.15   Gains of p-PERC, TOPCon, HJT vs. p-PERC monofacial at CEA INES

Table 5.2   Cumulative bifacial and power temperature coefficient gains

	Chambery	Malaga	Calama
Cumulative bifacial gain	9.66%	8.83%	12.74%
Cumulative power temp. Coeff gain	1.05%	1.91%	2.18%
Total	10.81%	10.91%	15.19%

The number of modules is adjusted in between 16 and 18 modules to accommodate the DC voltage of 1,000 V and main electrical characteristics of the system used to run the PVSYST simulations are reported. Four module technologies represented by first class representative commercial modules and virtual modules directly derived from some of these references by changing the nominal bifacial ratio are added to isolate a single evolution parameter, power temperature coefficient, or bifacial ratio.

Simulations are run for three different locations with main operating characteristics and GPS coordinates reported in Table 5.2.

The results of the simulation are shown below in a graphical form.

For the three locations, most of the gain in producible comes from bifaciality, which starts at 0.0 with monofacial p-PERC modules to end at 0.9 for a virtual HJT module with an optimized rear side, while stepwise producible gains from better power temperature coefficient (from $-0.34\%/°C$ for p-PERC to $-0.24\%/°C$ for n-HJT) come in second. The picture is clearer in the (Figure 5.15) reporting separate gains from bifaciality and from power temperature coefficient estimated for Chambery (INES site).

The cumulative gain due to power temperature coefficient gradually improving from mainstream p-PERC to the best passivated n-type HJT technology is about 1% while combined bifaciality gains represent 9.5%, about 1% per 10 points of bifaciality.

The figure evolves slightly when going to sunnier (more direct irradiation) and warmer areas like Malaga with temperature induced gain increased by 0.9% and bifaciality gain decreased by 0.8%. In desert areas with higher ratio DNI/GHI, moderate average air temperature and increased albedo values like in Calama (measured albedo at Plataforma Solar del Desierto de Atacama (PSDA) from ATAMOSTEC: 0.4), the overall gain increased mainly due to bifaciality effect. Note that the average air temperature used for simulation is biased because of a larger difference between day and night temperatures in desert areas.

5.5   Field performance of p-type and n-type bifacial systems

Three PV technologies have been tested since the end of 2019 on the Atacama Desert Solar Platform PSDA by AtamosTec Institute. P-type PERC, n-type PERT, and n-type HJT bifacial modules are implemented in small PV systems on three

Figure 5.16   Aerial view of PSDA (courtesy of ATAMOSTEC)

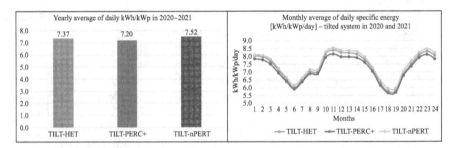

Figure 5.17   Specific energies on fixed-tilted structure

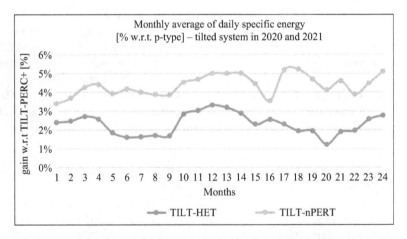

Figure 5.18   Monthly gains of n-type versus p-type on fixed-tilted structure

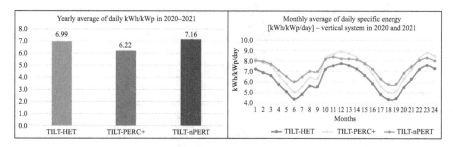

Figure 5.19 Specific energies on vertical structure

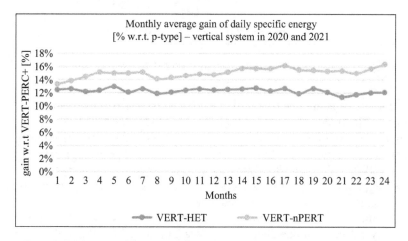

Figure 5.20 Monthly gains of n-type vs. p-type on vertical structure

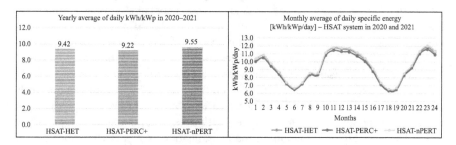

Figure 5.21 Specific energies on tracking structure

different structures: fixed-tilted, east/west vertical, and horizontal-axis tracker (HSAT) (Figure 5.16).

As previously said, the albedo value (0.4) is favorable to bifacial modules. The bifaciality factors of p-type PERC, n-type PERT and n-type HJT modules are, respectively, 0.60, 0.82, and 0.88.

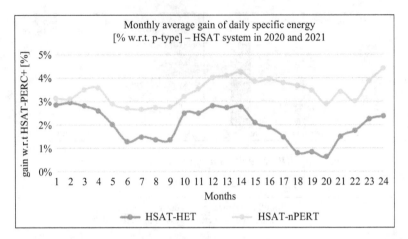

Figure 5.22   Monthly gains of n-type vs. p-type on tracking structure

On each structure, the layout of modules and the electrical system architecture are similar in order to have fair comparisons of module technologies.

On the fixed-tilted structure, i.e. facing North (Atacama desert is in the Southern hemisphere) with 20° inclination, results over 2 years of operation show a gain on kWh/kWp production for n-type technologies. Gain of n-type PERT is 4.4% and gain of n-type HJT is 2.4%. These gains are higher in Summer time as irradiation is higher (Figures 5.17 and 5.18).

On the fixed-vertical structure, i.e. facing east/west with 90° inclination, results over 2 years of operation also show a gain on kWh/kWp production for n-type technologies. Gain of n-type PERT is 15.1% and gain of n-type HJT is 12.4% (Figures 5.19 and 5.20). The gain is much higher in this configuration as 100% of the potential of each face of the module is used successively in the morning and afternoon; therefore, a high bifaciality factor makes the difference. Both gains are rather constant over the 2 years of operation.

This ratio looks rather constant because the difference between back/front irradiation ratios is itself constant between n-type with regard to p-type.

On the tracker structure, i.e. facing sun from east to west with variable incli-nation, results over 2 years of operation also show a gain on kWh/kWp production for n-type technologies. Gain of n-type PERT is 3.5% and gain of n-type HJT is 2.1%. They are higher in summer time as irradiation is higher (Figures 5.21 and 5.22).

The result of this 2-year monitoring in the desert of Atacama shows a clear gain in performance for the n-type technologies of a few percent in fixed-tilted and tracking configurations, and a more significant gain in the vertical configuration.

If we look at the surface energy indicator (i.e. kWh/m^2 of modules), conclu-sions are similar, with gains slightly higher and n-type HJT most performing technology due to differences in efficiency (Figure 5.23).

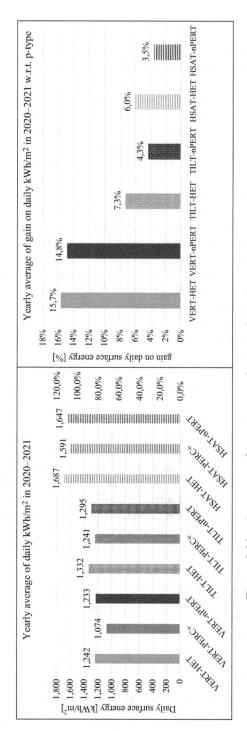

Figure 5.23   Average surface energy and gains of n-type vs. p-type on all structures

5.6 Conclusions

The energy transition becomes a reality with recent planning for the massive integration of photovoltaics in France and the 100 GW installed by 2050 or more than 200 GW installed in Germany by 2030. It would be thus lead to a global reduction of more than 20% of CO_2 with a worldwide installed capacity equal to 8 TW by 2050. The massive integration of photovoltaics is driven by the cost and performances of n-types cells assembled into modules. Photovoltaic systems will be open to a large scope of applications. Even if large ground solar plants remain the main driver, specific applications are rising. The expansion of photovoltaics is linked with the exposition to harsh environments, which may degrade drastically the performances and impact LCOE. The challenges will remain focused on durability of systems with the understanding of degradation mechanisms, the return of experience for the optimization of dedicated designs regarding the modules but also inverters. O&M management, losses evaluation, and detection are keys with the detection of failures with advanced monitoring and diagnostic tools with the collection of data in a way that the so-called big data management will aim to provide adapted solutions to address the predictive maintenance and qualify modules for optimized repowering and second life management. Optimal design of systems with current, new, and future technologies (bifacial nowadays and eco-designed modules in a near future) is a challenge with the multiplicity of usages. The design must takes into account the optimization of energy performances but also adapt the solutions to match with environmental footprint with biodiversity and landscape preservations, with the integration on degraded lands with specific constraints (linear vertical PV systems, solar road, etc.), co-activity (floating, agrivoltaic). Additionally, the provision of solutions aiming to reduce the BOS by the raising of operating voltage from low to medium voltages could be a driver to improve the BOS costs by reducing the number and section of cables, reducing the current rating of semi-conductors embedded into inverters [18]. This is particularly true for GW scale power plants where optimal design should not be constrained by the limitation of low-voltage technologies. For that particular case, standards should be revised and open the market to medium-voltage products with PV modules, inverters, cables, and protections specifically designed.

References

[1]   International Technology Roadmap or Photovoltaic, 2019 Results, VDMA, April 2020.
[2]   IHS Markit 2020.
[3]   Market Research PVT. LTD.
[4]   L. G. Alves Rodrigues and G. Perez. A 200 kW three-level flying capacitor inverter using Si/SiC based devices for photovoltaic applications. In: *PCIM Europe Digital Days 2021; International Exhibition and Conference for Power Electronics, Intelligent Motion, Renewable Energy and Energy Management*, 2021, pp. 1–8.

[5] Power Sic Materials, Devices and Applications 2020 Report, Yole Development, 2020.

[6] Power GaN 2019: Epitaxy, Devices, Applications & Technology Trends Report, Yole Development, 2019.

[7] T. Kimoto. SiC technologies for future energy electronics. In: *Technical Digests of 2010 Symp. on VLSI Technology* (Honolulu, 2010), pp. 9–14, doi: 10.1109/VLSIT.2010.5556137.

[8] T. P. Chow and R. Tyagi. Wide bandgap compound semiconductors for superior high-voltage power devices. In: *5th International Symposium on Power Semiconductor Devices and IC's, 1993*.

[9] V. S. Nguyen, S. Catellani, A. Bier, J. Martin, H. Zara, and J. Aime. A compact high-efficiency GaN based 400W solar micro inverter in ZVS operation. In: *PCIM Europe Digital Days 2020; International Exhibition and Conference for Power Electronics, Intelligent Motion, Renewable Energy and Energy Management*, 2020, pp. 1–7.

[10] Y. Cai, A. J. Forsyth, and R. Todd. Impact of GaN HEMT dynamic on-state resistance on converter performance. In: *2017 IEEE Applied Power Electronics Conference and Exposition (APEC)*, 2017, pp. 1689–1694, doi:10.1109/APEC.2017.7930926.

[11] N. Van Sang, A. Bier, R. Escoffier, S. Catellani, J. Martin, and C. Gillot, A high precision dynamic characterization bench with a current collapse measurement circuit for GaN HEMT operating at 175°C. In: *PCIM Europe Digital Days 2021; International Exhibition and Conference for Power Electronics, Intelligent Motion, Renewable Energy and Energy Management*, 2021, pp. 1–8.

[12] L. G. A. Rodrigues, L. Gabriel, and A. Rodrigues. Design and characterization of a three-phase current source inverter using 1.7kV SiC power devices for photovoltaic applications. In: *Electric Power*. Université Grenoble Alpes, 2019.

[13] Solid State Transformer Market – Global Industry Analysis, 2019, Zion Market Research.

[14] Off-Grid Solar: Market Trends Report 2020.

[15] D. Chudinzow, S. Nagel, J. Güsewell, and L. Eltrop, Vertical bifacial photovoltaics – A complementary technology for the European electricity supply? Applied Energy, Applied Energy 2020, 264, 114782.

[16] Optimising the Electrical Architecture of Linear Vertical PV Bifacial Plants, PVSEC 2020 Hervé Colin, Yadav Nepal, CEA INES.

[17] Trends in Photovoltaic Applications 2021, Report IEA-PVPS T1-41:2021.

[18] https://tigon-project.eu/

[19] T. Bieniek, G. Janczyk, A. Sitnik, and A. Messina. The "first and euRopEAn siC eigTh Inches pilOt line" – REACTION project as a driver for key European SiC Technologies focused on power electronics development. In: *TechConnect Briefs 2019*, pp. 256–259, TechConnect.org.

Chapter 6

Cost of ownership of n-type silicon solar cells and modules and life cycle analysis

Joris Libal[1] and Nouha Gazbour[2]

The ultimate scope of all technological development along the whole value chain of photovoltaic (PV) manufacturing is to further increase the deployment of PV systems in order to reach the targets set by the various international treaties aiming for the mitigation of global warming in terms of the contribution of PV-generated electricity to the overall global electricity generation. In order to further accelerate the PV capacity additions, the cost of PV-generated electricity (levelized cost of electricity, LCOE) must be further decreased. Even if at current module prices, in high irradiance locations, extremely low LCOE can be achieved, a further cost reduction will open up more and more applications for PV-generated electricity, such as production of hydrogen, ammonia or methane and will make PV-generated electricity the cheapest energy source even when including the cost of storage in batteries or by other means. In addition, of course, also in lower irradiance regions, the LCOE of PV will become more and more competitive compared to other methods of electricity generation.

Having this said, the investigation of new technologies to be introduced in any part of the PV value chain must always involve a techno-economic analysis in order to determine the impact of the new processes on the LCOE and in order to compare it with the respective current state-of-the art. As explained in detail in Section 6.2 of this chapter, calculating the LCOE mainly requires the calculation of the following two values: the overall lifecycle cost of the PV system and the overall amount of electricity generated during the lifetime of the PV system. The key element of the lifecycle cost, strongly determined by the technology choices, is the cost of the PV module. Accordingly, the first part of this chapter deals with the calculation of the cost of ownership of solar cells and modules considering three different n-type Si-based cell technologies (TOPCon, ZEBRA-IBC, and SHJ) as well as the p-type Si-based PERC as a benchmark representing the current industrial mainstream technology.

In the following part, using PVsyst simulations and taking into account the different technical characteristics of the investigated solar cell technologies and of

[1]ISC Konstanz, Konstanz, Germany
[2]CEA INES, Chambéry, France

the related PV modules, the energy yield of ground mounted utility scale PV systems will be calculated for three different locations—in three different climatic zones—within Europe.

A second topic that will gain significantly more importance than in the past is the environmental impact of the manufacturing and operation of PV system components. As, in order to reach the above-mentioned goals for the fight against global warming, a yearly PV module production capacity of 1,000 GWp/year will be reached at latest in 2030, the overall environmental impact of PV-generated electricity must be further reduced. Accordingly, in the same way as for the economic aspects mentioned above, development of new PV technologies must always keep an eye on the environmental impact resulting from the use of the new processes and technologies—potentially also involving the use of new raw materials.

In summary, R&D in the field of PV must assure that the resulting commercial products are sustainable from the economic as well as the ecological point of view.

6.1 Cost of ownership of n-type solar cells & modules

6.1.1 COO definition and assumptions

Total cost of ownership (TCO or COO) is a methodology that, in addition to the price of a purchase, incorporates other purchase-related costs. TCO can be defined as "the present value of all costs associated with a product, service, or capital equipment that are incurred over its expected life" [1].

Accordingly when applied to process steps used for the manufacturing for solar cells and modules, this method not only takes into account for the cost of the required raw materials but also for the following cost items:

- depreciation of the CAPEX for processing & utility equipment as well for the respective factory building area (including also cleanroom area);
- electricity, water, compressed air, and other utilities;
- labor;
- disposal;
- yield losses;
- facility operation, spare parts, and maintenance.

In order to correctly take into account for the above listed cost items and to be in the position to calculate the correct cost per piece (cell or module), a detailed analysis of the net throughput of each involved equipment has to be performed. Such an analysis, in turn, requires a complete and correct dataset about the uptime of the considered process equipment as well as about the overall yield (composed of electrical, optical, and mechanical yield). The bottom up COO calculation model for PV cells and modules used for the analysis presented in this chapter has been developed at ISC Konstanz. This model allows for the calculation of the COO of each single process step as well as of complete cell and module process flows. A detailed outline of the model can be found in [2]. The analysis in the present chapter will be limited to the solar cell processing and the module assembly step. For all required materials, including the silicon wafer, a market price is used. It has to be noted that while for ZEBRA and TOPCon as well as for p-type PERC technology all detailed information has been available to the authors, for SHJ, the level of detail of such available information has not been sufficient to apply the ISC COO model. In order to get some relevant—and comparable—COO results for SHJ as well, the COO information about SHJ cells contained in [3] has been used and the main assumptions used in the ISC model for PERC; TOPCon and ZEBRA COO calculations have been aligned with those reported in [3], while for others (such as e.g., the wafer cost) updated cost has been used wherever possible (see Tables 6.1–6.3).

The first part of the cost of ownership analysis consists in defining a scenario regarding the factory location as well as the yearly production capacity. In addition, realistic unit costs for all raw materials (chemicals and gases for cells, raw materials such as glass and Al frames for modules) have to be researched and to be implemented within the cost model. Regarding the key materials for cell processing—due to their high impact on the COO—which are the poly-Si as well as the pure silver, determining the cost of the Cz-Si wafer and the Ag-based screen printing pastes, respectively, a separate strategy has been followed: as these materials have on the one hand a high impact on the cell COO and are on the other hand very much fluctuating according to the respective market situation, one specific case had to be selected. For the case of the poly-Si, 15 USD/kg had been chosen as assumption, representing a value which is assumed to be valid—at least for the current mainstream Siemens purification process—for market situations where demand and supply are balanced. Regarding the Silver quotation, the cost of pure Ag, a cost of 560 USD/kg has been assumed in order to be in agreement with the assumptions used in [3] (Table 6.2).

Before having a look at the results of the COO calculations and analyze them, it is important to remind the scope these calculations. The main scope is—based on the fact that for all studied technologies the identical cost structure has been applied—to obtain an apple-to-apple comparison of the manufacturing cost of the upcoming n-type technologies and, in this way to quantify the relative cost gap between these technologies as well compared to the p-type PERC benchmark.

Table 6.1   Common assumptions for cell & module COO

Cell & module factory			
Factory location and capacity	Low-cost Asian region—5 GWp/year		
Electricity cost	0.06 USD/kWh		
Interest rate for debt (CAPEX financing)	2%		
Depreciation period	Process equipment	Facility equipment	Building
	5 years	10 years	20 years
Labor cost (4 teams for 24 h/7d production)	Operator	Technician	Supervisor/ engineer
Fully loaded salary (USD/year)	10,000	13,000	18,000

Table 6.2   Assumptions for cell COO

Cells

Price of pure Ag		560 USD/kg		
Polysilicon price		15 USD/kg		
Wafer format		M10—182 mm × 182 mm		
Wafer area		330.69 cm		
Wafer price	p-type Cz-Si	n-type Cz-Si		
	M10	M10		
	0.62	0.682		
	USD/wafer	USD/wafer		
Cell efficiency	**PERC**	**TOPCon**	**ZEBRA**	**SHJ**
	22.8%	23.5%	23.7%	24.0%

Table 6.3   Assumptions for module COO, bill of materials (BOM) include glass, encapsulants, ribbons for cell and string interconnection, junction box, aluminium frame, flux, silicone, packaging and label

Modules: 144 half-cut M10 cells with Al-frame

	Bifacial glass–glass modules with white reflector in cell interspaces			
Cell-to-module Pmpp loss	0% for PERC, TOPCon, ZEBRA-IBC and SHJ			
BOM cost	43.45 USD/module			
	PERC	TOPCon	ZEBRA	SHJ
Module dimensions	1,134 mm × 2,278 mm	1,134 mm × 2,278 mm	1,134 mm × 2,269 mm	1,134 mm × 2,278 mm
Module power	542 Wp	558 Wp	563 Wp	570 Wp
Module efficiency	21.0%	21.6%	21.9%	22.1%

The second scope is to obtain absolute values for the module COO of all 4 technologies that are—at least for those of the technologies that are in mass production on a GW-scale—quantitatively comparable with the average sales prices in the market at a given point of time where the key assumptions (mainly poly-Si price and Ag cost) are aligned with the respective market situation. Overhead cost such as cost for R&D and sales and administration has not been taken into account, as well as no profit margin is included in the cell and module COO.

Due to the fluctuating nature of the key cost assumptions and due to the fact that, depending on the specific scenario (factory location, size, import duties for raw materials, etc.), the value of each single input parameter may vary—some of them significantly—from those used for the calculations presented here, it is clear that in order to obtain correct absolute values of the cell and module COO (USD/Wp) for a given scenario, a dedicated calculation has to be performed for each specific scenario.

6.1.2 COO results for TOPCon, ZEBRA-IBC, and SHJ compared to PERC

The present section shows the results of the cell and module COO calculations for the up-coming n-type silicon solar cell technologies based on the assumptions listed in the previous Section 6.1 (Tables 6.1–6.3) and compares these values with the COO for PERC cells and modules obtained using the same cost model and the same cost structure.

Regarding the cell COO, an important metrics for an economic evaluation of the related process flow, is the transformation cost, which is the COO for the manufacturing of the solar cell without taking into account for the cost of the Si wafer. The cell transformation cost is interesting to consider as it allows to decouple the cell COO from the wafer price which is on the one hand, strongly fluctuating and on the other hand, differs between n-type and p-type.

Accordingly, in Figure 6.1, the cell transformation cost as a first intermediate result is shown as a cost per cell and—taking into account in addition also for the cell efficiency and consequently for the Pmpp (Wp) per cell—in Figure 6.2 as cost per Wp. Comparing the transformation cost of the four investigated cell technologies, the following key elements can be noted:

• All the three n-type technologies have a higher cost contribution from metal pastes compared to p-type PERC. This is due to the fact that, while PERC uses low-cost Al-paste on the rear side of the cell and Ag-pastes only on the front side, the n-type technologies currently still require Ag-paste on both sides of the cells.

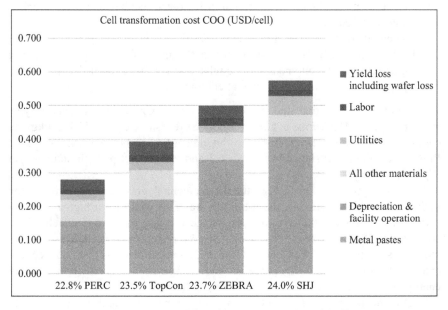

Figure 6.1   Cell transformation cost USD/cell

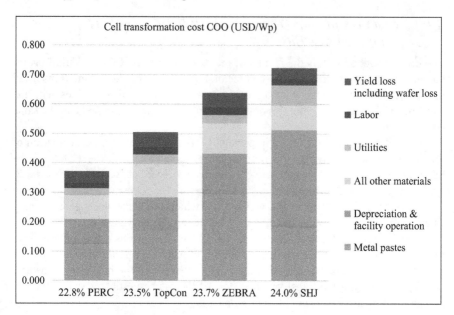

Figure 6.2   Cell transformation cost (USD/Wp)

- Compared to PERC, the cost of yield loss is higher for the three n-type technologies. This is mainly due to the higher cost for the n-type wafer as well as due to the higher Ag content per cell. In addition, TOPCon and ZEBRA require a higher number of process steps than PERC (each step having a yield which is lower than 100%).
- Compared to PERC, the depreciation for the CAPEX is slightly higher for TOPCon and ZEBRA-IBC and significantly higher for SHJ, partly due to a higher number of equipment and the related larger clean-room size, partly due to higher unit cost of certain special equipment.

Including, in a second step, also the cost for the Cz-Si wafer (p- and n-type, respectively) in M10 format assuming a poly-Si feedstock cost of 15 USD/kg, the complete COO for the manufacturing of the three n-type technologies as well as for PERC is obtained and shown in Figures 6.3 and 6.4 as cost per cell and cost per Wp, respectively.

As the higher Pmpp per cell of the n-type technologies partially mitigates the impact of the higher cost for the n-type Si wafer as well of the higher Ag consumption in terms of USD/Wp, the percentage increase of the n-type cell COOs compared to PERC is lower in terms of USD/Wp than for USD/cell. This effect is mostly visible for the 24% efficient SHJ technology and explains that—to a certain extent—the use of more costly processes and materials can be justified by a higher energy conversion efficiency of the resulting solar cells.

As it will be shown in the following, going to the next step of the value chain, the module assembly, the reward for higher cell efficiency becomes even higher.

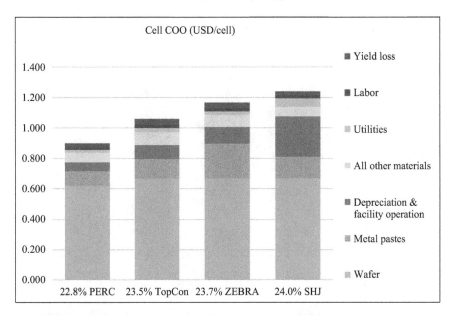

Figure 6.3   Total COO for solar cells (USD/cell)

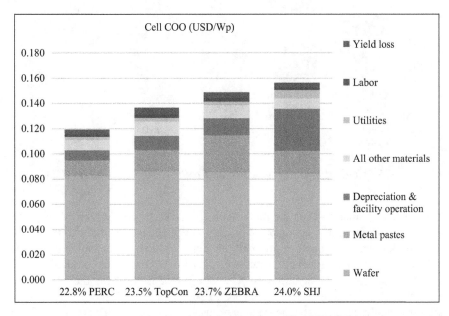

Figure 6.4   Total COO for solar cells (USD/Wp)

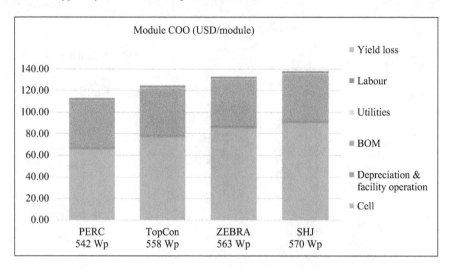

Figure 6.5   COO of modules with 144 half cells (M10 cell format) (USD/module)

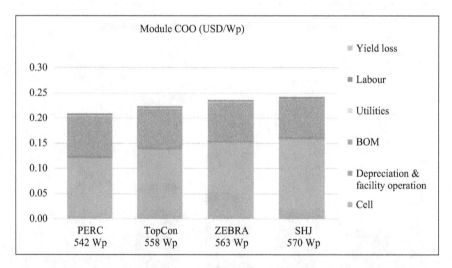

Figure 6.6   COO of modules with 144 half cells (M10 cell format) (USD/Wp)

Assuming the same cost for the module bill of materials (BOM) for all four cell technologies represents a good approximation of the actual industrial reality. As the only back-contact technology among the here considered technologies is the ZEBRA-IBC technology which features a contact design which is compatible with the same standard tabbing/stringing technology that is used for the both sided contacted cell technologies (PERC, TOPCon, SHJ), the module processing cost for all four modules results to be the same in terms of USD/module. The resulting COO values for all module technologies are shown in Figures 6.5 and 6.6.

6.2 LCOE

6.2.1 LCOE definition

As mentioned in the introduction of this chapter, for most applications of PV systems, the LCOE is the key criteria used for the selection of PV module technology, system configuration, and geographical location of the PV system. LCOE is "a widely used metric that aims to include, on the one hand, the complete cost related to the construction and operation of a PV system and on the other hand, all factors that have an impact on the total electricity generated (in kWh) during the lifetime of the PV system" (from [4], where also a more detailed description of the LCOE concept can be found). It can be represented as follows:

$$LCOE = \frac{total\ life\ cycle\ cost}{total\ lifetime\ electricity\ generation} \tag{6.1}$$

Furthermore, the LCOE concept includes also the financing conditions and takes into account the fact that money that will be spent in the future has a lower value than money that is spent today, which is one of the fundamentals of financial mathematics and is described by the concept of "Net Present Value." Accordingly, the LCOE is defined as the energy price (e.g., USD/kWh) for which the Net Present Value of the total project cost is zero. In other words, the LCOE is the averaged ("levelized") energy price (euro/kWh) over the complete lifecycle of the project for which the project reaches the financial break-even.

Taking into account the above mentioned concepts, the LCOE can be calculated as follows:

$$LCOE = \sum_{t=1}^{N} \frac{(I_t + O_t)/(1 + d)^t}{E_t/(1 + d)^t} \tag{6.2}$$

where d is the real discount rate (without inflation), I_t is the repayment for debt and equity—in order to cover the cost of purchase of PV modules and all other system components as well as of the installation—in the year t, O_t is the expenses for O&M (operation and maintenance) in the year t and E_t is the energy generated in year t.

In order to be in the position to calculate the LCOE according to (6.2) for a given scenario, a set of assumptions and input data is required. An overview of the base assumptions as well as of the scenarios studied in the present chapter is given in the following section.

6.2.2 Assumptions for LCOE calculations for TOPCon, ZEBRA-IBC, and SHJ compared to p-type PERC

As the scope of the present analysis is to investigate the potential of all considered n-type cell technologies for achieving a lower LCOE than the PERC mainstream technology, the scenario of a utility scale ground mounted system has been chosen. This is because for this application in particular, the LCOE is the crucial criteria regarding the PV system project owner's decision between several technological options.

Although for such applications, bifacial horizontal single axis tracking has been shown to enable the lowest LCOE, for simplicity, in the present study, bifacial fixed tilt has been selected for the system configuration, mostly in terms of determination of the cost of the balance of system components.

In addition, in order to illustrate the impact of irradiance and ambient temperature throughout the year on the LCOE for the various technologies, a sensitivity analysis has been conducted by selecting three different geographic locations within Europe, situated in three different climates. The assumptions regarding system configuration and some main climatic parameters of the three installation sites are shown in Table 6.4.

For these three scenarios and based on the technological and economic assumptions listed in Table 6.5, for all four cell technologies, the monofacial specific energy yield (kWh/kWp) before degradation has been simulated using PVsyst 6.88 for each of the three system locations. The solar cell characteristics with the most significant impact on the specific energy yield is considered to be the temperature coefficient for Pmpp, which describes the percentage reduction of Pmpp per °C temperature increase. Accordingly, in order to exclude any impact of potential differences in module design on the energy yield, and thus, in order to obtain a pure comparison of the impact of cell technology, the technical characteristics of the modules based on the four cell technologies have been implemented in the PVsyst simulations in the following way:

- for the PERC module, the data contained in the .pan-file (file used by PVsyst to store all relevant technical characteristics of a given PV module type) of a recent commercially available 144 half cells (M10) module has been selected;
- the .pan-files for the three n-type PV modules have been created based on the dataset contained in the .pan-file of this PERC module by solely changing the temperature coefficient of Pmpp to the value that applies to the respective cell technology.

According to this procedure, the difference in the specific energy yield resulting for the various cell technologies is determined uniquely by different thermal losses and not e.g., by different IAM (incident angle modifier, the incidence angle dependent

Table 6.4   Assumptions for the system configuration and for the three system locations (temperature and GHI from Meteonorm 7.2)

	Copenhagen	Konstanz	Malaga
Ground coverage ratio		0.4	
Fixed tilt of modules		30°	
Yearly global horizontal irradiance (GHI)	977.3	1,180.5	1,829
Global irradiance in module plane (GTI)	1,147	1,348.6	2,091.6
Average ambient temperature	9.11	10.36	18.58

Table 6.5   Technological and economic assumptions for LCOE calculations

	PERC	TOPCon	ZEBRA-IBC	SHJ
System lifetime		25 years		
WACC, discount rate		8%		
Area-related BOS cost (module area)		47.15 USD/m^2		
Power-related BOS cost		0.196 USD/Wp		
O&M cost		1% of PV system CAPEX		
Inverter lifetime		25 years		
Inverter replacement cost		35		
		USD/kWp		
Temperature coefficient Pmpp (%/°C)	−0.35	−0.30	−0.30	−0.25
Bifacial factor of module	0.7	0.8	0.7	0.9
Bifacial gain	12%	13.7%	12%	15.4%
Initial degradation	2%	1%	1%	1%
Yearly degradation		0.4%		
PV module cost USD/Wp	0.209	0.224	0.236	0.242
Module area (with frame) m^2	2.58	2.58	2.57[1]	2.58
Module power Pmpp	542 Wp	558 Wp	563 Wp	570 Wp
Module efficiency	21%	21.6%	21.9%	22.2%
Total cost of installed system (including PV modules)	0.622	0.630	0.640	0.644

[1]Reduced module length for IBC module due gapless interconnection of back-contact cells.

reflectivity—depending mostly on the properties of the module glass) or by different efficiencies at low irradiance levels (see as an example in Figure 6.7). The later point could cause an inaccuracy of the present study only in case that for one or several of the studied cell technologies it should be impossible (or very costly) by design, to obtain a shunt resistivity that is sufficiently high to avoid measurable efficiency losses at low irradiance.

Starting from the initial specific monofacial energy yield obtained from the PVsyst simulations described above, the initial specific bifacial energy yield has been calculated based on the assumption of a ground albedo of around 30% assumed to result in a bifacial energy yield gain of 12% of PV system based on the PERC modules (Table 6.5). The initial specific bifacial energy yield for the four technologies at the three locations has been used to calculate the respective values for the actual specific energy yield for year 1 until year 25 (system lifetime has been assumed to be 25 years), taking into account for the initial and yearly degradation rates indicated in Table 6.5.

The yearly expenses from year 1 to year 25 have been calculated as the net present values of the sum of the respective annuity of the debt (interest rate is the WACC) plus the yearly O&M cost. In addition, the cost for replacement of the inverters has been added up to the total expenses according to the inverter lifetime (15 years).

Using these intermediate results, the LCOE has been calculated according to (6.2) and the results are summarized in the following section.

Loss diagram over the whole year

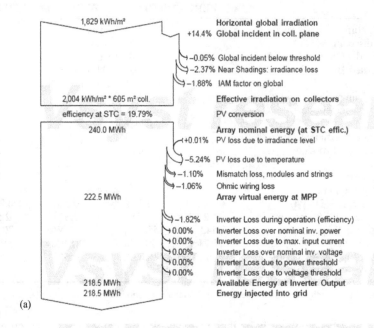

(a)

Loss diagram over the whole year

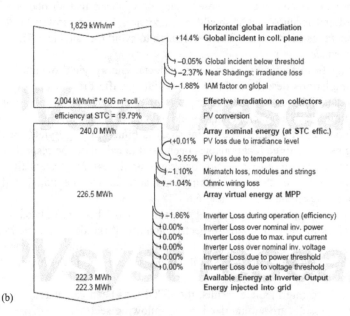

(b)

Figure 6.7   Comparison of the electrical losses at system level for PERC (a) vs.
SHJ (b) for a system located (loss diagram from PVsyst simulation
report for PV systems located in Malaga)

6.2.3 LCOE results for TOPCon, ZEBRA-IBC, and SHJ compared to PERC

The present section summarizes the results of the LCOE calculations that have been performed within the present study based on the assumptions described in the previous section of the chapter.

The results of the PVsyst simulations of the energy yield for the four technologies when implemented in PV systems located in three different geographic locations are summarized in Table 6.6.

Table 6.6   PVsyst simulation results for specific energy yield of the four technologies in ground mounted fixed tilt bifacial PV systems installed in three locations

Copenhagen/Denmark	PERC	TOPCon	ZEBRA-IBC	SHJ
Yearly global horizontal irradiance (GHI) (kWh/m^2)			977.3	
Global irradiance in module plane (GTI) (kWh/m^2)			1,147	
Average ambient temperature (°C)			9.11	
PVsyst simulation results				
Performance ratio (PR)	0.888	0.890	0.890	0.891
Initial specific monofacial yield (kWh/kWp)	1,018.5	1,021	1,021	1,022
Bifacial gain	12%	13.7%	12%	15.4%
Initial specific bifacial yield (kWh/kWp)	1,140.8	1,161.0	1,143.5	1,179.7
Konstanz/Germany	**PERC**	**TOPCon**	**ZEBRA-IBC**	**SHJ**
Yearly global horizontal irradiance (GHI) (kWh/m^2)			1,180.5	
Global irradiance in module plane (GTI) (kWh/m^2)			1,348.6	
Average ambient temperature (°C)			10.36	
PVsyst simulation results				
PR	0.890	0.893	0.893	0.895
Initial specific monofacial yield (kWh/kWp)	1,200	1,204	1,204	1,207
Bifacial gain	12%	13.7%	12%	15.4%
Initial specific bifacial yield (kWh/kWp)	1,344.0	1,369.1	1,348.5	1,393
Malaga/Spain	**PERC**	**TOPCon**	**ZEBRA-IBC**	**SHJ**
Yearly global horizontal irradiance (GHI) (kWh/m^2)			1,829	
Global irradiance in module plane (GTI) (kWh/m^2)			2,091.6	
Average ambient temperature (°C)			18.58	
PVsyst simulation results				
PR	0.873	0.882	0.882	0.889
Initial specific monofacial yield (kWh/kWp)	1,825	1,844	1,844	1,858
Bifacial gain	12%	13.7%	12%	15.4%
Initial specific bifacial yield (kWh/kWp)	2,044	2,096.9	2,065.3	2,144.7

Observing the dependency of the PR from cell technology (related to the temperature coefficient of Pmpp) and system location (ambient temperature and GHI), it can be concluded that for the hotter climate with a high irradiance (Malaga), the advantage of SHJ—the technology with the lowest temperature coefficient—is much more visible than for the climate with lower ambient temperature and lower irradiance (Copenhagen). Furthermore, for each location, the performance ratio is increasing with decreasing temperature coefficient of the respective cell technologies. While, when considering one specific cell technology, it shows the highest PR for the coolest climate.

These energy yield simulation results, taking into account for the assumptions summarized in Table 6.5, lead to the values of LCOE shown in Figure 6.8.

When analyzing these LCOE results, it has to be kept in mind that they are valid based on the following main assumptions that have been used here:

- fixed tilt, ground mounted utility scale system: rather low area related BOS cost, which reduces the LCOE benefit that can be obtained through high module efficiency (in particular ZEBRA and SHJ). Compared to the present scenario, scenarios with higher area-related BOS cost (such as e.g., roof-top applications) will increase the LCOE benefit for ZEBRA and SHJ;
- bifacial system with rather high (30%) ground albedo—a lower albedo reduces the benefit of modules with high bifacial factor (TOPCon and SHJ);
- medium-high WACC of 8% (values down to 4% are reported for certain projects); lower WACC leads to lower negative impact on LCOE of higher cost modules (ZEBRRA and SHJ), while higher WACC values are beneficial for low-cost modules (PERC and TOPCon).

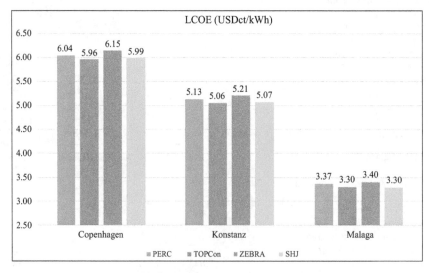

Figure 6.8   Results of LCOE calculations for the four cell technologies based on the module COO results from Figure 6.2

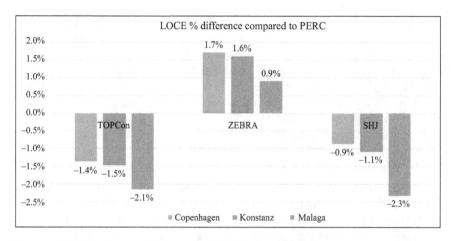

Figure 6.9 Percentage difference of LCOE results for the three n-type technologies compared to PERC

Accordingly, for different scenarios implying a different set of assumptions, the LCOE ranking can differ from the ranking obtained for the scenarios investigated in the present study.

Having this said, the most visible dependency when looking at the LCOE results shown in Figure 6.8 is the much lower LCOE for the high irradiance location with hot climate (Malaga) for all cell technologies compared to the installation sites in Central and Northern Europe with cooler climate and lower irradiance (Konstanz and Copenhagen). The percentage comparison of the LCOE of the n-type technologies with the LOCE of the respective PERC based PV systems is shown in Figure 6.9 and allows for an easier quantitative analysis.

Having a closer look at Figure 6.9, shows, on the one hand, that for ground mounted utility scale application—based on the current module efficiency and module cost as well as with the assumption of a rather high ground albedo of 30%—the ZEBRA-IBC technology shows a higher LCOE than PERC for all three locations, while—under the same conditions—TOPCon and SHJ result in a lower LCOE compared to the PERC benchmark.

When looking at a comparison between TOPCon and SHJ, TOPCon shows a lower LCOE than SHJ for two out of the three locations, in particular the gap between TOPCon and SHJ is decreasing with increasing temperature and irradiance; showing the potential of SHJ to reach lowest LCOE for installation sites with hotter climate and higher irradiance.

6.3 Environmental impact analysis

This section will focus on the environmental impact, via a Life Cycle Assessment (LCA), of the different N-type technologies addressed in this chapter in the perspective of moving towards eco-design approaches in this sector. The comparison between the technologies will be at the beginning at the module level to identify the critical points in the value chain of a PV module. The comparison will then be expanded to the system level with a focus on the carbon footprint and the resources depletion criteria in different location. This analysis is followed by a sensitivity analysis to identify the critical parameters that need to be optimized to improve the environmental profile of a PV system and reach the net zero emissions of the PV technologies in the future.

6.3.1   Methodology of LCA

Among the environmental analysis tools developed, LCA offers the most reliable and comprehensive approach to environmental assessment of products. It is a standardized assessment method (ISO 14040 and ISO 14044) that allows "the evaluation of the potential environmental impacts of a product system throughout its life cycle" (ISO, 2006a) [5]. This method is articulated in four distinct but interdependent phases, as illustrated in Figure 6.10.

Objectives of the study

The environmental assessment was structured in three tasks as followed:

- Objective 1: LCA at the module level by comparing the environmental profile of different technologies with identification of the critical points in the value chain.
- Objective 2: LCA at the system level by comparing the environmental profile of different technologies in different location.

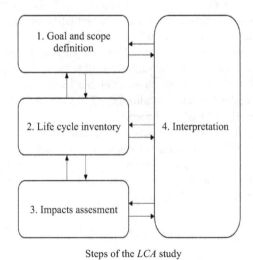

Steps of the *LCA* study

Figure 6.10   Steps of the LCA methodology [5]

● Objective 3: Sensitivity analysis to identify critical parameters that need to be optimized to reach the net zero emissions of the PV technologies.

The tools and the data used for the analysis are described in Table 6.7.

Environmental impact categories

The LCA approach aims to make the assessment as objective as possible by accounting for quantifiable and measurable material flows. The impact assessment step has the aim of translating the consumption and discharges identified during the inventory into environmental impacts (greenhouse effect, destruction of the ozone layer, acidification, eutrophication, toxicity, etc.). For this, these flows are classified in the different impact categories to which they contribute. These sets of flows are then characterized from indicators in environmental impacts.

The environmental categories, selected in this study, are recommended in the Product Environmental Footprint [6]. It describes a common way proposed by the European Commission (EC) of measuring environmental performance.

These environmental impacts could be divided in four categories: (A) impacts on the eco-system, (B) impacts on the Human health, (C) impacts on the water damage, and (D) impacts on the resource depletion (Figure 6.11).

Quality of the data

The reliability of the LCA results relies mainly on the quality of the data collected for the inventory modeling. It is therefore important to have recent inventory data that best represents current PV systems. The data collection is therefore based on primary data at an industrial scale. The data was also supplemented by the use of the Ecoinvent database, along with calculations and estimates specific to the study. This data collection was done in such a way as to meet the quality requirements set by Weidema and Wesnoes [7] and which were later adopted by the ISO 14040 standard [8].

● Time factor: Collected from early 2021 to Q2 2022, the data correspond to the current PV industry.
● Representativeness: Through their experience, their equipment and their partnerships, CEA and ISC experts are constantly working with the industrial sector, which guarantees a good representativeness of the data.
● Consistency: The data collection and analysis method was designed to be as uniform as possible for each elementary process evaluated.

Table 6.7   Tools and methods of the LCA study

Software	SimaPro 9.3
Data base	Ecoinvent V3.8
Data source	CEA/ISC & Ecoinvent
Evaluation method	EF3.0 adapted method
Impacts categories	All the listed impact categories in Figure 6.11

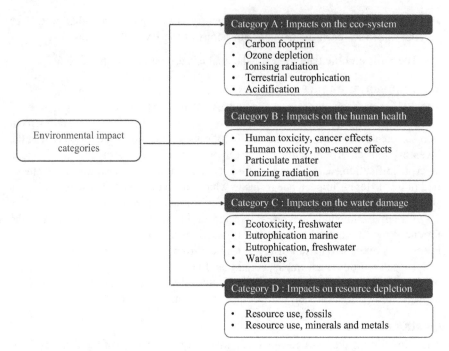

Figure 6.11   List of the environmental impact categories

6.3.2   LCA of PV technologies at the module level

This section will focus on the analysis of the environmental impacts of the n-type technologies at the module level. It takes into account only the manufacturing stage of the PV module (Cradle to gate) as described in Figure 6.12. As currently the large majority of PV modules are produced there, it was assumed that all the components of the value chain are manufactured in China and the PV module is installed in Europe. Since the installation stage was not considered in this part, the functional unit applied for the evaluation is "1 kWp of PV modules."

The assumptions of the PV technologies performance are summarized in Table 6.8 and are detailed in Section 6.2.

The results of the environmental comparison are described in Figure 6.13. On all the environmental impacts categories, it is noted a reduction that vary between 3% and 5% compared to the PERC technology. This reduction is mainly due to the increase of the module efficiency with IBC ZEBRA and HJT technologies.

To go deeper in the analysis and identify the critical points in the value chain of the PV module, the share of each components on the impact was analyzed with the focus on the carbon footprint, the human toxicity, and the resource depletion impacts.

The results are presented in Figure 6.14. For the "carbon footprint" and "resource use, fossils impacts" indicators, the production of the polysilicon and the ingots represents more than 50% of the impact. This is mainly caused by the large

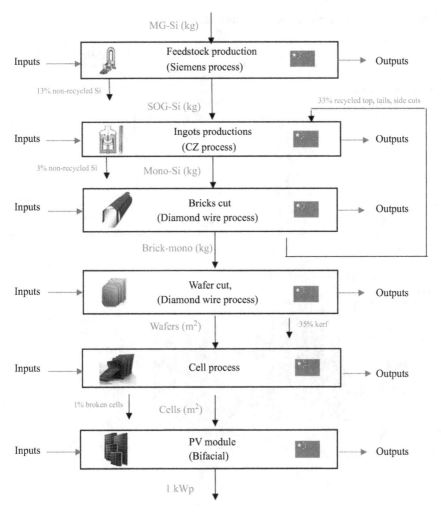

Figure 6.12 System boundary at the module level

amount of electricity, produced from the Chinese coal mix and consumed during this process. The aluminum frame represents around 10% of the impact, which is non-negligible. Frameless module could be an alternative to reduce the carbon footprint of the PV modules.

For "the human toxicity, cancer effects" impact, the aluminum frame represents more than 90% of the impact. It is caused mainly by the process of extraction of the primary aluminum

The "resource use, minerals and fossils" impact is expressed on kg Sb éq. The indicator takes into account the extraction rate and the available stocks in relation to the antimony reference. As shown in Figure 6.14, the cell and module steps

Table 6.8   Assumptions of the PV technologies performance

	PERC	TOPCON	IBC ZEBRA	HJT
Wafer size	M10	M10	M10	M10
Wafer thickness (μm)	170	170	170	170
Wafer area (m^2)	0.033	0.033	0.033	0.033
Cell efficiency	22.80%	23.50%	23.7%	24.0%
Cell number	72	72	72	72
Module area (m^2)	2.58	2.58	2.57	2.58
Module power (Wp)	542	558	563	570
Module power (Wp/m^2)	210	216	219	221
Glass thickness (mm)	2	2	2	2
Module type	Bifacial	Bifacial	Bifacial	Bifacial
Encapsulant type	POE	POE	POE	POE
Frame	Yes	Yes	Yes	Yes

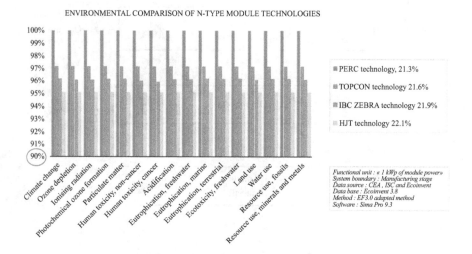

Figure 6.13   Environmental comparison of N-type technologies at the module level

represent more than 90% of the resource use, minerals, and fossils. The impact is mainly caused by the consumption of the silver at the cell level and the cooper for the ribbons at the module level.

Therefore, the high efficiency of the n-type technologies is a beneficial alternative to improve the environmental profile of the PV module. However, adopting new alternatives such as shifting to produce the silicon in countries with low carbon electricity mix, reducing the consumption of the silver and cooper are necessary to reach the net zero emissions of PV modules.

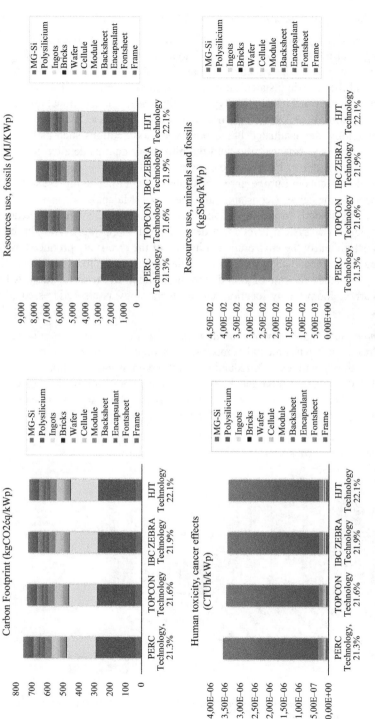

Figure 6.14 Distribution of the impact by components for different impact categories

6.3.3   *LCA of PV technologies at the system level*

This section will focus on the analysis of the environmental impacts of the n-type technologies at the system level. It takes into account the manufacturing stage and the installation stage. The installation stage includes the mounting structure, the inverter, and the electric installation in the LCA modeling. Given that the modules are bifacial, the defined installation type is utility scale. The functional unit applied for the evaluation is "1 kWh of electricity produced." The electricity produced by each technology was evaluated based on several assumptions such as the degradation rate, the performance ratio, and the bifaciality gain as detailed in Sections 6.2 and 6.3. In order to have a global vision about the environmental impact of photovoltaic systems in Europe, three locations were defined to represent the north (Copenhagen), the center (Konstanz), and the south (Malaga).

The analysis starts by a comparison of the environmental impact of the defined technologies at the system level for a defined location (Centre). The aim of this analysis is to identify the environmental impact of the electricity produced by each technology. The second analysis focuses on the impact of the location for a defined technology to study the rate of gain due to the location of the installation. A focus on the carbon footprint by location and by technology is also addressed in this analysis.

- *Comparison of the environmental impacts at the system level*: Figure 6.15 addresses an environmental comparison between the defined n-type technologies at the system level. A reduction on all the environmental impacts is noted with IBC Zebra Topcon and HJT technologies compared to the PERC technology. This reduction vary from 3% to 10%. This variation is due to the variation of the electricity produced for each technology depending on their

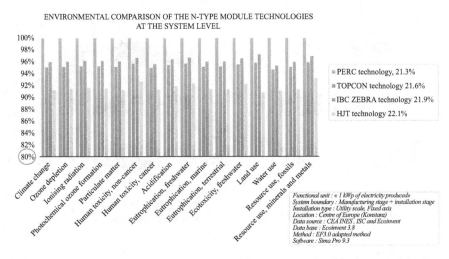

Figure 6.15   Environmental comparison of the n-type technologies with PERC at the system level

system parameters. For example, the bifaciality gain of HJT (15.4%) is higher than the PERC one (12%) leading therefore to a higher electricity production and lower environmental impacts. With the same bifaciality gain (12%), the IBC ZEBRA is more profitable from an environmental point of view than the PERC technology. This gain is due to the higher module efficiency of the IBC ZEBRA compared to the PERC. The environmental impact of the IBC ZEBRA technology is higher that the TOPCON one (+2% à 3%) because of the lower bifaciality gain of the IBC ZEBRA (12%) compared to the TOPCON one (13.7%) associated with a small delta of the module efficiency (21.9% vs. 21.6%).

- *Impact of the location on the environmental impacts*: The impact of the installation location on the environmental impacts is described in Figure 6.16. The results show a reduction on all the environmental impacts of 15% and 55% with a center and south installation, respectively, compared to a PV installation in the North. This reduction is due mainly to the higher solar irradiance in the south compared to the north leading therefore to a higher *energy yield* for a defined PV technology.

- *Carbon footprint by technology and by location*: Figure 6.17 describes the carbon footprint impact on gCO2éq/kWh for the different technologies in the three locations described above. Results show that the carbon footprint value decreases with the location installation and the technology efficiency. It varies from 46 to 42 gCO2éq/kWh in the North of Europe, 40 to 36 gCO2éq/kWh in the center, and from 26 to 23 gCO2éq/kWh in the South of Europe, which lead it to be competitive to other renewable sources of energy such as hydraulic and wind energies with 6 gCO2éq/kWh and 32 gCO2éq/kWh, respectively [9].

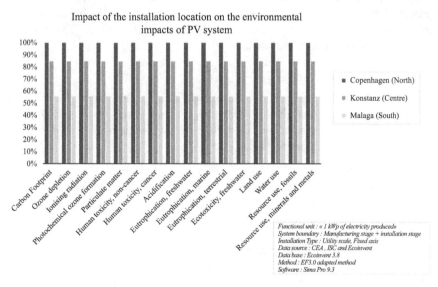

Figure 6.16   Impact of the installation location on the environmental impacts of PV system

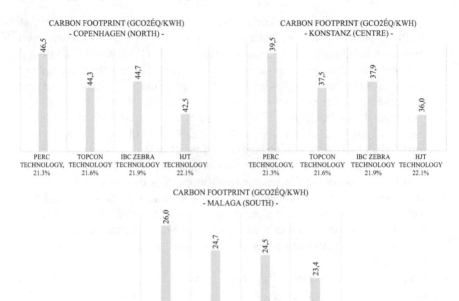

Figure 6.17   Carbon footprint on gCO2éq/kWh by location and by technology

Therefore, the specific energy yield (kWh/kWp) of the PV system is a key element to improve the environmental impact of the PV energy compared to the other sources of energies. It depends mainly on the solar irradiance, the bifaciality gain, and temperature coeffecient. Optimizing these parameters is the best way toward the net zero emissions of PV system.

6.3.4   Sensitivity analysis

Based on a sensitivity analysis, this section will address alternatives to reduce the environmental impacts of the PV technologies and make the PV energy the "greener" one by excellence. Two alternatives in alignment with the PV roadmap industry [10, 11] are suggested in this study:

- The wafer thickness.
- The location of the production of the PV components (polysilicon, ingots, wafer, etc.).

The wafer thickness

To evaluate the impact of the reduction of the wafer thickness, several scenarios of the wafer thickness are studied for a defined technology (reference) as described in Table 6.9.

Results show a reduction on the environmental impacts with the reduction of the wafer thickness (Figure 6.18). This reduction varies with the environmental

Table 6.9 Wafer thickness assumptions for the different scenarios

	Reference	**Scenario 1**	**Scenario 2**	**Scenario 3**
Wafer size	M10	M10	M10	M10
Wafer thickness (μm)	170	150	130	120
Wafer area (m^2)	0.033	0.033	0.033	0.033
Cell efficiency	22.80%	22.80%	22.80%	22.80%
Cell number	72	72	72	72
Module area (m^2)	2.55	2.55	2.55	2.55
Module power (Wp)	542	542	542	542
Module power (Wp/m^2)	213	213	213	213
Glass thickness (mm)	2	2	2	2
Module type	Bifacial	Bifacial	Bifacial	Bifacial
Encapsulant type	POE	POE	POE	POE
Frame	Yes	Yes	Yes	Yes

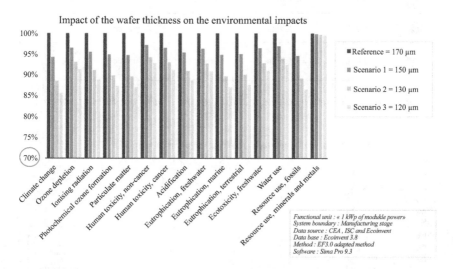

Figure 6.18 Impact of the wafer thickness on the environmental impacts

indicator. It is around 15% for the carbon footprint and less than 1% for the resources use, minerals, and metals indicator. In fact, the reduction of the wafer thickness aims to a reduction of the consumption of the silicon so an optimization of the polysilicon and ingots steps in the value chain. Therefore, the wafer thickness reduction improves the environmental impacts where the contribution of the polysilicon and ingots are the higher ones in the value chain.

The manufacturing location of the PV components

As detailed in Section 6.3.2, the main contributor to the carbon footprint impact is the Chinese electricity mainly produced from coal. Therefore, shifting the production of PV components to European countries with low carbon electricity mix could be an alternative to improve the carbon footprint of the PV module.

Figure 6.19 describes different scenarios of the location of the production of the PV components. These scenarios are established based on the European PV market in 2022 [12].

As described in Figure 6.20, producing the polysilicon and ingots in Norway allows a reduction of almost 70% of the carbon footprint (scenario 3). This reduction is due mainly to the low carbon footprint of 1 kWh of Norway Electricity

		Reference	Scenario 1	Scenario 2	Scenario 3
	Polysilicon	China	China	Germany	Norway
	Ingots	China	China	Norway	Norway
	Wafer	China	China	Norway	Norway
	Cells	China	France	France	France
	Module	China	France	France	France

Figure 6.19   Scenarios of the location production of the PV components

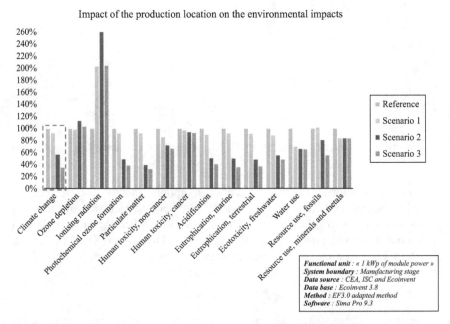

Figure 6.20   Impact of the production location on the environmental impacts

mix (28 gCO2éq/kWh vs. 1,024 gCO2éq/kWh of the Chinese electricity). Producing the cells and modules in France offers also a reduction of the carbon footprint but increase other indicator such as "ionizing radiation" because of the nuclear energy in the France mix.

In general terms, boosting the European PV market could be an important alternatives toward the net zero emissions of PV systems.

6.4 Conclusions

In the present chapter, important factors related to the economic viability as well as of the environmental impact of three upcoming industrially mature n-type Si solar cell technologies have been investigated on a cell, module, and system level and compared to the current mainstream p-type PERC technology.

The first part shows an in-depth analysis of cost-of-ownership on cell and module level of the n-type technologies and extends the analysis to system level by calculating the LCOE for a ground mounted bifacial system on three different locations and climatic zones. For the scenario considered here—a ground mounted utility scale system—the TOPCon and SHJ show a lower LCOE than the PERC benchmark. Further technological developments—in particular reduction of the cost gap between n- and p-type wafer as well as the reduction of the Ag consumption for all n-type technologies—will further improve the competitiveness of all three n-type technologies.

The second part shows an overview of the environmental profile of the three n-type PV technologies throughout their life cycle and compare the results with PERC. In this sector, innovation appears to be an essential condition to further improve the sustainability of PV modules. The results of the present study show that the high efficiency of the n-type technologies as well as their system performance (bifaciality gain, degradation rate, etc.) offers a large potential to improve the environmental profile of PV system and make the PV energy "even greener" as it already is.

References

[1] M. Monczka, R.B. Handfield, and L.C. Giunipero. *Purchasing and Supply Chain Management*, 6th ed., Boston, MA: Cengage Learning, 2016.

[2] F. Eiermann. HTWG Konstanz, Master Thesis, *Bottom-up COO-Modell für Solarzellentechnologien*, 2021.

[3] B.L. Smith. *Photovoltaic (PV) Module Technologies: 2020 Benchmark Costs and Technology Evolution Framework Results*, Golden, CO: National Renewable Energy Laboratory, 2021, NREL/TP-7A40-78173.

[4] J. Libal and R. Kopecek. *Bifacial Photovoltaics: Technology, Applications and Economics Bificial Photovoltaics*, UK: Institution of Engineering and Technology, doi: 10.1049/PBPO107E.

[5] ISO 14040, 2006a. « Environmental Management e Life Cycle Assessment e Principles and Framework».

[6]   F. Wyss, R. Frischknecht, M. Wild Scolten, and Ph. Stolz. « PEF Screening Report of Electricity from Photovoltaic Panels in the Context of the EU Product Environmental Footprint Category Rules (PEFCR) Pilots», 2015.

[7]   B.P. Weidema and M.S. Wesnæs. Data quality management for life cycle inventories—an example of using data quality indicators. *Journal of Cleaner Production*, 1996;4(3):167–174, doi: 10.1016/S0959-6526(96)00043-1.

[8]   European Commission. Joint Research Centre. Institute for Environment and Sustainability. *International Reference Life Cycle Data System (ILCD) Handbook: General Guide for Life Cycle Assessment: Detailed Guidance*, LU: Publications Office, 2010. Accessed October 13, 2021, https://data.europa.eu/doi/10.2788/38479

[9]   "ecoinvent – ecoinvent," https://ecoinvent.org/

[10] ITRPV, 2022. «International Technology Roadmap for Photovoltaic for Photovoltaic», Results 2021 Inc., 12 edition.

[11] IEA 2022, Executive Summary – Solar PV Global Supply Chains – Analysis – IEA, https://www.iea.org/reports/solar-pv-global-supply-chains/executive-summary

[12] F. Ise and P.P. GmbH. Photovoltaics Report, p. 52.

Chapter 7

Future of n-type PV

*Amran Al-Ashouri[1], Mathieu Boccard[2], Can Han[3],
Olindo Isabella[3], Eike Köhnen[1], Lars Korte[1],
Paul Procel[3] and Guangtao Yang[3]*

7.1 Introduction

As already discussed in the first chapter of this book, the ultimate efficiency limits
of today's PV workhorse, the passivated emitter and rear cell (PERC) technology,
are in sight. At the same time, the rate of global solar deployment continues to
increase significantly for many leading markets—171 GW_p of PV have been
installed globally in 2021, and projections see 209 GW_p to be installed in 2022 and
231 GW_p in 2023 [1]. These two elements create ample opportunity for technologies with higher efficiency potential to increase their market share. Overarching
topics in these "beyond PERC" technologies are:

1. Alternative "passivating contacts" technologies. The most prominent of these
    are the tunnel oxide passivated contact (TOPCon)/polycrystalline silicon on
    oxide (POLO) and the silicon heterojunction (SHJ) technology, which already
    today are in the market—and SHJ has been there for quite a while, with the
    first Sanyo HIT modules being deployed in the late 1990s. In TOPCon/POLO,
    a thin silicon oxide is used as a passivating interlayer between the c-Si wafer
    and the highly doped polycrystalline silicon contact layers (cf. also Chapter 3).
    In SHJ, a few nm of undoped hydrogenated amorphous silicon, a-Si:H, takes
    the role of the passivating layer, and doped a-Si:H thin films form the carrier
    selective contacts. Both TOPCon/POLO and SHJ have already demonstrated
    efficiencies above 26%. Going beyond these technologies, alternative materi-
    als, or stacks thereof, are investigated in order to mitigate the few remaining
    losses, e.g. to achieve higher transparency of the top contacts in the blue/UV

[1]Department Perovskite Tandem Solar Cells, Helmholtz-Zentrum Berlin für Materialien und Energie
GmbH (HZB), Germany
[2]Institute of Microengineering (IMT), Photovoltaics and Thin-Film Electronics Laboratory (PV-Lab),
Ecole Polytechnique Federale de Lausanne (EPFL), Switzerland
[3]Photovoltaic Materials and Devices Group, Electrical Sustainable Energy Department, Delft University
of Technology, The Netherlands

and near IR spectral range, or aiming at potentially more cost-effective production processes such as sputtering instead of plasma-enhanced chemical vapor deposition (PECVD), or even wet-chemical in lieu of vacuum processing.

2. Tandem and multijunction cells. Even with further improvements in contact layer transparency etc., the practical limit of solar cells based on a single c-Si absorber are in sight, cf. Section 7.3. In order to surpass this limit, multi-junction cells can be used, as already demonstrated over the last decades in applications with a different cost structure such as concentrator PV and space applications. For the utility and rooftop market at the TW scale, a different technology will be required, and it can be expected that it will be silicon-based. In Section 7.5, we discuss options for Si-based tandems.

In most of these technologies, n-type wafers have been used to demonstrate highest efficiencies. The reasons for this are summarized in Chapters 2 and 3 on materials and cell concepts; briefly, the benefit of n-type wafers is rooted in the basic physics of point defects in silicon: Most transition metal point defects, as well as silicon surface states have larger electron- than hole capture cross-sections. Therefore, for the same type and concentration of impurities in the crystal, the minority carrier lifetime, i.e. the lifetime of holes in n-type c-Si, is higher than the lifetime of electrons in p-type wafers. This yields a higher potential in both, V_{oc} and fill factor of solar cells on (n)c-Si. Furthermore, conventional boron-doped (p)c-Si is prone to light-induced degradation (LID) through boron–oxygen and/or boron–iron complex formation [2]. This degradation mechanism is not present in (n)c-Si, which makes even relatively oxygen-rich Czochralski wafers a viable option for the high efficiency cells discussed in the following.

7.2   Carrier-selective passivated contacts

As we will see in the following sections, all of today's as well as, most likely, future very high efficiency silicon solar cells are based on so-called carrier selective passivated contacts (CSPCs). What are the differences of such CSPCs to the diffused junctions used traditionally in photovoltaics?

In very general terms, a solar cell under illumination can be understood as a device where electrons and holes are generated by the impinging photons in an absorber (in our case, the silicon wafer), and these charges are then extracted selectively through two contacts according to their polarity, cf. Figure 7.1(a). Such selectivity for just one carrier type is achieved through asymmetric conductivities for the two carrier types. In Figure 7.1(b), this idea is depicted for an electron-selective contact: It has a high conductivity for electrons, σ_e, much greater than the conductivity for holes, σ_h. Therefore, a large current of electrons toward the contact can be sustained with just a small driving force, i.e. a small gradient of the quasi-Fermi level of electrons (dashed blue line). However, with a small σ_h, the hole current to this contact will also be small even with a large gradient of the quasi-Fermi level of holes (dashed red line). In traditional diffused junction cells, this selectivity, i.e. large difference of conductivities, is achieved through doping, cf. Figure 7.1(c). This works

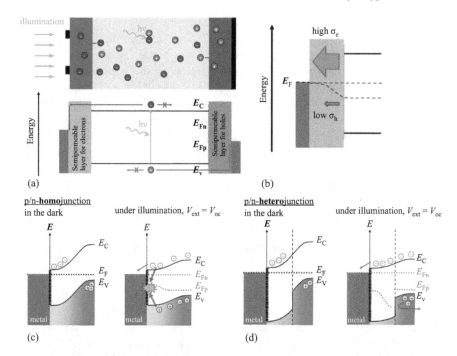

Figure 7.1 (a) Schematic of a generic solar cell structure with semipermeable layers which selectively extract photogenerated electrons (left) and holes (right) to the contacts. Bottom: simplified band diagram. (b) Generic band diagram of a semipermeable layer with high conductivity σ_e for electrons, but low conductivity σ_h for holes. (c) Band diagram of an n/p junction, and (d) of an electron-selective heterojunction in the dark and under illumination. Black dashes at the metal/semiconductor interface represent interfacial defects

reasonably well in the dark, where a high density of electrons close to the contact ensures high electron conductivity, whereas the hole density is low, yielding a low conductivity for holes.* However, under illumination, the density of both electrons and holes is greatly enhanced, in high injection conditions even surpassing the doping-induced density of majority carriers in the bulk of the wafer (holes, for the case of p-type c-Si sketched here). The conductivity is the product of charge carrier mobility, μ_i, and concentration, n_i: $\sigma_i = \pm q\, n_i\, \mu_i$, where the index i represents electrons (e) or holes (h) and q is the elementary charge. Thus, when the densities of

---

*This discussion is inspired by the work of Würfel *et al.* [3], and it is different from the usual textbook arguments involving electrical fields in the p/n junction, which are said to repel and attract opposite charge types. In the framework of this theory, the electrical fields in such junctions are merely a by-product of the different carrier densities, thus conductivities, achieved by doping. See e.g. Ref. [4] for further implications of this approach to understanding carrier selectivity.

electrons and holes become similar under illumination, conductivities towards the contacts also become similar in diffused junctions since the mobilities μ_p and μ_n in silicon differ by less than half an order of magnitude. Therefore, Shockley–Read–Hall (SRH) recombination across defects at the semiconductor/metal interface will become very efficient, since both electrons and holes are then present at the interface in large quantities. This strong recombination limits the quasi-Fermi level splitting in the wafer bulk, thus the V_{oc} of the solar cell. However, if the diffused junction is replaced by a heterostructure, i.e. a second semiconductor deposited on the c-Si wafer that gives rise to an asymmetric band offset, this problem can be resolved, cf. Figure 7.1(d): As sketched here, the holes see a large barrier that they cannot overcome even when the bands are flattened out under illumination. Thus, the concentration of holes at the recombination active interface is greatly reduced, and so is the SRH recombination. This is why such contacts, where the recombination-active interface is separated from the absorber by a semipermeable layer, are called "passivated". On the other hand, electrons are not hindered (in this ideal case) by an additional barrier and thus can move freely to the contact to be extracted. This is the difference between passivated surfaces/interfaces (such as c-Si/SiO$_2$, c-Si/AlO$_x$, etc.[†]) and carrier-selective passivated contacts (CSPCs): the latter still allow to extract one charge carrier type, while the former block both carrier types.

In the following sections, we will see how this general concept of CSPCs is implemented both in already industrialized as well as in future high efficiency cell concepts. In practical devices, the passivating CSCs often consist of double-layer stacks, where the layer adjacent to the wafer serves to passivate the c-Si surface and a second, highly doped and thus highly conductive layer improves carrier selectivity. In addition, a third layer—often a transparent conductive oxide (TCO)—may be required to provide sufficient lateral conductivity to the contact fingers.

7.2.1 Silicon heterojunction solar cells

Silicon heterojunction (SHJ) solar cells hold the world record of power conversion efficiency (PCE) among single-junction solar cells, with 26.81% and 26.7% as front-back contact and interdigitated back-contact devices, respectively [5–9]. The key feature of SHJ cells is a particularly high open-circuit voltage (V_{oc}) due to the excellent passivation provided by hydrogenated amorphous silicon stacks [10] deposited on high-quality monocrystalline wafers. SHJ cells also show a lower temperature coefficient than diffused-junction devices, enhancing their energy yield in the field as compared to classical Si cells [11]. From a manufacturing point of view, the fabrication of HJT solar cells requires only unstructured blanket layer depositions (except the metal contact grid) and a reduced number of process steps compared to their mainstream counterparts, yielding a relatively low CAPEX [11]. Due to the combination of these advantageous features, especially the superior energy yield and

---

[†]Note that SiO$_2$, etc. are also used in CSCs, e.g. in TopCon/POLO type junctions. However, for this purpose, they are made ultrathin, of the order of 1 nm, so that charges can cross the barrier by quantum-mechanical tunneling.

simple fabrication process, SHJ solar cells are expected to continuously gain market share in the coming years [12]. For the same reasons, they are also considered an excellent candidate for the bottom cell of Si/perovskite tandem devices [13].

The progression in best SHJ cell efficiencies over the last 30 years is shown in Figure 7.2(b). Many aspects of the SHJ device concept will be discussed only briefly in the following. For further reading, several review articles [11,14–16] and a monograph [17] on the same subject are recommended.

Silicon heterojunction device structure and history: A schematic layer stack and a band diagram for a both sides contacted SHJ cell on n-type wafer and with p/n junction on the front is depicted in Figure 7.2(a). The key feature of SHJ solar cells are stacks of a few nm thin hydrogenated amorphous silicon (a-Si:H) films deposited on the crystalline silicon absorber. They are typically grown by plasma-enhanced chemical vapor deposition (PECVD) at temperatures around 200°C. The hydrogen, which is present in the a-Si:H at a level of a few percent, passivates both internal defects of the amorphous silicon network, and most of the remaining dangling bonds at the a-Si:H/c-Si interface. Device grade a-Si:H has a band gap of the order of 1.7 eV, with asymmetric band offsets to c-Si [18]. As can be seen from the band diagram in Figure 7.2(a), the amorphous/crystalline silicon heterojunction is thus not ideally suited for charge carrier extraction from the crystalline silicon absorber: Instead of providing a high band offset, thus a high barrier, for one carrier type and a vanishing one (no band offset) for the other, both carrier types "see" barriers. Furthermore, these barriers are the same for both the electron and hole contacts. This is why the additional doped layers are required: They define the

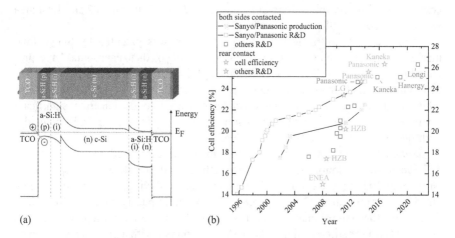

(a)   (b)

Figure 7.2   (a) Schematic of a front-back contacted silicon heterojunction solar cell with p/n junction on the front side (illuminated from the left side), and the corresponding band diagram. (b) Development of silicon heterojunction cell efficiencies over the last 30 years. All both sides contacted cells reported after the year 2000 are >100 cm², while rear contact devices are small (1 or 4 cm²) as well as large area.

direction of the current flow by providing a high conductivity for the desired carrier polarity. The a-Si:H/c-Si junction will thus function as a passivating CSC which separates the contacts (where the density of electronic defects is high) spatially and electronically from the c-Si absorber. The concept of amorphous/crystalline silicon heterojunctions was introduced by Fuhs *et al.* in the 1970s using a direct hetero-junction between doped a-Si:H and c-Si [19]. In the 1990s, the company Sanyo introduced an additional, undoped, thus nominally intrinsic (i) a-Si:H buffer layer between the doped a-Si:H films and the c-Si wafer [20]. Due to the absence of dopants, the (i) a-Si:H interlayer has much reduced defect densities and can be fine-tuned to minimize a-Si:/c-Si interface recombination, which enabled SHJ cell V_{oc}s above 700 mV. Further process steps in the fabrication of SHJ cells comprise: (i) the deposition of TCO layers, typically sputter-deposited indium-tin oxide or other indium-based oxides, on top of the a-Si:H. The TCO serves as antireflection coat-ing and at the same time enables lateral carrier transport to the contact grid fingers. (ii) Screen-printing of front and rear side metallizations, using low-temperature curable silver pastes, and (iii) a final curing step for the paste, which serves at the same time to anneal defects present in the a-Si:H either from the initial PECVD or the subsequent TCO sputtering. Overall, the cell process is based on comparatively simple full-area deposition steps and screen-printed contacts. This is in contrast to conventional c-Si cells with diffused emitters where efficiencies well above 20% can only be reached using local contacting schemes requiring additional patterning. Furthermore, since all process steps occur at temperatures around or below 200°C, SHJ cell fabrication has a low thermal budget. In contrast to conventional diffusion processes that lead to wafer bowing and increased breakage, the SHJ process is compatible with very thin wafers: for example, a 24.7% cell with 102 cm^2 total area was fabricated on a 98 μm thin wafer [21].

The use of an intrinsic passivating interlayer was patented by Sanyo (and marketed as HIT[®] technology). Over 20 years, Sanyo, then Panasonic, have pro-duced more than 500MW/a of HIT modules [22], i.e. a cumulated SHJ production of around 10 GW. Inspecting Figure 7.2(b), it is obvious that around 2010, also other institutes and PV manufacturers have started to demonstrate significant improvements in cell efficiency. Interestingly, this steep increase in PCEs coincides with the expiration of the (i) a-Si:H patent.

State of the art: As mentioned above, the highest reported power conversion efficiency for a double-side contacted silicon cell to date was reported in November 2022 by the Chinese PV company LONGi, with a 26.81% SHJ cell on an M6 wafer (cell area 274 cm^2) [5,8]. The excellent passivation of c-Si surface defects by the heterojunction also results in the highest reported open circuit voltage (V_{OC}) for silicon solar cells of 751 mV, on par with a previously achieved V_{oc} result, also with SHJ, by Panasonic [23]. Instead of doped amorphous silicon layers, at least for the n-type contact, nanocrystalline films are used in this SHJ. This probably also contributes to the efficient carrier extraction demonstrated by the extremely high fill factor of 86.1%, as well as to the high current density j_{sc} of 41.45 mA/cm^2.

The j_{sc} is the only individual parameter where interdigitated back side contacted (IBC) SHJ can outperform their double-side contacted siblings. Until recently, IBC SHJ cells also held the overall world record for silicon solar cells, with a PCE of 26.7% (V_{oc} 738 mV, j_{sc} 42.65 mA/cm^2, FF 84.9%; designated measurement area: 79 cm^2), realized by Kaneka [6,7]. As also discussed in Ref. [7], the "practical limit" for SHJ-based solar cells is estimated at around 27.1%,[3] probably to be achieved with a back contacted cell due to the further reduced optical absorption losses in this geometry. Thus, the present record results are only a few tenths of a percent from this "practical limit."

Future R&D requirements: According to a market analysis performed at the end of 2020 [24], about 20 companies had expressed interest in starting SHJ manufacturing, and at least four (REC, GS Solar, Tongwei, Hevel) have started production already. Moving forward, the following points are likely to be key to reaching highest efficiencies and cost competitiveness. A more in-depth discussion of some of these points can be found e.g. in the review by Haschke *et al.* [11]. Note that module technology aspects are not considered here.

(i) As discussed e.g. in [15], the availability of silicon wafers with very high carrier lifetime is crucial not only to achieve a high V_{oc}, but even more so for realizing ultimate fill factors: If recombination at bulk and interface defects can be suppressed to a level where Auger recombination dominates the carrier lifetime also at the maximum power point, diode ideality factors below 1 become possible. This translates into fill factors approaching the ideal value of ~89% (assuming only Auger recombination and no series resistance [15]).

To improve the SHJ cell's optics, thus the j_{sc}, a more transparent front side is desired. Several aspects need to be considered, which can also have considerable influence on production costs:

(ii) Light absorbed in the doped amorphous silicon layers does not contribute to the SHJ's photocurrent, since minority carrier lifetimes in these layers are very low. To reduce this parasitic absorption, alternative Si-based doped layers with wider band gaps and/or lower absorption coefficients have been explored, such as amorphous or nanocrystalline silicon carbide, a-/nc-SiC [25,26], or nanocrystalline silicon oxide, nc-SiO$_x$ [27,28].[4] Furthermore, it was found that p-doped a-Si:H films have inferior optical and electrical properties (lower doping efficiency; higher defect densities, thus parasitic sub-bandgap absorption; smaller band gap) than intrinsic and n-type films. Therefore, it is beneficial to place the (p)a-Si:H (or (p) nc-Si:H) film on the SHJ cell's rear side. On an n-type c-Si wafer, this creates a rear p/n junction cell with inverted polarity as compared to Figure 7.2(a). Note that such devices are often called "rear emitter" cells, but this is actually a misnomer [29].

---

[3]Note that a "practical limit" FF of 85.3% was estimated for the 27.1% "practical efficiency limit." Taking the 86.6% of the double-side contacted record SHJ, this would scale to slightly higher "practical efficiency limit" of 27.5%. However, it is unclear whether such a high FF can also be achieved in an interdigitated rear contact geometry, where the contact area has to be shared by both contacts and the series resistance is thus likely to be higher.

[4]Historically, such films were labeled as "microcrystalline," due to the visibility of features on the μm scale e.g. in SEM images. The actual crystallites have sizes in the nm scale, though.

(iii) Placing the p/n junction on the cell's rear side has additional benefits for lateral charge transport towards the grid fingers on the front side of the cell: with an n-doped a-Si:H (or nc-SiO:H) film on the front, an electron accumulation layer is formed. This helps in transporting the photogenerated current toward the grid fingers, and the front side TCO can therefore have a higher sheet resistance [30]. This enables to use more oxidic versions of the "standard" TCO, i.e. indium tin oxide (ITO), which show less free carrier absorption (FCA) in the near IR range due to reduced carrier densities, thus improving the near-IR EQE and overall j_{sc}. Other In-based TCOs such as InO:H [31] show improved electrical properties (increased electron mobilities), allowing for low carrier concentration and high IR transparency while maintaining high conductivity. However, all In-based TCOs share the same cost risks: The $\sim$80 nm thick ITO films which are typically deposited on the front and back of SHJ cells consume about 3.5 g of Indium/m^2 of modules, costing $\sim$1 \$/m^2, or below 0.5 ct/W for a 335 W, 60 cell module [11]. However, the In price is relatively volatile, it has been fluctuating by a factor of 2 over the last 10 years. To mitigate this cost risk, alternative In-free TCOs such as aluminium-doped zinc oxide (ZnO:Al) can be attractive, especially for the SHJ cell's front side in combination with the rear p/n junction geometry. Indeed, rear junction SHJ with ZnO:Al and identical performance as the ITO references as well as good stability under damp-heat tests have been demonstrated recently [32,33]; combinations of In-containing and In-free TCOs could also be an attractive option to reduce overall In consumption [34]. The topic of alternative TCOs also has a direct connection to alternative carrier-selective contacts as discussed in Section 7.4.

(iv) Another cost driver in SHJ cells is the contact metallization: due to the low annealing temperatures of only $\sim$200°C allowed for SHJ due to the presence of the a-Si:H films, low-curing-temperature silver screen printing pastes need to be used. The conductivities of such pastes have been greatly improved over the last 10 years, approaching the specific bulk resistivity of Ag ([11], Figure 7.10). Still, alternative concepts for cell contacting, such as copper plating, reviewed in Ref. [35] or the so-called SmartWire® technology [36] are interesting. Ref. [37] presents an overview over these diverse options.

(vi) Finally, the remaining parasitic light absorption in the front contact formed by the silicon layer stack together with the TCO, as well as the shading by the metal grid can be avoided altogether by using highly transparent passivation layers such as SiO_x/SiN_x on the front and placing both contacts on the rear side of the cell. A geometry of alternating p- and n-type stripe contacts is used in the already mentioned interdigitated back contact silicon heterojunction (IBC-SHJ) cell concept. As discussed above, IBC-SHJ hold the record for the highest solar cell efficiency in a single junction silicon cell to date, of 26.7% [7]. While it appears that such IBC-SHJ cells will probably be the devices to demonstrate ultimate efficiencies also in future, practical issues remain especially in structuring the a-Si:H, TCO and metal films of the IBC, which might also lead a cost penalty [38]. However, notable progress has been made in simplifying these structuring steps by using shadow masks during silicon thin film and TCO deposition, requiring just a single alignment step to define the n and p contact areas [39]. Note, that the IBC geometry with two contacts at the cell's back also opens up entirely new options also for tandem solar cells, enabling three-terminal tandem configurations [40] with enhanced energy yield also under conditions with pronounced spectral mismatch, cf. Section 7.5.

7.2.2   TOPCon/POLO

History: The theoretical efficiency of the PERC cell, which dominates the current PV market, is limited to 24.5% due to the high metallization induced recombination. To further push the c-Si cell efficiency to its limit, different passivating contact concepts have been introduced, which are able to bring c-Si cell efficiencies closer to their ultimate limit around 29% [41]. Besides the silicon heterojunction and doping free passivating contacts, passivating contacts based on poly-Si on ultra-thin silicon oxide have also proven to enable record high efficiencies, with the so-called tunneling oxide passivating contact (TOPCon) [42] and poly-Si on oxide (POLO) solar cell structures [43]. The poly-Si/silicon oxide concept was first introduced into the PV field with the so-called semi-insulating poly-Si (SIPOS) hetero-contact by Yablonovitch *et al.*, which led to a V_{OC} of 720 mV in a solar cell test structure [44]. The first solar cell device with this contact structure showed a cell efficiency of 12% [45]. Recently, together with the demonstration of record efficiencies of c-Si solar cells with passivating contacts, the SiO_x/poly-Si as one of the most promising contact structures has drawn extensive attention from PV research institutes and industry.

Working principle: The poly-Si-based passivating contact structure consists of a highly doped poly-Si on an ultra-thin silicon oxide (SiO_x) layer, see Figure 7.3(a). The ultra-thin SiO_x layer plays a role as chemical passivation layer for the c-Si surface, but also builds a potential barrier for the carriers' transport through it, see Figure 7.3(b). Owing to a much higher doping level within the poly-Si layer than that in the c-Si bulk, see Figure 7.3(c,d), an electrical field is built up at the c-Si interface, which enables field effect passivation of the c-Si surface and carrier selective collection, see Figure 7.3(a). The doping profile at the interface is a crucial factor that influences the electrical field passivation quality, see the inset tables in Figure 7.3(c,d), as well as the transport of the majority carriers: the optimization of the doping tail inside the c-Si surface region induces an efficient separation of carriers inside the c-Si bulk before they reach the c-Si surface with its high defect density D_{it}. On the other hand, by alloying the poly-Si with oxygen [46–48], carbon [49,50], and nitrogen [51], the modified electronic properties of the formed layer can eventually improve the carrier selective collection.

Passivation/hydrogenation: Besides the chemical passivation induced by the ultra-thin SiO_x and the field effect passivation due to the high doping within the poly-Si layer, hydrogen passivation of the c-Si surface defects is another key factor that contributes to the passivation quality of this passivating contact structure. The hydrogen passivation process is also referred to as hydrogenation, which is achieved by introducing hydrogen atoms to the c-Si surface to saturate the Si dangling bonds. Hydrogenation is normally done by diffusing hydrogen species from a solid [52–54], gas [52], or ionized [55] hydrogen source at a high temperature between 400°C and 850°C. The accumulation of hydrogen atoms at the c-Si/SiO_x/poly-Si interfaces [53] minimizes the c-Si surface defect density,

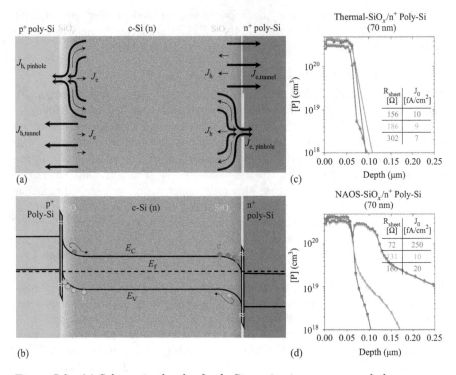

Figure 7.3 *(a) Schematic sketch of poly-Si passivating contact and charge transport into n^+ and p^+ poly-Si based on tunneling and pinholes structure. (b) Band diagram of the n^+ and p^+ poly-Si passivated contacts on n-type c-Si wafer. The doping profiles of phosphorous, P, in n^+ poly-Si passivating contacts prepared with (c) thermal-SiO$_x$ and (d) NAOS-SiO$_x$ as interfacial layer. Replot based on Ref. [56]*

therefore reducing the c-Si surface recombination velocity. The quality of passivation is quantified by the recombination current density J_0. The lower the J_0 value the better the passivation.

Carrier collection: The type of carrier extracted at the contact is determined by the doping type of the poly-Si layer. The carrier selectivity for holes and electrons, $S_{10,h}$ and $S_{10,e}$ respectively, has been discussed in the literature [57]. They influence the recombination parameter J_0 and the difficulty of carrier transport through the contact stack, quantified by the contact resistivity ρ_C. This means that the carrier selectivity optimization of the poly-Si passivating contact stands for the minimization of the aforementioned J_0 at the c-Si surface, together with minimization of ρ_C. For ρ_C, besides influences from the doping level and doping profile of the contact structure, the quality and thickness of the thin-SiO$_x$ layer are also dominating factors. Especially the SiO$_x$ layer thickness is a crucial parameter that determines the carrier transport mechanisms, by tunneling (for SiO$_x$ thickness < 1.5 nm) or through pinholes (for thicker SiO_x layers), see Figure 7.3(a). The pinholes within in the thick SiO_x layers are usually formed by high-temperature annealing processes during the activation/diffusion of the dopants within the poly-Si layer, which also leads to locally thinned SiO_x regions [58,59].

Preparation technology: The fabrication of the poly-Si-based passivation contacts mainly consists in the following five steps: [1] formation of the ultra-thin SiO_x layer, [2] poly-Si layer deposition, [3] doping process, [4] high-temperature annealing and [5] hydrogenation process. Different process technologies were explored to study the influences of material properties on the performance of the contact structures. For example, the ultra-thin SiO_x layer has been prepared by thermal-oxidation [60], oxygen-plasma oxide [51], chemical-oxidation with nitric acid [61], sulfuric acid [62], hydrogen peroxide [63], ozone-based oxides [64], and ALD prepared SiO_x [65]. The poly-Si layers can be deposited by PECVD [66], LPCVD [67], and PVD [68]. In practice, depending on the used technologies for each step, a few of the aforementioned steps can be integrated into one single step. For example, it is possible to integrate the first two or even three steps into one single process step by using a thermal/plasma-SiO_x layer, which is followed by intrinsic/in-situ doped a-Si layer deposition with LPCVD or PVD technology [51,69–71].

Solar cell structure and efficiency evolution: The application of poly-Si-based passivating contacts in c-Si solar cells can be grouped into front/back contact (FBC) solar cells and interdigitated back-contact (IBC) solar cells. A FBC cell can feature one poly-Si contact, for example, the TOPCon [72] and iTOPCon [71] cell, or two poly-Si contacts with localized poly-Si contact regions [73] or on the full surface area [74]. An IBC solar cell can also feature one or two contacts with poly-Si/SiO_x/c-Si junctions. The cell sketches are shown in Figure 7.4 (left).

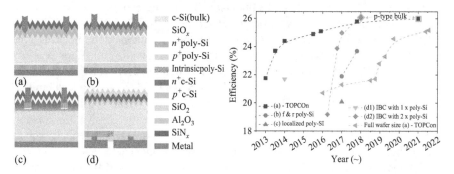

Figure 7.4 Left: Sketch examples of solar cells that deploy poly-Si CSPC structure. (a) One polarization TOPCon cell, (b) two polarizations on full area poly-poly FBC, (c) localized poly-Si contact poly-finger FBC cell, and (d) interdigitated back contacted solar cells poly-Si IBC. Re-sketch based on reference [16]. Right: The efficiency evolution of n-type solar cells with different structures, shown on the left, fabricated with poly-Si-based passivating contacts. Data taken from Refs. [42,49,56,60,61,71–73,75–87]

The efficiency of n-type solar cells with one or two poly-Si contacts progresses rapidly, for both laboratory scale and commercial scale cells, along with the optimization of the poly-Si passivating contact and a deeper understanding of the physics background. In Figure 7.4 (right), we present the efficiency evolution of n-type solar cells with different structures fabricated with poly-Si-based passivating contacts, from < 20% laboratory scale efficiency in 2013 to >25% full wafer size commercial scale cell efficiency in 2021. Note that by 2022 the highest TOPCon cell efficiency is 26.4% as announced by Jinko with a n-type large area solar cell [193].

7.2.3   Bifacial cells

7.2.3.1   Bifacial solar cell development

The idea behind bifacial solar cell designs is to make them light sensitive on both sides of the substrate, such that the absorption of sunlight can be maximized and the energy yield can ultimately be improved. The history of bifacial cell concepts starts as early as 1960 [88]. However, it was only around 1980 that scientists realized that a bifacial solar cell could also be sensitive to the natural albedo [89], i.e., reflected and scattered light from the surroundings [90]. Since the 1990s, the University of New South Wales (UNSW) has been developing bifacial metallization alternatives to achieve high efficiency bifacial solar cells [91–93]. Figure 7.5 summarizes the efficiency evolution of c-Si based bifacial solar cells as reported in literature. Besides, there are newly developed concepts such as thin-film or tandem-based bifacial solar cells [94–101], which are beyond the scope of this book. From Figure 7.5, one can see that (i) with respect to *p*-type cell technologies, *n*-type cells are gaining more momentum. This could be attributed to their higher efficiency potential, little LID and less sensitivity to degradation upon high-temperature processing [102]. (ii) Regarding the bifacial cell structure evolution, bifacial buried contact solar cells (BCSCs) were proposed and developed by UNSW in the 1990s [91,103]. However, such BCSCs suffered from high recombination losses, and the controllability in processing the grooves and in the metallization steps was not sufficient [104]. In the past decade, BCSC [105] was replaced by PERx technologies such as PERC (passivated emitter and rear contact), PERT (passivated emitter, rear totally diffused), and PERL (passivated emitter, rear locally diffused). Furthermore, bifacial c-Si solar cells with carrier selective passivating contacts (CSPCs) are becoming more attractive due to the high efficiency potential, such as silicon heterojunction (SHJ), and Tunnel Oxide Passivated Contact (TOPCon). So far, the highest bifacial cell efficiency is 25.4% as reported by Jinko with an n-TOPCon solar cell structure [105].

7.2.3.2   Bifacial solar cells in the PV market

Figure 7.6(a) displays the status of bifacial cells in the PV market. PERC is currently dominating the market, due to its relatively high cell efficiency and low production cost. It is noteworthy that the cost of bifacial PERC cells has been close to that of monofacial cells [89]. Moreover, one should take into consideration the energy yield gain due to bifacial cell use [121]. The bifaciality factor (which is the

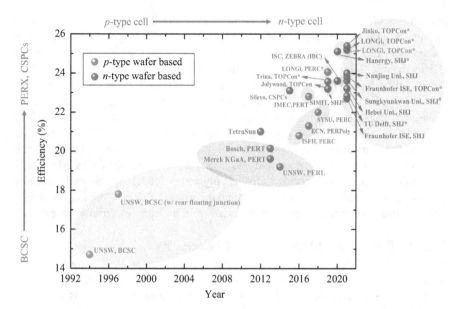

Figure 7.5 *The efficiency evolution of c-Si-based bifacial solar cells as reported in literature [71,87,91,92,106–120], in which the bluish background indicates the buried contact solar cells (BCSCs); the yellowish background denotes the PERX cells, and the greenish background depicts c-Si solar cells with carrier selective passivating contacts (CSPCs). Note: unless indicated with with a "#," the cell performance was measured with single-side illumination; The "*" implies that the data was certified from a third party institution*

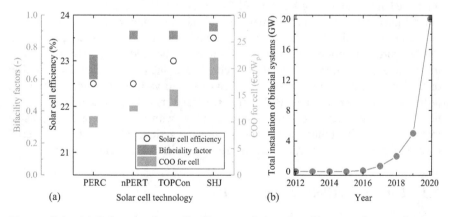

Figure 7.6 *(a) Bifacial solar cell efficiency, bifaciality factor, and cost of ownership (COO) as a function of solar cell technologies and (b) accumulated capacity of bifacial PV systems. The data were obtained from [102]*

ratio of the one-side efficiency to the efficiency under illumination from the other side under standard test conditions) of PERC cell is in the range of 0.60–0.75. For comparison, c-Si cells with CSPCs (such as TOPCon and SHJ) show both higher efficiency and bifacility factors, but the costs of such cells are also higher. Figure 7.6(b) depicts the accumulated capacity of bifacial PV systems. Historically, the world's first bifacial PV module was launched in 2000 by Sanyo (now Panasonic), based on its proprietary heterojunction with intrinsic thin layer (HIT) design [122]. However, it was only after 2010 that the first >1 MW$_p$ system was built in Japan by a private investor with bifacial nPERT modules [102]. In the following 10 years, with more and more players involved (such as MegaCell, Sunpreme, Yingli, SolarWorld, LONGi, Jolywood) [102,123], bifacial modules became more readily available. In particular, bifacial PV is becoming bankable [89,102] and is expected to play an important role in future electricity generation.

7.2.3.3   Prospects and challenges of bifacial PV

After decades of development, bifacial cells are now able to deliver high efficiencies above 25% (single-side illumination), bifacial modules have become available on the market, and bifacial PV is becoming bankable. Besides, an International Electrotechnical Commission (IEC) specification to measure and label bifacial PV products is available since 2019 [124]. This represents an important step forward toward standardization. Furthermore, different energy yield simulation tools have been developed which could take bifaciality into account [90]. However, when more detailed and precise simulations are required, more advanced programs remain to be developed and deployed [102]. On the other hand, next-generation PV technology remains to be considered. As PERC is coming to its efficiency limit, novel cell concepts need to be implemented in production lines. Both TOPCon and SHJ are promising technologies. TOPCon could be easily integrated from upgrading existing PERC lines, while SHJ implementation may face higher Capex (capital expenditure) due to equipment incompatibility. However, comprehensive long-term cost analyses still need to be performed. Furthermore, the following items still need to be addressed: (i) continuous improvements of different bifacial PV technologies, including c-Si based, thin film or tandem-based; (ii) integration of bifacial PV into various applications, such as utility-scale PV, agrivoltaics, BIPV; (iii) advanced metallization and interconnection designs, in order to maximize power output with minimal use of expensive materials; and (iv) other innovations at module/system level, such as material choice of transparent cover, tracker/tilt design in bifacial module installation.

7.3   Ultimate efficiency in c-Si PV

Silicon is an element (material) that is abundant in earth and features a bandgap of 1.1 eV, close to the optimal bandgap for AM1.5 spectrum for PV applications according to the Shockley–Queisser (SQ) limit for a single junction. However, owing to the indirect band gap of Si, silicon solar cells are not able to reach this limit. The indirect bandgap reduces the probability for both absorption and radiative recombination, resulting in a relatively weak absorption and the dominance of Auger recombination, respectively. Considering such drawbacks, the limiting efficiency has been calculated to 29.43% for 110 μm thick un-doped silicon [125].

To overcome the low absorption, thicker silicon wafers are required. In addition, c-Si solar cells demand advanced light management techniques to trap most of the light inside the absorber. Thus, these light trapping methods are focused on reducing optical losses by, for instance, using nano-texturing [126] or multi-anti-reflective coatings (ARC) in the front side together with distributed Bragg reflectors [127] on the rear side. However, the implementation of such upgrades in c-Si solar cells implies modifications at interfaces and thus might entail new technological challenges for mitigating recombination and improving the transport of carriers toward electrodes.

In general, recombination mechanisms limit the conversion efficiency. Important improvements have been reported in mitigating SRH recombination due to defects at interfaces and also inside the absorber bulk. In principle, metallic electrodes were in direct contact to the absorber bulk yielding highly defective c-Si/metal interfaces with strong recombination. The development and imple-mentation of the so-called passivating contacts allowed to overcome the limitation of recombination at interfaces. Similarly, important technology improvements have been developed to mitigate defects in the bulk and thus reducing recombination [128,129]. Indeed, such high lifetime materials expose the interesting behavior of intrinsic recombination in c-Si. Thus, relevant calibration of the semi-empirical models of intrinsic recombination have been reported lately [130–134]. Interestingly, the theoretical limiting efficiency calculated with updated Auger parameterizations exhibits slight variations in the range of 29.4–29.6% [134].

It is well known that Auger recombination increases with the doping, sug-gesting that minimal recombination is achieved for un-doped silicon. However, intrinsic recombination also depends on the injection of free carriers, that becomes relevant for relatively low doping. Indeed, considering the AM 1.5 spectrum and maximum power conditions, a slight increase of intrinsic recombination is calcu-lated until 10^{15} cm^{-3} doping as reported by Richter *et al.* [125] (see Figure 7.7). For higher doping above 10^{15} cm^{-3} the intrinsic recombination significantly increases (see red arrow in figure), while a slight decrease is calculated for lower doping (see green arrow in Figure 7.7). This means that the practical limiting efficiency could be enabled for solar cells using wafers with resistivity above 10 Ω*cm with high quality (lifetime) and using passivating contacts.

The limiting efficiency was calculated for around 100 μm thick silicon wafer [125] (Figure 7.8). Such a value is result of the optimal trade-off between absorbed light and intrinsic recombination in a 1D device model [125,134]. Note that both mechanisms increase with the thickness. In fact, intrinsic recombination increases with the thickness due to the increment of the path for collecting carriers that becomes optimal in the simplified 1D picture. However, in real solar cells, the collecting path becomes 2D and even 3D depending on the design and architecture as front and back contact, or interdigitated back contact solar cells. Indeed, small contact sizes and distance between contacts are preferred for improving transport of

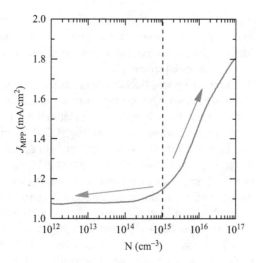

Figure 7.7   Intrinsic loss current density at maximum power point J$_{MPP}$ as a function of the doping concentration N for ideal 110 μm thick n-type solar cells under illumination with the AM 1.5 spectrum and 25°C. Green arrow indicates a slight J$_{MPP}$ decrease for N < 10^{15} cm^{-3}. Red arrow indicates a significant J$_{MPP}$ increase for N > 10^{15} cm^{-3}. J$_{MPP}$ values are digitized from [125]

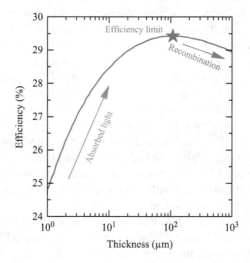

Figure 7.8   Efficiency as a function of the thickness for undoped c-Si cells under AM 1.5 illumination and 25°C. Green arrow indicates the increase of efficiency due to the improvement of absorbed light. Red arrow indicates the efficiency decrease due to intrinsic recombination. The efficiency limit is achieved for a thickness with the best trade-off: absorbed light vs recombination. Efficiency values are digitized from [125]

carriers. Therefore, the contact size/width and patterning resolution become relevant to minimize the path for collecting carriers and accordingly is a technological limitation for achieving the theoretical limiting efficiency.

The gap between the theoretical and practical efficiency limit for c-Si solar cells is expected to be reduced by: (i) using optimal light management at front and rear interface, (ii) using optimal passivation at interfaces (passivating contacts), (iii) using high-quality wafers with resistivity above 10 Ω cm and with thickness around 110 μm, iv) enabling smaller contacts and features.

7.4   Alternative carrier-selective contacts

7.4.1   *Metal oxide and alkali metal–halogen compounds*

Whereas direct metal contacts on silicon are straightforward and enable to use a simple window-layer stack with excellent transparency, the application of passivating contacts using doped silicon leads to complex processing and parasitically absorbing layers. There has therefore been since 2014 a resurgence of research aiming at applying metal compounds as carrier-selective contacts to silicon solar cells. Research has so far targeted mainly transparent materials, i.e. with a bandgap typically over 3 eV, with a few exceptions for use on the rear side of the device.

The most widely studied material families are the metal-oxide and the alkali (or alkaline-earth) metal–halogen compounds. The main material systems employed as selective contacts are shown in Figure 7.9. Using sub-stoichiometric molybdenum trioxide capped with ITO as a full-area hole-selective contact on the front side of the device, an efficiency of 23.5% was reached in 2019 [135] and further augmented to 23.8% in 2022 [136]. Conversely, using a titanium dioxide-lithium fluoride stack capped with aluminium as a partial-area electron-selective contact on the rear-side of the device, an efficiency of 23.1% was reached in 2018 [137]. These results were at that time very promising considering the little history for these approaches.

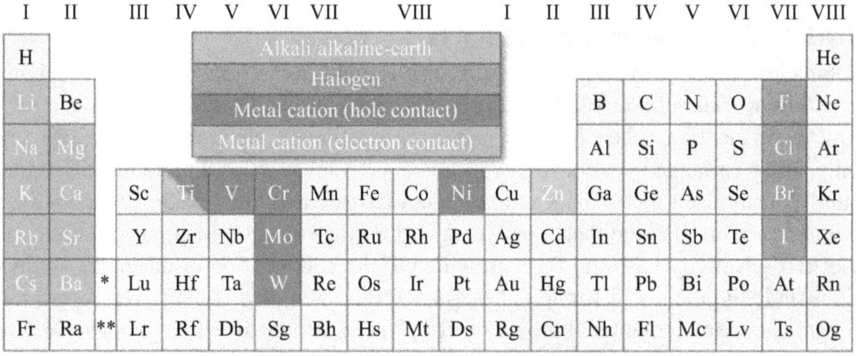

I	II	III	IV	V	VI	VII	VIII	I	II	III	IV	V	VI	VII	VIII
H															He
Li	Be									B	C	N	O	F	Ne
Na	Mg									Al	Si	P	S	Cl	Ar
K	Ca	Sc	Ti	V	Cr	Mn	Fe	Co	Ni	Cu	Zn	Ga	Ge	As	Se	Br	Kr
Rb	Sr	Y	Zr	Nb	Mo	Tc	Ru	Rh	Pd	Ag	Cd	In	Sn	Sb	Te	I	Xe
Cs	Ba	*	Lu	Hf	Ta	W	Re	Os	Ir	Pt	Au	Hg	Tl	Pb	Bi	Po	At	Rn
Fr	Ra	**	Lr	Rf	Db	Sg	Bh	Hs	Mt	Ds	Rg	Cn	Nh	Fl	Mc	Lv	Ts	Og

Legend:
- Alkali/alkaline-earth
- Halogen
- Metal cation (hole contact)
- Metal cation (electron contact)

Figure 7.9   Chemical elements used in carrier-selective contacts based on metal oxide and alkali metal–halogen compounds

7.4.1.1 Hole-selective contacts

High-workfunction metal oxides using a fully oxidized metal cation from column VI of the periodic table (CrO_3, MoO_3, and WO_3) are the most studied materials, together with V_2O_5. When stoichiometric, they typically are insulators with a bandgap over 3 eV and a work function over 6 eV. Nevertheless, they naturally tend to be sub-stoichiometric, which reduces slightly their workfunction while enabling conductivity. When deposited on silicon, this high workfunction induces a strong bending of the conduction and valence bands, leading to accumulation for a p-type wafer (inversion for an n-type wafer) thus inducing hole selectivity. Thermal evaporation is the most widely used deposition technique; reactive sputtering so far yielded underwhelming performances, yet remarkable performance was obtained with V_2O_5 deposited by ALD [138]. Since they do not provide efficient surface passivation, an additional passivation layer (most typically amorphous silicon) is inserted between the metal oxide and the silicon wafer.

Deposition conditions as well as pre- or post-deposition treatments prove to be very critical to the performance of the contact. MoO_3 was shown to be sensitive to reduction, which typically can be caused by hydrogen effusion from the passivating amorphous silicon layer. Preventing such effusion is thus critical, e.g. by reducing the H content of the a-Si layer prior to MoO_3 deposition, or by maintaining the thermal budget following MoO_3 deposition below 150°C. Additionally, plasma treatments prior to MoO_3 deposition were shown to enable an increase in performance together with a reduction of the required MoO_3 thickness [139]. Even more striking is the case of titanium dioxide, traditionally used as electron-selective contact as discussed in the following, but which can behave as a hole-selective contact when processed with dedicated approaches [140].

Several approaches were therefore investigated to implement metal-oxide hole-selective layers, but without a clear demonstration of improved performance compared to traditional contacts. Parasitic light absorption, and significant contact resistance impede the overall solar cell performance. Besides, the need for a dedicated passivation layer makes processing similarly complicated as for standard doped-silicon-based contacts.

7.4.1.2 Electron-selective contacts

A broader variety of materials were used as electron-selective contacts, mostly relying on a low workfunction. Alkali or alkaline-earth metal halides typically provide such low work function. They however function as efficient electron-selective contacts only when combined with a low-workfunction metal such as aluminum or magnesium. Mostly used metal-oxides are titanium dioxide and zinc oxide, also combined with a low-workfunction metal, or even a lithium-fluoride/aluminium stack. Although electron selectivity has long been attributed to the combination of adequate conduction band alignment and strong valence-band mismatch, this is being reassessed following the aforementioned discovery that TiO_2 can also be an efficient hole-selective contact. Part of the selectivity is therefore thought to stem from the electrode (or electrode stack). The absence of transparent electrodes with a low workfunction prevents to use these materials on the front side of the device.

A higher level of passivation was shown when using metal-oxides as electron-selective contacts than in the hole-selective contact case. This probably stems from the use of deposition techniques including some hydrogen, such as ALD or CVD. Nevertheless, dedicated passivation layers, such as a-Si:H or SiO_x:H are generally included for higher passivation. Performance stability is an active research topic, with some of the best-performing strategies degrading rapidly even without strong external stress [141]. Furthermore, best-performing contacts require an elaborate stack of layers with nanometric thicknesses, with similar or higher complexity compared to traditional contacts.

Alternative materials being evaluated include metal nitrides such as tantalum nitride or titanium nitride. Using the latter, working solar-cell devices were demonstrated with a single TiN layer as electron-selective contact. Efficiency was however not as high as when using more complex stacks.

7.4.1.3 Perspectives

Multiple strategies were investigated to implement non-silicon-based contacts in silicon solar cells. Remarkable performance was demonstrated in a short time-frame. Nevertheless, similarly to traditional contact strategies, high performance comes at the cost of relatively high complexity. There is nevertheless still a wide range of candidate materials that has not yet been widely investigated. Most studied materials so far were inspired from the organic semiconductor field (photovoltaics as well as light-emitting diodes). The field of wide-band-gap semiconductors remains to be investigated in detail, and the very promising results recently achieved with nanocrystalline silicon carbide suggest that this is highly relevant [26]. Indeed, the combination of a wide band gap and efficient p- or n-type doping is anticipated to be key in enabling a given material to act as an efficient carrier-selective contact in silicon solar cells. In this framework, the perovskite-oxide, III–V or II–VI families include very attractive materials. With these, dedicated substrates and growth techniques are however required to ensure sufficient material quality, making a direct use as contact in silicon devices unpractical. In parallel, numerous novel semiconductors are being predicted and experimentally demonstrated, widening further the possibilities. To this regard, ternary nitrides offer underexplored space with promising properties such as dopability or defect tolerance. Considering the variety of materials available, high-throughput experimentation would likely be appropriate to scan the para-meter space efficiently. Nevertheless, their efficient implementation as contacts in silicon solar cells is unlikely to be an easy task, making dedicated effort likely required to reach high performance. A fine balance between rapid exploration and detailed investigation will thus have to be found should such alternative approach be envisioned. That is worthwhile though since unveiling an easy-to-process contact material with improved transparency compared to doped silicon together with similar passivation, selectivity, and reliability would enable solar cells in mass production to reach efficiency values up to 27%, making this a game-changer in the photovoltaics industry.

7.4.2   Organic carrier selective contacts

Besides inorganic carrier-selective contacts, substantial efforts have been put on researching the organic route since the first Si/organic solar cell reported in 1990 by Sailor *et al.* [142]. That device employed poly-$(CH_3)_3$Si-cyclooctatetraene as hole-selective contact, but later other materials were tested, such as poly[2-methoxy-5-(2-ethylhexyloxy)-1,4-phenylenevinylene] (MEH-PPV) [143], poly(3-hexylthiophene) (P3HT) [144], and poly(3,4-ethylenedioxythiophene) (PEDOT) [145]. Especially the last has received attention for its appealing basic properties such as optical transparency in its conducting state, high stability, moderate band gap, and low redox potential [146,147]. This material can be doped with polystyrene sulfonate (PSS) to form PEDOT:PSS, whose augmented conductivity is crucial in high performance Si/PEDOT:PSS solar cells [16]. So far, the highest conversion efficiency achieved in c-Si PV technology by means of an organic hole-selective contact (PEDOT:PSS) is 20.6% [148]. The architecture of such a device, dubbed *BackPEDOT*, essentially simplifies the rear side of the PERC architecture, and thus embodies the promise of a low-CAPEX organic route with FF $> 80\%$ and j_{sc} close to 39 mA/cm^2.

Passing to the electron-selective contacts, Buckminsterfullerene (C_{60}) [149] and Lewis base polymers [150] have been successfully integrated in c-Si solar cells. In the first case, C_{60} was doped with tetrabutyl ammonium iodide (C_{60}:TBAI) and deployed at the front side of a simple, double side flat c-Si solar cell, yielding an efficiency of 8.43%. In the latter case, branched polyethylenimine (b-PEI) as well as other Lewis base polymers were tested on double side textured c-Si hetero-junction cells endowed with standard i/p hole-selective contact stack based on a-Si:H at the front side and a-Si:H surface passivation at the rear side. The device with 3-nm thick b-PEI film yielded an efficiency of 19.4%, mostly supported by state-of-the-art $V_{oc} = 720$ mV. As strong donor of electrons, b-PEI induces strong downward bending in c-Si, thus favoring electron transport.

Leveraging existing processes typical of PERC or FBC heterojunction archi-tectures (e.g. diffusion, PECVD a-Si:H, sputtering, screen printing, etc.), organic charge-selective contacts hold the potential to simplify and/or make cheaper their fabrication. However, further advancement in this route demands research efforts not only for improving the conversion efficiency of organic/c-Si solar cells but also their stability against temperatures beyond 200°C, UV light, and oxygen [16].

7.5   Silicon-based tandem cells

As outlined above, silicon single-junction solar cells are currently dominating the photovoltaics market and are slowly reaching their practical limit in module effi-ciency. For a further drop in the levelized cost of solar energy, aiming for higher power conversion efficiency is an obvious path. Today, the only experimental demonstration of overcoming the single-junction detailed balance limit has been achieved by multijunction solar cell technology. Multijunction cells tackle the shortcoming that incident photons with an energy above bandgap lead to a dissipation of their excess energy into the lattice of the absorber as heat (thermalization).

By adding additional solar cells with higher bandgaps, the potential of the high-energy photons is harvested more effectively [151,152]. The simplest realization of this concept is a tandem solar cell. Here, a combination of two subcells shares the solar spectrum (see Figure 7.10) and, upon optimal choice of the bandgaps, this leads to higher power conversion efficiencies.

The two subcells can be combined in three ways (see Figure 7.11): (1) monolithic stacking, where the top cell is processed on top of the bottom cell. This leads to two-terminal (2T) tandems. Both cells then share the top and bottom electrodes, and they are interconnected by a recombination layer. Ideally, this interlayer behaves like an ohmic contact with a low (vertical) series resistance. Due

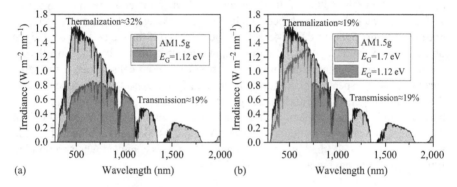

(a)   (b)

Figure 7.10   Fraction of the AM1.5g spectrum which can be used in a single-junction solar cell with a bandgap of 1.12 eV such as silicon (A) or in a multijunction solar cell where materials with bandgaps of 1.7 eV and 1.12 eV are combined (B). In this specific single-junction solar cell, around 19% of the incoming power is lost due to transmission (photons with energies lower than the lowest bandgap) whereas 32% is lost due to thermalization. In tandem solar cells, high energy photons are absorbed in the wide bandgap top cell, which reduces the amount of thermalization. For this specific combination of bandgaps, the losses due to thermalization are reduced to 19%

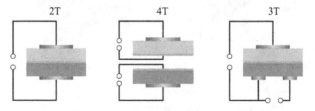

Figure 7.11   Schematics of different tandem architectures, where the blue subcell is the top cell and the orange cell represents the bottom cell: 2T, 4T, and 3T

to the series connection, the voltages of the subcells add up and the photocurrent of the device is limited by the lower one of the photocurrents generated in the two subcells. (2) Mechanical stacking resulting in four-terminal (4T) tandems, where both subcells are independent from each other and the top cell acts as an optical filter for the bottom cell. (3) Three-terminal (3T) interconnection combines the characteristics of both 2T and 4T tandems, in that the device stack is monolithic, but the two subcells can still be controlled independently [40].

Figure 7.12 shows the detailed balance efficiency limit of 2T and 4T tandem solar cells as a function of the top- and bottom cell bandgap under AM1.5g illumination (calculated according to [151]). The independent operation of the subcells in the 4T architecture enables a broad window where high efficiencies can be achieved. The highest PCE of 46.1% is achieved when materials with bandgaps of 1.73 eV and 0.93 eV are combined. If silicon with a bandgap of 1.12 eV is used as a bottom cell, the highest PCE of 45.2% can be achieved if the top cell has a bandgap of 1.82 eV. In the 2T architecture, the output current is limited by the subcell with the lowest current generation. Thus, in the best case, both cells generate the same current at their respective maximum power point (same J_{MPP} for both subcells). As the amount of generated current strongly depends on the subcell bandgap, only a narrow window exists where high PCEs can be achieved. The highest PCE of 45.7% is achieved when bandgaps of 1.60 eV and 0.94 eV are combined. If silicon is used as bottom cell, the highest efficiency of 45.0% is achieved when combining it with a material having a bandgap of 1.73 eV.

7.5.1   Perovskite-silicon tandems

As seen in Figure 7.12, a good tandem partner for Si cells would have a bandgap of around 1.7 eV [153]. Ideally, the cost of the top cell should be minimal, such that the efficiency gain leads to a reduced levelized cost of electricity. On top, the

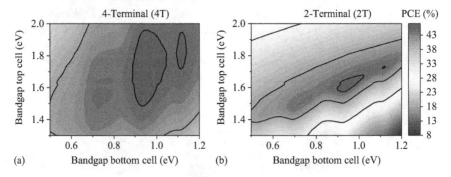

Figure 7.12   Detailed balance limit of tandem solar cells in 4T (A) and 2T (B) architecture as a function of the top- and bottom cell bandgap. The black lines indicate a PCE increment of 5% (i.e. 5%, 10%, 15%, etc.). The detailed balance limit is calculated according to [151] using the AM1.5g spectrum

processing complexity of the top cell should be minimal, to facilitate integration into existing Si cell production lines with a minimal number of barriers for the factories. Metal-halide perovskite semiconductors have the potential to combine all these aspects: favorable optoelectronic properties at low processing temperatures of around 100°C [154], low sub-bandgap absorption [155], a wide variety of fabrication methods and a variable bandgap. Their general compositional formula is ABX_3, where A denotes a monovalent cation (usually mixed organic/inorganic atoms or small molecules), B is a metal (mostly Pb, Sn or a mixture), and X denotes a halide anion (I, Br, Cl or a mixture thereof). For the A site, a wide variety of combinations can be found in literature [13]. For highly efficient perovskite/Si tandem solar cells, the organic cations formamidinium, methylammonium, and caesium have been used in mixtures, while the bandgap (optimized at 1.66–1.68 eV [156,157]) has been mostly controlled by the iodide-to-bromide ratio in the composition [158].

The first functional 2T perovskite-silicon tandem solar cell was published by Mailoa *et al.* in 2015. The combination of a standard n-type back surface field (BSF) silicon cell with a perovskite ($MAPbI_3$) top cell in n–i–p configuration (where the electron-selective layer was deposited first, followed by the intrinsic perovskite and then the hole extraction layer) enabled a PCE of 13.7%. Later in 2015, Albrecht *et al.* used SnO_2 deposited via atomic layer deposition (ALD) as the perovskite cell's n-contact, which was deposited at temperatures below 120°C. This allowed to fabricate tandem solar cells with silicon heterojunctions (SHJ) as bottom cells. The so fabricated tandem solar cell improved the PCE to a stabilized value of 18.1%.

Since this achievement, the majority of highly efficient (>25%) tandem solar cells were fabricated on SHJ solar cells on (n)c-Si wafers. Furthermore, following a publication by Bush *et al.*, in which the perovskite is deposited on the hole-selective layer and the electron contact contained a SnO_2 buffer layer [159], most perovskite-silicon tandems employ this "inverted" p–i–n perovskite cell architecture.

Usually, silicon bottom cells have textured surfaces to reduce reflection losses. However, the several micrometre high pyramids are not compatible with solution processing of the perovskite, which is an order of magnitude thinner than the height of these pyramids. Sahli *et al.* presented a conformal deposition method for perovskites, which enabled the utilization of silicon bottom cells with textured front sides [160]. The reduced reflection enabled high photogenerated current densities in the subcells and with that a high j_{SC} in the tandem solar cell. The certified 25.2% were published in 2018. Afterwards, Al-Ashouri and Köhnen *et al.* optimized the perovskite composition, introduced new hole-selective materials, namely self-assembled monolayers (SAMs) and carried out further optical optimizations, thereby increasing the power conversion efficiency to 29.2% [161,162]. The optical cell design was further optimized with a nanotextured interconnection layer by Tockhorn and Sutter *et al.*, enabling a certified efficiency of 29.8% [163,164]. In 2022, EPFL and CSEM announced a certified efficiency of 31.3% on silicon with textured front side and 30.9% using a planarized silicon front side. Later that year, HZB achieved a certified efficiency of 32.5% on a planar silicon front side by

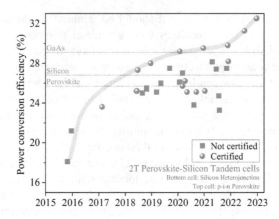

Figure 7.13   *Best power conversion efficiencies of two-terminal perovskite-silicon tandem solar cells. The red line is a guide to the eye. The gray horizontal lines are for reference and indicate record efficiencies of other, single junciton technologies (non-concentrator) under global AM1.5 spectrum: GaAs (thin film cell) with 29.1%, silicon (crystalline cell) with 26.7% and perovskite (thin film) with 25.7% [5,8]*

optimizing interfaces and optical modifications at the front side of the perovskite cell [194]. These values not just surpass the highest single-junction solar cell (GaAs, 29.1%) and even the (Auger-corrected) detailed-balance limit of silicon single-junction solar cells but approach the highest overall 2T tandem efficiency of 32.9% for GaInP/GaAs [10.1002/pip.3646]. These high efficiencies were achieved mostly on n-type silicon.

This impressively rapid improvement of the PCE for perovskite-silicon tandem solar cells is yet unique among all solar cell technologies. A plot of the time evolution of best power conversion efficiencies is depicted in Figure 7.13. With increasing PCE, the scientific focus in the community shifts more toward increasing the stability of the tandem cells, dealing with the sensitivity of perovskite layers against heat and moisture and also investigating the behaviour under real-world conditions [157,165]. Furthermore, the implementation of "drop-in tandem upgrades" for existing PERx technologies can be expected to become an important topic, as discussed by Messmer *et al.* based on simulations [166]. First such proof-of-concept perovskite/PERC/POLO tandems have been realized just recently in a collaboration between Helmholtz-Zentrum Berlin and ISFH [167].

7.5.2   III–V/Silicon tandems

While perovskite-silicon solar cells are today's shooting star, when it comes to tandem solar cells, III–V/silicon tandems have a much longer history starting in the 1990s. Today, III–V/Si cells also hold the efficiency record for tandem cells on silicon under the unconcentrated AM1.5g spectrum, with 32.8% in a mechanically stacked 4T GaAs/Si tandem [168]. Furthermore, 2T wafer bonded as well as 4T mechanically stacked triple-junction cells—GaInP/GaInAsP//Si and GaInP/GaAs//Si stacks, respectively—have been demonstrated with identical efficiencies of 35.9% [5,168,169].

In these cells, knowledge gained in the fabrication of III–V compound semiconductor-based multijunction cells for space and concentrator PV applications is transferred for fabricating tandems and triple junction cells on silicon. Major advantages of the III–V semiconductors are (i) their direct band gap, thus the possibility to use thin film absorbers, and (ii) the wide range of band gaps that can be covered by variations in stoichiometry, especially when not only binary alloys such as GaAs, but also ternary and even quaternary alloys are considered [170]. For example, alloying Ga with indium (In) and arsenic (As) or phosphorous (P), a band gap range from ~1.4 to ~2.3 eV can be covered. This range extends even well beyond the most interesting region for both 2T and 4T tandems on silicon bottom cells, cf. Figure 7.12.

Historically, much of the development of III–V-based tandems on wafers was carried out on gallium arsenide (GaAs) and germanium (Ge) substrates. However, the shortage and relatively high cost of Ge makes this a difficult material for large-scale deployment of PV. Therefore, the focus has shifted to using silicon as an alternative substrate; it is worth mentioning that this is also an interesting field of research for the microelectronics industry [171].

As with perovskite-Si tandems, also III–V-Si devices can be broadly classified into 2T and 4T devices, and they can be fabricated either by direct growth of the III–V films on the Si wafer, or by mechanical stacking.

For fabricating 2T tandems by direct growth of III–V on Si, yielding 2T tandems, the challenge lies in achieving high quality of material both at the growth interface and in the III–V film's bulk. This requires the epitaxial growth of an antiphase-free nucleation layer, using MBE and/or MOCVD techniques. Typically, GaP is grown since it is lattice-matched to silicon with a suitable surface reconstruction. Epitaxy could be achieved with sufficient quality [172,173], but special offcut and polished Si wafers had to be used. Thus, this is approach is not compatible with usual saw damage etched PV-grade silicon, much less with textured wafers. Continuing with the growth of the actual top cell absorber layer, a second challenge arises: to achieve suitable band gaps, the ternary alloys GaAsP or GaInP should be used (e.g., $GaAs_{0.8}P_{0.2}$ has a band gap of ~1.66eV). However, the lattice constants of these materials are larger than that of Si by several percent. In addition, the thermal expansion coefficients of the materials differ significantly. Thus, such heteroepitaxial growth will induce the formation of dislocations, which are detrimental to the electronic quality. To tackle this issue, the so-called metamorphic epitaxy, which is key to record efficiencies in III–V multijunction technology, has been applied for III–V on Si growth. However, the device performances achieved so far are rather moderate, at 23.1% for a GaAsP/Si 2T tandem cell [5,174] and 25.9% for a GaInP/GaAs/Si monolithic 2T triple junction cell [175]. In the latter cell, a graded $Al_xGa_{1-x}As_yP_{1-y}$ buffer layer with a gradual increase of the band gap was used for the metamorphic transition on the Si wafer. Still, the efficiency for this quite complicated device stays below the Si single junction record.

The limitations in the direct growth approach render bonding of III–V on Si for 2T tandems an interesting option: Here, the III–V top and Si bottom cells are fabricated separately, alleviating problems due to process incompatibilities, cross contamination, etc. For 2T tandem cells, the III–V cells are then attached to their Si partner by mechanical bonding, and the support used in III–V growth is removed. The bonding step at the same time forms an electrical series interconnection; in literature, the bonding interface is often denoted with a double slash, "//," to distinguish it from interfaces formed by layer growth. The main challenges for the bonding approach are the need for very flat/polished interfaces, incompatible with texture, and the development of various process steps for: (i) surface cleaning to remove any residual oxides or contaminations and promote adhesion; (ii) homogeneous, well-aligned bonding also on large areas; (ii) re-use of the wafer on which the III–V cells have been grown. If suitable, cost efficient solutions can be developed, this technology has significant potential: while in tandem cells, moderate efficiencies of 21.1% have been demonstrated using AlGaAs//Si [176], a wafer-bonded GaInP/GaInAsP//Si 2T triple junction cell has shown an efficiency of 35.9% (j_{sc} 13.1 mA/cm², V_{oc} 3.25 V, FF 84.3%) [169]. Among others, a TOPCon bottom cell, double antireflective coating and a nanostructured diffractive rear-side grating have been used to maximize performance. This is today's world record efficiency for such devices, on par with a mechanically stacked 4T triple junction device.

As an alternative to bonding, the so-called "smart stack" [177] has been developed: it uses transparent conductive adhesives for subcell interconnection into 2T III–V-Si multijunction cells and has recently shown interesting potential, with up to 30.8% efficient InGaP/AlGaAs//Si triple junction cells in 2T configuration [178].

Another classical option is to process both subcells into full cells, with two terminals each, and then form a mechanical stack using transparent, refractive index-matched and electrically insulating glue. This yields III–V on Si 4T tandems. The same advantages and caveats as for such 4T tandems with other combinations of absorber materials apply, i.e. briefly: potentially better energy yield due to better resilience toward non-optimal illumination spectra vs. additional parasitic absorption in the two additional contact stacks consisting of laterally conductive layers and metallization grids. Despite the latter disadvantage, a 35.9% efficient GaInP/GaAs//Si triple junction 4T cell could be fabricated. The other parameters were: V_{oc} 2.52 and 0.681 V in the III–V top and Si bottom cell, respectively, j_{sc} 13.6 and 11.0 mA/cm², FF 87.5 and 78.5% [168]. In the same publication, a 32.8% 4T tandem cell with GaInP and Si absorbers was demonstrated. Furthermore, the paper contains a cost analysis which reveals the biggest challenge in III–V/Si technology: the use of costly GaAs wafer substrates and MOVBE epitaxy processes for the III–V solar cell stack. Thus, techniques for more than 100 substrate reuses as well as faster deposition techniques, such as high-growth-rate MOVPE, close-space vapor transport and hydride vapor phase epitaxy (HVPE) need to be developed further to make III–V on Si multijunctions a viable option.

7.6 Projections

The energy transition from fossil fuels to renewable energy is one of the biggest challenges mankind faces nowadays. The last two decades have seen a continued exponential growth of renewable energy technologies such as wind and solar [179], contributing to the energy mix of several countries in appreciable percentages [180]. This development mitigates climate change, supports the UN sustainable development goals [181], and is driven by the electrification of society. In this age of information and digitalization, electricity is the most versatile energy carrier, serving the built environment, industrial processes, electrical mobility, and sustainable farming while enabling smart data connection.

The electrical energy system of today transforms the chemical energy of fossil fuels into thermal, mechanical, and ultimately electrical energy. This system is centralized with a few power plants that unidirectionally distribute electricity from very high-voltage levels down to medium and low-voltage levels. For its reliance on fossil fuels and related CO_2 emissions, today's electrical energy system is one of the major causes of climate change. Most of the cumulative installed power of wind and solar can be thus far ascribed to GW-scale onshore wind [182] and PV plants [183] that are co-existing with fossil fuel and nuclear power plants. It is undeniable that PV modules based on p-type c-Si technology have played a crucial role in such growth, benefitting from an industrial maturity that has steadily increased the conversion efficiency while driving down the costs. Owing to this success, the power plant level LCOoE of solar electrical energy is today the lowest in many countries [184–186].

The generation of electricity from renewable energy sources can be achieved also at medium and low-voltage levels, that is, closer to local consumption. While the installation of onshore wind turbines is not always practical for issues related to wildlife safety, societal acceptance, and ecological impact [187], PV systems can be massively integrated in both urban and open environments with much lower societal and economic barriers. By 2050, more than 65% of the ever-growing global population will live in urban areas [188] resulting in two major challenges. First, cities will need more energy to satisfy their population's needs and, second, nearby rural areas will be primarily used for food production. In this respect, the generation of green electricity by means of PV systems will compete in both cities and rural areas with land scarcity.

PV systems based on n-type c-Si solar cell architectures will tackle these challenges by ensuring a continued increase in conversion efficiency by single-junction technology and, later, by embodying the ideal bottom sub-cell technology for future tandem PV modules. That is, given the same form factor and environmental conditions, n-type PV modules will exhibit higher annual energy density (kW h/m^2) than their current p-type counterparts, for which either smaller form factors can be devised with similar energy output to current modules or standard form factors can be produced enabling intra-module customization and thus new applications. Owing to the higher conversion efficiency, the production in volumes of such n-type modules will further drive down the c-Si PV learning curve. At the same time, as n-type solar cells enable several new applications, e.g. environment-integration (EIPV), urban-integration (UIPV), and vehicle-integration (VIPV), customized n-type PV modules will also be produced with economic return.

Therefore, in future cities, the electricity-driven energy system will be fed by high-performance n-type PV systems that can pave every surface in the environment providing useful green electricity for the sustainable electrification of society. Also, shade-resilient modules [189] integrated with storage of electrical or thermal energy and communication capabilities will constitute the so-called *PV-based intelligent energy agents*. These will cover the whole conversion chain from photons to electrons to bits, marking the advent of the photovoltatronics age [190]. At the same time, as n-type PV systems can deliver higher performance than current p-type based ones, more powerful PV plants occupying less space will be realized for a more granular green electricity generation in the vicinity of cities. Finally, diverse and powerful EIPV systems, such as floating PV or agri-PV systems, will be installed competing less for land or water otherwise needed for other utilizations.

With more than 700 GW$_p$ cumulatively installed in 2020 [191] and, according to projections [192], more than 900 GW$_p$ cumulatively installed in 2021, PV systems installations are well on track to break the TW$_p$ scale by 2022. These formidable numbers are due to PERX/iTOPCon architectures, that dominate the market with 80% share [12], and the arising of silicon heterojunction (HTJ) technology, whose market share is expected to grow well above 15% by 2030.

7.7 Summary

In this chapter, we have reviewed candidates for further enhancement of cell efficiencies beyond those of today's mainstream PERC cells, with a focus on technological aspects rather than, e.g. cost. Regarding silicon single junctions, the prevalent theme is the use of carrier-selective passivating contacts, CSPCs. Of these, silicon heterojunction and polysilicon-on-silicon oxide (TOPCon/POLO) are most advanced and have enabled record high efficiencies above and close to 26%, respectively, on n-type silicon wafers. Further important topics are bifacial cell designs, which can be applied to different PV technologies. Single-side efficiencies above 25% have been achieved on bifacial TOPCon and bifacial SHJ solar cells. With proven bankability, bifacial PV products can be expected to gain more momentum in future development. In contrast, contacts based on metal compounds have yielded remarkable results in the last decade, yet failing to clearly evidence a significant advantage compared to the ones based on silicon. Further research is needed to unravel the material combination that would enable the long-awaited ultimate passivating contact for Si solar cells.

The second major topic are tandem and multijunction cells. This is the technology to move beyond the ultimate efficiency barrier of 29.4% for silicon PV and indeed, efficiencies well above 29% have been demonstrated in the lab for Si-based tandems. We have reviewed the current state of the art in lead halide perovskite-silicon tandems as well as III–V/silicon tandems. The former have reached a record PCE of 32.5% in monolithically integrated 2-terminal tandems, while III–V/Si 2T tandems currently stand at 23.4%. However, in III–V-Si devices, the number of absorbers has already been increased further, to three: in triple junction III–V/III–V/Si cells, PCEs of 35.9% have been realized with both 2T and 4T architectures.

With a substantially higher cost for the III–V technology as compared to perovskites, but still inferior long-term stability in perovskites, as well as challenges in upscaling for both technologies, it remains to be seen which one of these technologies will gain an advantage. It should be mentioned that an important difference between reported silicon single junction and tandem/multijunction record devices is the cell area: while the single junction Si record devices have "industrial-size" active areas of several tens of cm^2 or even full wafers, record tandem cells are lab-scale 1–4 cm^2. Thus, up-scaling of tandem cells will remain an important topic in the near future.

At any rate, it can be expected that the exponential growth of PV as well as the diversity of applications (utility, rooftop and BIPV, agri-PV, etc.) will create ample opportunity for the market entry of quite a few of the mentioned technologies, and even for entirely new concepts such as three-terminal tandems or, at the module level, integrated PV and storage systems.

References

[1] Snapshot of Global PV Markets 2021. IEA-PVPS; 2021, p. 22. [cited 2022 Feb 28]. Available from: https://iea-pvps.org/snapshot-reports/snapshot-2021/
[2] Niewelt T, Schön J, Warta W, Glunz SW, and Schubert MC. Degradation of crystalline silicon due to boron–oxygen defects. *IEEE J Photovolt.* 2017; 7(1):383–98.
[3] Wurfel U, Cuevas A, and Wurfel P. Charge carrier separation in solar cells. *IEEE J Photovolt.* 2015;5(1):461–9.
[4] Rau U and Kirchartz T. Charge carrier collection and contact selectivity in solar cells. *Adv Mater Interfaces.* 2019;6:1900252.
[5] Green MA, Dunlop ED, Siefer G, *et al.* Solar cell efficiency tables (version 61). *Prog Photovolt Res Appl.* 2023;31(1):3–16.
[6] Yamamoto K, Yoshikawa K, Uzu H, and Adachi D. High-efficiency heterojunction crystalline Si solar cells. *Jpn J Appl Phys.* 2018;57(8S3): 08RB20.
[7] Yoshikawa K, Kawasaki H, Yoshida W, *et al.* Silicon heterojunction solar cell with interdigitated back contacts for a photoconversion efficiency over 26%. *Nat Energy.* 2017;2(5):17032.
[8] At 26.81%, LONGi sets a new world record efficiency for silicon solar cells. [cited 2023 Jan 05]. Available from: https://www.longi.com/en/news/propelling-the-transformation/
[9] LONGi announces new world record cell efficiencies. [cited 2023 Jan 05]. Available from: https://www.longi.com/en/news/new-record-cell-for-hjt/
[10] Taguchi M, Terakawa A, Maruyama E, and Tanaka M. Obtaining a higher Voc in HIT cells. *Prog Photovolt Res Appl.* 2005;13(6):481–8.
[11] Haschke J, Dupre O, Boccard M, and Ballif C. Silicon heterojunction solar cells: recent technological development and practical aspects – from lab to industry. *Sol Energy Mater Sol Cells.* 2018;187:140–53.

[12] ITRPV Roadmap, 12 ed. VDMA, 2021.

[13] Jošt M, Kegelmann L, Korte L, and Albrecht S. Monolithic perovskite tandem solar cells: a review of the present status and advanced characterization methods toward 30% efficiency. *Adv Energy Mater.* 2020;10(26): 1904102.

[14] De Wolf S, Descoeudres A, Holman ZC, and Ballif C. High-efficiency silicon heterojunction solar cells: a review. *Green.* 2012;2:7–24.

[15] Battaglia C, Cuevas A, and Wolf SD. High-efficiency crystalline silicon solar cells: status and perspectives. *Energy Environ Sci.* 2016;9(5):1552–76.

[16] Liu Y, Li Y, Wu Y, *et al.* High-efficiency silicon heterojunction solar cells: materials, devices and applications. *Mater Sci Eng R Rep.* 2020;142:100579.

[17] Sark WGJHM, van Korte L, and Roca F, editors. *Physics and Technology of Amorphous-Crystalline Heterostructure Silicon Solar Cells*, 2011th ed. New York, NY: Springer; 2011. 596 p.

[18] Korte L and Schmidt M. Doping type and thickness dependence of band offsets at the amorphous/crystalline silicon heterojunction. *J Appl Phys.* 2011;109(6):063714.

[19] Fuhs W, Niemann K, and Stuke J. Heterojunctions of amorphous silicon and silicon single crystals. In: *Tetrahedrally Bonded Amorphous Semiconductors.* Yorktown Heights, 1974, pp. 345–50 (American Institute of Physics Conference Proceedings; vol. 20).

[20] Tanaka M, Taguchi M, Matsuyama T, *et al.* Development of new a-Si/c-Si heterojunction solar cells: ACJ-HIT (artificially constructed junction-heterojunction with intrinsic thin-layer). *Jpn J Appl Phys.* 1992;31:3518–22.

[21] Yano A, Tohoda S, Matsuyama K, *et al.* 24.7% Record efficiency HIT[®] solar cell on thin silicon wafer. In: *28th Eur Photovolt Sol Energy Conf Exhib.*, 2013 Nov 22, pp. 748–51.

[22] Taguchi M. Review—Development history of high efficiency silicon heterojunction solar cell: from discovery to practical use. *ECS J Solid State Sci Technol.* 2021;10(2):025002.

[23] Masuko K, Shigematsu M, Hashiguchi T, *et al.* Achievement of more than 25% conversion efficiency with crystalline silicon heterojunction solar cell. *IEEE J Photovolt.* 2014;4(6):1433–5.

[24] Chunduri SK and Schmela M. *Heterojunction Solar Technology*, 2020 ed. Düsseldorf, Germany: TaiyangNews, 2020. Available from: https://taiyang-news.info/reports/heterojunction-solar-technology-2020-report/

[25] Boccard M and Holman ZC. Amorphous silicon carbide passivating layers for crystalline-silicon-based heterojunction solar cells. *J Appl Phys.* 2015; 118(6):065704.

[26] Köhler M, Pomaska M, Procel P, *et al.* A silicon carbide-based highly transparent passivating contact for crystalline silicon solar cells approaching efficiencies of 24%. *Nat Energy.* 2021;6(5):529–37.

[27] Mazzarella L, Kirner S, Stannowski B, Korte L, Rech B, and Schlatmann R. p-type microcrystalline silicon oxide emitter for silicon heterojunction

solar cells allowing current densities above 40 mA/cm^2. *Appl Phys Lett.* 2015;106(2):023902.

[28] Seif JP, Descoeudres A, Filipič M, *et al.* Amorphous silicon oxide window layers for high-efficiency silicon heterojunction solar cells. *J Appl Phys.* 2014;115(2):024502.

[29] Cuevas A and Yan D. Misconceptions and misnomers in solar cells. *IEEE J Photovolt.* 2013;3(2):916–23.

[30] Bivour M, Schröer S, Hermle M, and Glunz SW. Silicon heterojunction rear emitter solar cells: less restrictions on the optoelectrical properties of front side TCOs. *Sol Energy Mater Sol Cells.* 2014;122:120–9.

[31] Koida T, Ueno Y, and Shibata H. In$_2$O$_3$-based transparent conducting oxide films with high electron mobility fabricated at low process temperatures. *Phys Status Solidi A.* 2018;215(7):1700506.

[32] Meza D, Cruz A, Morales-Vilches AB, Korte L, and Stannowski B. Aluminum-doped zinc oxide as front electrode for rear emitter silicon heterojunction solar cells with high efficiency. *Appl Sci.* 2019;9(5):862.

[33] Morales-Vilches AB, Cruz A, Pingel S, *et al.* ITO-free silicon heterojunction solar cells with ZnO:Al/SiO$_2$ front electrodes reaching a conversion efficiency of 23%. *IEEE J Photovolt.* 2019;9(1):34–9.

[34] Barraud L, Holman ZC, Badel N, *et al.* Hydrogen-doped indium oxide/indium tin oxide bilayers for high-efficiency silicon heterojunction solar cells. *Sol Energy Mater Sol Cells.* 2013;115:151–6.

[35] Yu J, Li J, Zhao Y, *et al.* Copper metallization of electrodes for silicon heterojunction solar cells: process, reliability and challenges. *Sol Energy Mater Sol Cells.* 2021;224:110993.

[36] Yao Y, Papet P, Hermans J, *et al.* Module integration of solar cells with diverse metallization schemes enabled by SmartWire Connection Technology. In: *2015 IEEE 42nd Photovoltaic Specialist Conference (PVSC)*, 2015, pp. 1–5.

[37] Geissbühler J, Faes A, Lachowicz C, Ballif C, and Despeisse M. Metallization techniques and interconnection schemes for high efficiency silicon heterojunction PV. *PV Tech.* 2017;37:61–9.

[38] Louwen A, van Sark W, Schropp R, and Faaij A. A cost roadmap for silicon heterojunction solar cells. *Sol Energy Mater Sol Cells.* 2016;147: 295–314.

[39] Tomasi A, Paviet-Salomon B, Jeangros Q, *et al.* Simple processing of back-contacted silicon heterojunction solar cells using selective-area crystalline growth. *Nat Energy.* 2017;2(5):1–8.

[40] Tockhorn P, Wagner P, Kegelmann L, *et al.* Three-terminal perovskite/silicon tandem solar cells with top and interdigitated rear contacts. *ACS Appl Energy Mater.* 2020;3(2):1381–92.

[41] Long W, Yin S, Peng F, *et al.* On the limiting efficiency for silicon heterojunction solar cells. *Sol Energy Mater Sol Cells.* 2021;231:111291.

[42] Richter A, Müller R, Benick J, *et al.* Design rules for high-efficiency both-sides-contacted silicon solar cells with balanced charge carrier transport and recombination losses. *Nat Energy.* 2021;6(4):429–38.

[43] Haase F, Hollemann C, Schäfer S, *et al.* Laser contact openings for local poly-Si-metal contacts enabling 26.1%-efficient POLO-IBC solar cells. *Sol Energy Mater Sol Cells.* 2018;186:184–93.

[44] Yablonovitch E, Gmitter T, Swanson RM, and Kwark YH. A 720 mV open circuit voltage SiOx:c-Si:SiOx double heterostructure solar cell. *Appl Phys Lett.* 1985;47(11):1211–3.

[45] Lindholm FA, Neugroschel A, Arienzo M, and Ilies PA. Heavily doped polysilicon-contact solar. *IEEE Electron Device Lett.* 1985;6(7):363–5.

[46] Yang G, Zhang Y, Procel P, Weeber A, Isabella O, and Zeman M. Poly-Si (O)x passivating contacts for high-efficiency c-Si IBC solar cells. *Energy Proc.* 2017;124:392–9.

[47] Yang G, Guo P, Procel P, Weeber A, Isabella O, and Zeman M. Poly-crystalline silicon-oxide films as carrier-selective passivating contacts for c-Si solar cells. *Appl Phys Lett.* 2018;112(19):1–6.

[48] Stuckelberger J, Nogay G, Wyss P, *et al.* Passivating electron contact based on highly crystalline nanostructured silicon oxide layers for silicon solar cells. *Sol Energy Mater Sol Cells.* 2016;158:2–10.

[49] Nogay G, Stuckelberger J, Wyss P, *et al.* Interplay of annealing temperature and doping in hole selective rear contacts based on silicon-rich silicon-carbide thin films. *Sol Energy Mater Sol Cells.* 2017;173:18–24.

[50] Nogay G, Stuckelberger J, Wyss P, *et al.* Silicon-rich silicon carbide hole-selective rear contacts for crystalline-silicon-based solar cells. *ACS Appl Mater Interfaces.* 2016;8(51):35660–7.

[51] Yang Q, Liu Z, Lin Y, *et al.* Passivating contact with phosphorus-doped polycrystalline silicon-nitride with an excellent implied open-circuit voltage of 745 mv and its application in 23.88% efficiency TOPCon solar cells. *Sol RRL.* 2021;5(11):2100644.

[52] Yang G, Van de Loo B, Stodolny M, *et al.* Passivation enhancement of poly-Si carrier-selective contacts by applying ALD Al_2O_3 capping layers. *IEEE J Photovolt.* 2021;1–8.

[53] van de Loo BWH, Macco B, Schnabel M, *et al.* On the hydrogenation of poly-Si passivating contacts by Al2O3 and SiNx thin films. *Sol Energy Mater Sol Cells.* 2020;215:110592.

[54] Reichel, C. Müller, R. Feldmann, F. Richter, A. Hermle, M, and Glunz SW. Interdigitated back contact silicon solar cells featuring ion-implanted poly-Si/SiOx passivating contacts. In: *33th Eur Photovolt Sol Energy Conf Exhib EU PVSEC,* 2017;

[55] Polzin JI, Feldmann F, Steinhauser B, Hermle M, and Glunz SW. Study on the interfacial oxide in passivating contacts. *AIP Conf Proc.* 2019;2147(1):040016.

[56] Geerligs BLJ, Stodolny MK, Wu Y, *et al.* LPCVD polysilicon passivating contact. In: *Workshop Cryst Silicon Sol Cells Modul Mater Process,* 2016, 28–31.

[57] Schmidt J, Peibst R, and Brendel R. Surface passivation of crystalline silicon solar cells: present and future. *Sol Energy Mater Sol Cells.* 2018;187:39–54.

[58] Steinkemper H, Feldmann F, Bivour M, and Hermle M. Numerical simulation of carrier-selective electron contacts featuring tunnel oxides. *IEEE J Photovolt.* 2015;5(5):1348–56.

[59] Peibst R, Romer U, Larionova Y, *et al.* Working principle of carrier selective poly-Si/c-Si junctions: Is tunnelling the whole story? *Sol Energy Mater Sol Cells.* 2016;158:60–7.

[60] Rienäcker M, Merkle A, Römer U, *et al.* Recombination behavior of photolithography-free back junction back contact solar cells with carrier-selective polysilicon on oxide junctions for both polarities. *Energy Proc.* 2016:412–8.

[61] Yang G, Ingenito A, Van Hameren N, Isabella O, and Zeman M. Design and application of ion-implanted polySi passivating contacts for interdigitated back contact c-Si solar cells. *Appl Phys Lett.* 2016;108(3):033903.

[62] Upadhyaya AD, Ok Y, Chang E, *et al.* Ion-implanted screen-printed n-type solar cell with tunnel oxide passivated back contact. *IEEE J Photovolt.* 2016;6(1):153–8.

[63] Kim H, Bae S, Ji K-sun, *et al.* Passivation properties of tunnel oxide layer in passivated contact silicon solar cells. *Appl Surf Sci.* 2017;409:140–8.

[64] Moldovan A, Feldmann F, Zimmer M, Rentsch J, Benick J, Hermle M. Tunnel oxide passivated carrier-selective contacts based on ultra-thin SiO_2 layers. *Sol Energy Mater Sol Cells.* 2015;142:123–7.

[65] Lozac'h M, Nunomura S, and Matsubara K. Double-sided TOPCon solar cells on textured wafer with ALD SiOx layer. *Sol Energy Mater Sol Cells.* 2020;207:110357.

[66] Stuckelberger J, Yan D, Phang SP, Samundsett C, and Macdonald D. Industrial solar cells featuring carrier selective front contacts. Presented at 36th Eur Photovolt Sol Energy Conf Exhib. 2019;

[67] Ding Z, Yan D, Stuckelberger J, *et al.* Phosphorus-doped polycrystalline silicon passivating contacts via spin-on doping. *Sol Energy Mater Sol Cells.* 2020;221:110902.

[68] Tutsch L, Feldmann F, Polzin J, *et al.* Implementing transparent conducting oxides by DC sputtering on ultrathin SiOx/poly-Si passivating contacts. *Sol Energy Mater Sol Cells*, 2019;200:109960.

[69] Hollemann C, Haase F, Schäfer S, Krügener J, Brendel R, and Peibst R. 26.1%-efficient POLO-IBC cells: quantification of electrical and optical loss mechanisms. *Prog Photovolt Res Appl.* 2019;27(11):950–8.

[70] Wu W, Bao J, Ma L, *et al.* Development of industrial n-type bifacial topcon solar cells and modules Weiliang. In: *36th Eur Photovolt Sol Energy Conf Exhib Dev.*, 2019, October, pp. 2–5.

[71] Chen Y, Chen D, Liu C, *et al.* Mass production of industrial tunnel oxide passivated contacts (i-TOPCon) silicon solar cells with average efficiency over 23% and modules over 345 W. *Prog Photovolt Res Appl.* 2019;27 (10):827–34.

[72]   Feldmann F, Bivour M, Reichel C, *et al.* Tunnel oxide passivated rear contact for large area n-type front junction silicon solar cells providing excellent carrier selectivity. *Sol Energy Mater Sol Cells.* 2014;131(Part A):46–50.

[73]   Ingenito A, Limodio G, Procel P, *et al.* Silicon solar cell architecture with front selective and rear full area ion-implanted passivating contacts. *Sol RRL.* 2017;1(7):1700040.

[74]   Ingenito A, Nogay G, Stuckelberger J, *et al.* Phosphorous-doped silicon carbide as front-side full-area passivating contact for double-side contacted c-Si solar cells. *IEEE J Photovolt.* 2019;9(2):346–54.

[75]   Chen J. The industrial application of n-type bifacial passivating-contact technology. In: *9th Int Conf Cryst Silicon Photovolt,* 2019.

[76]   Feldmann F, Bivour M, Reicherl C, Hermle H, and Glunz SW. A passivated rear contact for high-efficiency. In: *Proc 29th Eur Photovolt Sol Energy Conf Paris Fr,* 2013.

[77]   Glunz SW, Feldmann F, Richter A, *et al.* The irresistible charm of a simple current flow pattern – 25% with a solar cell featuring a full-area back contact. In: *Proc 31st Eur Photovolt Sol Energy Conf Exibition,* 2015, pp. 259–63.

[78]   Hermle M, Feldmann F, Eisenlohr J, *et al.* Approaching efficiencies above 25% with both sides-contacted silicon solar cells. In: *2015 IEEE 42nd Photovolt Spec Conf PVSC 2015,* 2015, pp. 8–10.

[79]   Krügener J, Haase F, Rienäcker M, Brendel R, Osten HJ, and Peibst R. Improvement of the SRH bulk lifetime upon formation of n-type POLO junctions for 25% efficient Si solar cells. *Sol Energy Mater Sol Cells.* 2017;173:85–91.

[80]   Morales-Vilches AB, Larionova Y, *et al.* ZnO:Al/a-SiOx front contact for polycrystalline-silicon-on-oxide (POLO) solar cells. *AIP Conf Proc.* 2018;1999(1):040016.

[81]   Nandakumar N, Rodriguez J, Kluge T, *et al.* Approaching 23% with large-area monoPoly cells using screen-printed and fired rear passivating contacts fabricated by inline PECVD. *Prog Photovolt Res Appl.* 2019; 27(2):107–12.

[82]   Nandakumar N, Rodriguez J, Kluge T, *et al.* 21.6% monoPoly TM cells with in-situ interfacial oxide and poly-Si layers deposited by inline PECVD. In: *2018 IEEE 7th World Conf Photovolt Energy Convers WCPEC 2018 – Jt Conf 45th IEEE PVSC 28th PVSEC 34th EU PVSEC,* 2018, 2048–51.

[83]   Reichel C, Müller R, Feldmann F, Richter A, Hermle M, and Glunz SW. Influence of the transition region between p- and n-type polycrystalline silicon passivating contacts on the performance of interdigitated back contact silicon solar cells. *J Appl Phys.* 2017;122(18):184502.

[84]   Richter A, Benick J, Müller R, *et al.* Tunnel oxide passivating electron contacts as full-area rear emitter of high-efficiency p-type silicon solar cells. *Prog Photovolt Res Appl.* 2018;26(8):579–86.

[85] Richter A, Benick J, Feldmann F, Fell A, Hermle M, and Glunz SW. n-Type Si solar cells with passivating electron contact: identifying sources for efficiency limitations by wafer thickness and resistivity variation. *Sol Energy Mater Sol Cells.* 2017;173:96–105.

[86] Römer U, Merkle A, Peibst R, *et al.* Ion-implanted poly-Si/c-Si junctions as a back-surface field in back-junction back-contacted solar cells. In: *Proc 29th Eur Photovolt Sol Energy Conf.*, 2014, October, pp. 1107–10.

[87] Stodolny MK, Anker J, Geerligs BL, *et al.* Material properties of LPCVD processed n-type polysilicon passivating contacts and its application in PERPoly industrial bifacial solar cells. *Energy Proc.* 2017;124:635–42.

[88] Cuevas A. The early history of bifacial solar cells. In: *Proceedings of the European Photovoltaic Solar Energy Conference*, Munich: WIP, 2005, pp. 801–5.

[89] Chunduri SK and Schmela M. Bifacial Solar Technology 2021 Edition – Part 1: Cells & Modules Report, 2021.

[90] Liang TS, Pravettoni M, Deline C, *et al.* A review of crystalline silicon bifacial photovoltaic performance characterisation and simulation. *Energy Environ Sci.* 2019;12(1):116–48.

[91] Ebong AU, Taouk M, and Wenham SR. A low cost metallization scheme for double sided buried contact silicon solar cells. *Sol Energy Mater Sol Cells.* 1994;31(4):499–507.

[92] Wang X, Allen V, Vais V, *et al.* Laser-doped metal-plated bifacial silicon solar cells. *Sol Energy Mater Sol Cells.* 2014;131:37–45.

[93] Chang Y-C, Wang S, Li S, Rong D, and Ji J. *Applications of simultaneously bifacial plating technology in bifacial solar cells. In: EU PVSEC 2021*, 2021, online.

[94] Fu F, Feurer T, Jager T, *et al.* Low-temperature-processed efficient semi-transparent planar perovskite solar cells for bifacial and tandem applications. *Nat Commun.* 2015;6:8932.

[95] Dupré O, Tuomiranta A, Jeangros Q, Boccard M, Alet P, and Ballif C. Design rules to fully benefit from bifaciality in two-terminal perovskite/silicon tandem solar cells. *IEEE J Photovolt.* 2020;1–8.

[96] De Bastiani M, Mirabelli AJ, Hou Y, *et al.* Efficient bifacial monolithic perovskite/silicon tandem solar cells via bandgap engineering. *Nat Energy.* 2021;1–9.

[97] Du D, Gao C, Zhang D, *et al.* Low-cost strategy for high-efficiency bifacial perovskite/c-Si tandem solar cells. *Sol RRL.* 2021;n/a(n/a).

[98] Kim S, Trinh TT, Park J, *et al.* Over 30% efficiency bifacial 4-terminal perovskite-heterojunction silicon tandem solar cells with spectral albedo. *Sci Rep.* 2021;11(1):15524.

[99] Chantana J, Kawano Y, Nishimura T, Mavlonov A, and Minemoto T. Optimized bandgaps of top and bottom subcells for bifacial two-terminal tandem solar cells under different back irradiances. *Sol Energy.* 2021;220:163–74.

[100]   Khan SN, Ge S, Gu E, *et al.* Bifacial Cu2ZnSn(S,Se)4 thin film solar cell based on molecular ink and rapid thermal processing. *Adv Mater Interfaces.* 2021;8:2100971.

[101]   Nishimura T, Chantana J, Mavlonov A, Kawano Y, Masuda T, and Minemoto T. Device design for high-performance bifacial Cu(In,Ga)Se2 solar cells under front and rear illuminations. *Sol Energy.* 2021;218:76–84.

[102]   Kopecek R and Libal J. Bifacial photovoltaics 2021: status, opportunities and challenges. *Energies.* 2021;14(8):2076.

[103]   Green MA, Wenham SR, Honsberg CB, and Hogg D. Transfer of buried contact cell laboratory sequences into commercial production. *Sol Energy Mater Sol Cells.* 1994;34(1–4):83–9.

[104]   Ebong AU. Double Sided Buried Contact Silicon Solar Cells, 1994.

[105]   Shenvekar S. N-Type TOPCon a Sign of the Times – Jinko Solar, 2021.

[106]   Bhambhani A. CPVT Confirms LONGi Produced First Bifacial Monocrystalline Silicon PERC Solar Cell Exceeding 24% on Commercial Wafer Size, 2019.

[107]   Barth S, Doll O, Koehler I, *et al.* 19.4 Efficient bifacial solar cell with spin-on boron diffusion. *Energy Proc.* 2013;38:410–5.

[108]   Dullweber T, Kranz C, Peibst R, *et al.* PERC+: industrial PERC solar cells with rear Al grid enabling bifaciality and reduced Al paste consumption. *Prog Photovolt Res Appl.* 2016;24(12):1487–98.

[109]   Ebong AU, Lee SH, Warta W, Honsberg CB, and Wenham SR. Characterization of high open-circuit voltage double sided buried contact (DSBC) silicon solar cells. *Sol Energy Mater Sol Cells.* 1997;45(3):283–99.

[110]   Heng J. B, Fu J, Kong B, *et al.* >23% High-efficiency tunnel oxide junction bifacial solar cell with electroplated Cu gridlines. *IEEE J Photovolt.* 2015; 5(1):82–6.

[111]   Longi Solar. Longi achieves 25.21% efficiency for TOPCon solar cell, 2021.

[112]   Pham DP, Lee S, Kim Y, and Yi J. Band-offset reduction for effective hole carrier collection in bifacial silicon heterojunction solar cells. *J Phys Chem Solids.* 2021;154:110059.

[113]   Schmiga C, Grubel B, Cimiotti G, *et al.* 23.8% efficient bifacial i-TOPCon silicon solar cells with <20 μm wide Ni/Cu/Ag-plated contact fingers. In: *Metallization and Interconnection Workshop*, 2021, Genk, Belgium.

[114]   Schultz-Wittmann O, De Ceuster D, Turner A, *et al.* Fine line copper based metallization for high efficiency crystalline silicon solar cells. In: *Proceedings of the 27th European Photovoltaic Solar Energy Conference.* München: WIP; 2012, pp. 596–9.

[115]   Böscke TS, Kania D, Helbig A, *et al.* Bifacial n-type cells with >20% front-side efficiency for industrial production. *IEEE J Photovolt.* 2013;3 (2):674–7.

[116]   Tous L, Russell R, Cornagliotti E, *et al.* 22.4% Bifacial n-PERT cells with Ni/Ag co-plated contacts and Voc~ 691 mV. *Energy Proc.* 2017;124: 922–9.

[117] Wang J, Xuan Y, Zheng L, Xu Y, and Yang L. Improvement of bifacial heterojunction silicon solar cells with light-trapping asymmetrical bifacial structures. *Sol Energy.* 2021;223:229–37.

[118] Wu W, Zhang Z, Zheng F, Lin W, Liang Z, and Shen H. Efficiency enhancement of bifacial PERC solar cells with laser-doped selective emitter and double-screen-printed Al grid. *Prog Photovolt Res Appl.* 2018;26(9):752–60.

[119] Yu B, Shi J, Li F, *et al.* Selective tunnel oxide passivated contact on the emitter of large-size n-type TOPCon bifacial solar cells. *J Alloys Compd.* 2021;870:159679.

[120] Yu J, Bian J, Duan W, *et al.* Tungsten doped indium oxide film: ready for bifacial copper metallization of silicon heterojunction solar cell. *Sol Energy Mater Sol Cells.* 2016;144:359–63.

[121] Rodríguez-Gallegos CD, Liu H, Gandhi O, *et al.* Global techno-economic performance of bifacial and tracking photovoltaic systems. *Joule.* 2020; 4(7):1514–41.

[122] Panasonic Solar. Panasonic Solar, 2018.

[123] Intersolar Europe: SolarWorld to Launch Glass-Glass Bifacial Modules, 2015.

[124] Liang TS, Pravettoni M, Singh JP, and Khoo YS. Meeting the requirements of IEC TS 60904-1-2 for single light source bifacial photovoltaic characterisation: evaluation of different back panel materials. *Eng Res Express.* 2020;2(1):015048.

[125] Richter A, Hermle M, and Glunz SW. Reassessment of the limiting efficiency for crystalline silicon solar cells. *IEEE J Photovolt.* 2013; 3(4):1184–91.

[126] Ingenito A, Isabella O, and Zeman M. Nano-cones on micro-pyramids: modulated surface textures for maximal spectral response and high-efficiency solar cells. *Prog Photovolt Res Appl.* 2015;23(11):1649–59.

[127] Ingenito A, Isabella O, and Zeman M. Experimental demonstration of 4n 2 classical absorption limit in nanotextured ultrathin solar cells with dielectric omnidirectional back reflector. *ACS Photonics.* 2014;1(3): 270–8.

[128] Niewelt T, Richter A, Kho TC, *et al.* Taking monocrystalline silicon to the ultimate lifetime limit. *Sol Energy Mater Sol Cells.* 2018;185:252–9.

[129] Steinhauser B, Niewelt T, Richter A, Eberle R, and Schubert MC. Extraordinarily high minority charge carrier lifetime observed in crystalline silicon. *Sol RRL.* 2021;5(11):2100605.

[130] Fell A, Niewelt T, Steinhauser B, Heinz FD, Schubert MC, and Glunz SW. Radiative recombination in silicon photovoltaics: modeling the influence of charge carrier densities and photon recycling. *Sol Energy Mater Sol Cells.* 2021;230:111198.

[131] Veith-Wolf BA, Schäfer S, Brendel R, and Schmidt J. Reassessment of intrinsic lifetime limit in n-type crystalline silicon and implication on maximum solar cell efficiency. *Sol Energy Mater Sol Cells.* 2018;186:194–9.

[132]   Black LE, Macdonald DH. On the quantification of Auger recombination in crystalline silicon. *Sol Energy Mater Sol Cells.* 2022 Jan;234(September 2021):111428.

[133]   Blinn JF. Models of light reflection for computer synthesized pictures. In: *Proceedings of the 4th Annual Conference on Computer Graphics and Interactive Techniques – SIGGRAPH '77.* New York, NY: ACM Press; 1977, pp. 192–8.

[134]   Niewelt T, Steinhauser B, Richter A, *et al.* Reassessment of the intrinsic bulk recombination in crystalline silicon. *Sol Energy Mater Sol Cells.* 2022;235:111467.

[135]   Dréon J, Jeangros Q, Cattin J, *et al.* 23.5%-Efficient silicon heterojunction silicon solar cell using molybdenum oxide as hole-selective contact. *Nano Energy.* 2020;70:104495.

[136]   PVMD Group, TU Delft. Certified measurement at ISFH CalTeC, Calmark: 002243, device area = 3.915 cm^2, η = (23.83 ± 0.29)%, 13.01.2022.

[137]   Bullock J, Wan Y, Hettick M, *et al.* Dopant-free partial rear contacts enabling 23% silicon solar cells. *Adv Energy Mater.* 2019;9:1803367.

[138]   Yang X, Xu H, Liu W, *et al.* Atomic layer deposition of vanadium oxide as hole-selective contact for crystalline silicon solar cells. *Adv Electron Mater.* 2020;6(8):2000467.

[139]   Mazzarella L, Alcañiz A, Procel P, *et al.* Strategy to mitigate the dipole interfacial states in (i)a-Si:H/MoOx passivating contacts solar cells. *Prog Photovolt Res Appl.* 2021;29(3):391–400.

[140]   Matsui T, Bivour M, Hermle M, and Sai H. Atomic-layer-deposited TiOx nanolayers function as efficient hole-selective passivating contacts in silicon solar cells. *ACS Appl Mater Interfaces.* 2020;12(44):49777–85.

[141]   Lin W, Boccard M, Zhong S, *et al.* Degradation mechanism and stability improvement of dopant-free ZnO/LiFx/Al electron nanocontacts in silicon heterojunction solar cells. *ACS Appl Nano Mater.* 2020;3(11): 11391–8.

[142]   Sailor MJ, Ginsburg EJ, Gorman CB, Kumar A, Grubbs RH, and Lewis NS. Thin films of n-Si/poly-(CH3)3Si-cyclooctatetraene: conducting-polymer solar cells and layered structures. *Science.* 1990;249(4973): 1146–9.

[143]   Gowrishankar V, Scully SR, McGehee MD, Wang Q, and Branz HM. Exciton splitting and carrier transport across the amorphous-silicon/polymer solar cell interface. *Appl Phys Lett.* 2006;89(25):252102.

[144]   Liu C-Y, Holman ZC, and Kortshagen UR. Hybrid solar cells from P3HT and silicon nanocrystals. *Nano Lett.* 2009;9(1):449–52.

[145]   Sun K, Zhang S, Li P, *et al.* Review on application of PEDOTs and PEDOT:PSS in energy conversion and storage devices. *J Mater Sci Mater Electron.* 2015;26(7):4438–62.

[146]   Groenendaal L, Zotti G, Aubert P-H, Waybright SM, Reynolds JR. Electrochemistry of poly(3,4-alkylenedioxythiophene) derivatives. *Adv Mater.* 2003;15(11):855–79.

[147] Heywang G and Jonas F. Poly(alkylenedioxythiophene)s—new, very stable conducting polymers. *Adv Mater.* 1992;4(2):116–8.

[148] Zielke D, Niehaves C, Lövenich W, Elschner A, Hörteis M, and Schmidt J. Organic-silicon solar cells exceeding 20% efficiency. *Energy Proc.* 2015;77:331–9.

[149] Yun MH, Kim JW, Park SY, Kim DS, Walker B, Kim JY.High-efficiency, hybrid Si/C60 heterojunction solar cells. *J Mater Chem A.* 2016 Oct 25; 4(42):16410–7.

[150] Ji W, Allen T, Yang X, Zeng G, De Wolf S, and Javey A. Polymeric electron-selective contact for crystalline silicon solar cells with an efficiency exceeding 19%. *ACS Energy Lett.* 2020;5(3):897–902.

[151] Vos AD. Detailed balance limit of the efficiency of tandem solar cells. *J Phys Appl Phys.* 1980;13(5):839–46.

[152] Brown AS and Green MA. Detailed balance limit for the series constrained two terminal tandem solar cell. *Phys E Low-Dimens Syst Nanostruct.* 2002;14(1–2):96–100.

[153] Futscher MH and Ehrler B. Efficiency limit of perovskite/si tandem solar cells. *ACS Energy Lett.* 2016;1(4):863–8.

[154] Steirer KX, Schulz P, Teeter G, *et al.* Defect tolerance in methylammonium lead triiodide perovskite. *ACS Energy Lett.* 2016;1(2):360–6.

[155] De Wolf S, Holovsky J, Moon S-J, *et al.* Organometallic halide perovskites: sharp optical absorption edge and its relation to photovoltaic performance. *J Phys Chem Lett.* 2014;5(6):1035–9.

[156] Jäger K, Korte L, Rech B, and Albrecht S. Numerical optical optimization of monolithic planar perovskite-silicon tandem solar cells with regular and inverted device architectures. *Opt Express.* 2017;25(12): A473.

[157] Aydin E, Allen TG, De Bastiani M, *et al.* Interplay between temperature and bandgap energies on the outdoor performance of perovskite/silicon tandem solar cells. *Nat Energy.* 2020 Nov;5(11):851–9.

[158] Unger EL, Kegelmann L, Suchan K, Sörell D, Korte L, and Albrecht S. Roadmap and roadblocks for the band gap tunability of metal halide perovskites. *J Mater Chem A.* 2017;5(23):11401–9.

[159] Bush KA, Palmstrom AF, Yu ZJ, *et al.* 23.6%-efficient monolithic perovskite/silicon tandem solar cells with improved stability. *Nat Energy.* 2017;2(4):17009.

[160] Sahli F, Werner J, Kamino BA, *et al.* Fully textured monolithic perovskite/ silicon tandem solar cells with 25.2% power conversion efficiency. *Nat Mater.* 2018;17(9):820–6.

[161] Al-Ashouri A, Magomedov A, Roß M, *et al.* Conformal monolayer contacts with lossless interfaces for perovskite single junction and monolithic tandem solar cells. *Energy Environ Sci.* 2019;12(11):3356–69.

[162] Al-Ashouri A, Köhnen E, Li B, *et al.* Monolithic perovskite/silicon tandem solar cell with >29% efficiency by enhanced hole extraction. *Science.* 2020;370(6522):1300–9.

[163]   Helmholtz-Zentrum Berlin für Materialien und Energie. World record again at HZB: Almost 30 % efficiency for next-generation tandem solar cells, 2021 [cited 2022 Feb 17]. Available from: https://www.helmholtz-berlin.de/pubbin/news_seite?nid=23248;sprache=en

[164]   NREL. Best Research-Cell Efficiency Chart. [cited 2022 Feb 28]. Available from: https://www.nrel.gov/pv/cell-efficiency.html

[165]   Jošt M, Lipovšek B, Glažar B, *et al.* Perovskite solar cells go outdoors: field testing and temperature effects on energy yield. *Adv Energy Mater.* 2020;10(25):2000454.

[166]   Messmer C, Schön J, Lohmüller S, *et al.* How to make PERC suitable for perovskite–silicon tandem solar cells: a simulation study. *Prog Photovolt Res Appl.* 2022;early view. [cited 2022 Jan 24]. Available from: https://onlinelibrary.wiley.com/doi/abs/10.1002/pip.3524

[167]   Mariotti S, Jäger K, Diederich M, *et al.* Monolithic perovskite/silicon tandem solar cells fabricated using industrial p-type polycrystalline silicon on oxide/passivated emitter and rear cell silicon bottom cell technology. *Sol RRL.* 2022;n/a:2101066.

[168]   Essig S, Allebé C, Remo T, *et al.* Raising the one-sun conversion efficiency of III–V/Si solar cells to 32.8% for two junctions and 35.9% for three junctions. *Nat Energy.* 2017;2(9):1–9.

[169]   Schygulla P, Müller R, Lackner D, *et al.* Two-terminal III–V//Si triple-junction solar cell with power conversion efficiency of 35.9 % atAM1.5g. *Prog Photovolt Res Appl.* ;n/a(n/a) [cited2021 Nov 12]. Available from: https://onlinelibrary.wiley.com/doi/abs/10.1002/pip.3503

[170]   Vurgaftman I, Meyer JR, and Ram-Mohan LR. Band parameters for III–V compound semiconductors and their alloys. *J Appl Phys.* 2001;89(11): 5815–75.

[171]   B. Bolkhovityanov Y and P. Pchelyakov O. III–V Compounds-on-Si: heterostructure fabrication, *application and prospects.* Open Nanosci J. 2009 Oct 27 [cited 2022 Feb 28];3(1). Available from: https://benthamopen.com/ABSTRACT/TONANOJ-3-20

[172]   Grassman TJ, Brenner MR, Rajagopalan S, *et al.* Control and elimination of nucleation-related defects in GaP/Si(001) heteroepitaxy. *Appl Phys Lett.* 2009;94(23):232106.

[173]   Warren EL, Kibbler AE, France RM, Norman AG, Stradins P, and McMahon WE. Growth of antiphase-domain-free GaP on Si substrates by metalorganic chemical vapor deposition using an in situ AsH3 surface preparation. *Appl Phys Lett.* 2015;107(8):082109.

[174]   Grassman TJ, Chmielewski DJ, Carnevale SD, Carlin JA, and Ringel SA. GaAs0.75P0.25/Si dual-junction solar cells grown by MBE and MOCVD. *IEEE J Photovolt.* 2016;6(1):326–31.

[175]   Feifel M, Lackner D, Schön J, *et al.* Epitaxial GaInP/GaAs/Si triple-junction solar cell with 25.9%AM1.5g efficiency enabled by transparent metamorphic $Al_xGa_{1-x}AsyP_{1-y}$ step-graded buffer structures. *Sol RRL.* 2021;5(5):2000763.

[176] Veinberg-Vidal E, Vauche L, Medjoubi K, *et al.* Characterization of dual-junction III–V on Si tandem solar cells with 23.7% efficiency under low concentration. *Prog Photovolt Res Appl.* 2019;27(7):652–61.

[177] Mizuno H, Makita K, and Matsubara K. Electrical and optical interconnection for mechanically stacked multi-junction solar cells mediated by metal nanoparticle arrays. *Appl Phys Lett.* 2012;101(19):191111.

[178] Makita K, Mizuno H, Tayagaki T, *et al.* III–V//Si multijunction solar cells with 30% efficiency using smart stack technology with Pd nanoparticle array. *Prog Photovolt Res Appl.* 2020;28(1):16–24.

[179] Victoria M, Haegel N, Peters IM, *et al.* Solar photovoltaics is ready to power a sustainable future. *Joule.* 2021;5(5):1041–56.

[180] Ritchie H and Roser M. Renewable Energy | Our World in Data. Our World in Data. 2020 [cited 2022 Feb 22]. Available from: https://ourworldindata.org/renewable-energy

[181] Sustainable Development Goals | United Nations Development Programme. United Nations Development Programme. [cited 2022 Feb 22]. Available from: https://www.undp.org/sustainable-development-goals

[182] Global Wind Report 2021. Global Wind Energy Council. 2021 [cited 2022 Feb 22]. Available from: https://gwec.net/global-wind-report-2021/

[183] Solar PV – Renewables 2020 – Analysis. IEA. [cited 2022 Feb 22]. Available from: https://www.iea.org/reports/renewables-2020/solar-pv

[184] Utility scale solar reaches LCOE of $0.028-$0.041/kWh in the US, Lazard finds. *PV Magazine International.* [cited 2022 Feb 22]. Available from: https://www.pv-magazine.com/2021/11/05/utility-scale-solar-reaches-lcoe-of-0-028-0-041-kwh-in-the-us-lazard-finds/

[185] Levelized Cost of Electricity: Renewables Clearly Superior to Conventional Power Plants Due to Rising CO2 Prices – Fraunhofer ISE. Fraunhofer Institute for Solar Energy Systems ISE. [cited 2022 Feb 22]. Available from: https://www.ise.fraunhofer.de/en/press-media/press-releases/2021/levelized-cost-of-electricity-renewables-clearly-superior-to-conventional-power-plants-due-to-rising-co2-prices.html

[186] Wade W. Power Plants Get More Expensive But Renewables Still Cheapest. Bloomberg.com. 2021 Dec 21 [cited 2022 Feb 22]; Available from: https://www.bloomberg.com/news/articles/2021-12-21/power-plants-get-more-expensive-but-renewables-still-cheapest

[187] Vuichard P, Broughel A, Wüstenhagen R, Tabi A, and Knauf J. Keep it local and bird-friendly: exploring the social acceptance of wind energy in Switzerland, Estonia, and Ukraine. *Energy Res Soc Sci.* 2022;88:102508.

[188] Ritchie H and Roser M. Urbanization | Our World in Data. Our World Data. 2018 Jun 13 [cited 2022 Feb 22]; Available from: https://ourworldindata.org/urbanization

[189] Calcabrini A, Muttillo M, Weegink R, Manganiello P, Zeman M, and Isabella O. A fully reconfigurable series-parallel photovoltaic module for higher energy yields in urban environments. *Renew Energy.* 2021;179:1–11.

[190] Ziar H, Manganiello P, Isabella O, and Zeman M. Photovoltatronics: intelligent PV-based devices for energy and information applications. *Energy Environ Sci.* 2021;14(1):106–26.

[191] Fraunhofer Institute for Solar Energy Systems, ISE. *Photovolt Report.* 2021;2021:50.

[192] Solar continues to break installation records, on track for Terawatt scale by 2022 – SolarPower Europe. SolarPower Europe. [cited 2022 Feb 22]. Available from: https://www.solarpowereurope.org/solar-continues-to-break-installation-records-on-track-for-terawatt-scale-by-2022/

[193] Jinko Solar. JinkoSolar's High-efficiency N-Type Monocrystalline Silicon Solar Cell Sets Our New Record with Maximum Conversion Efficiency of 26.4%, 2022. https://www.jinkosolar.com/en/site/newsdetail/1827

[194] World record back at HZB: Tandem solar cell achieves 32.5 percent efficiency. [cited 2023 Jan 05]. Available from: https://www.helmholtz-berlin.de/pubbin/news_seite?nid=24348&sprache=en&seitenid=1

Chapter 8

Summary

Delfina Muñoz[1]

8.1 Technological lessons learned from this book

The first solar cell fabricated by Bell Laboratories in 1954 was made of an n-type silicon wafer. Despite this head start, much of the research, development, and technological advances in solar photovoltaics (PV) were focused around p-type silicon wafers, thanks to their outer space performance. We have needed more than 60 years to come back to n-type and trust the early believers. A key point was to separate the space and the terrestrial residential and commercial solar markets to demonstrate that the final application is the driver for technology selection. So now, it is time for a change.

In the last few years, n-type cells have begun to accumulate market share due to all the advantages concerning performance, cost, manufacturing, and many other. In this sense, we believe that it is high time to compile this worldwide knowledge in a book allowing the community to understand why n-type are here and which challenges are next.

The main goal of this book was to review all the development of the last 20 years concerning n-type technologies and show the next steps to move to a photovoltaic terawatt era.

Obviously, the main change between p- and n-type solar cells relies on the silicon wafer. An extensive review on N-type silicon feedstock and wafers is done in Chapter 1 considering growth, wafering, and material engineering technologies. n-Type solar cells rely on high-quality Cz material to obtain very high performances avoiding any degradation. Current status is reported and main challenges to improve not only performance indicators but also cost issues are presented through wafer cost as still today one of the main drivers for the whole PV system. Even if some of the silicon improvements are common for p- and n-type wafers, n-type face dedicated items addressed today to improve production of n-type solar cells and modules.

The core of the book relies on Chapter 2 where n-type solar cells are described extensively. A technological review of the main n-type solar cells and their comparison to mainstream p-technologies is presented. It is important to mention that n-type solar cells architectures diverge from standard p-type cells with a variety of

[1]CEA, LITEN, INES, Université Grenoble Alpes, France

structures based on high and low temperature passivated contact approaches. All of them are presented and the state-of-the-art and challenges are reported. Even if major record devices of the last 10 years are obtained on n-type solar cells, industry has only started in 2020 to plan first Gigawatt factories based on n-type technology waiting also for dedicated equipment and facilities for such changes.

We cannot forget that the commercial product is the PV module that allows solar cells to perform in real conditions in the field. Chapter 3 is dedicated to module technology covering specific aspects for n-type technology. Here, an extensive review on module manufacturing and characterization tools is presented. Module technology has much evolved the last years to improve performance and reduce manufacturing costs at the R&D level but also in production. A variety of new architectures based on half, quarter cells, with interconnection improvements to reduce silver consumption but also to densify and optimize the energy produced by every m^2 is described in detail.

Two main topics are transversal to these two chapters: bifaciality and silver reduction. In both cases, there is a strong impact in levelized cost of energy (LCOE) (Chapter 7).

Bifaciality [1] is mainstream today for large area plants, enhancing energy produced by factors between 10% and 30% in function of the location and the technology. n-Type solar cells have bifaciality ratios higher than passivated emitter and rear contact (PERC) (intrinsically limited to $\approx 75\%$ [2]). They have already demonstrated $> 90\%$ for heterojunction technology.

Silver reduction is mandatory today for two reasons: cost and sustainability [3]. To address the massive penetration of PV in the market, a strong reduction in critical materials is needed and Silver plays the major role for all technologies. Reduction of silver plays also a crucial role on the module design, either for interconnection schemes but also for reliability issues making encapsulation requirements stronger.

The module is the photovoltaic product, but at the end, we need a whole system to generate energy. Chapter 4 gives an overview of a photovoltaic system. There is worth to mention that even if p- and n-type modules do not affect the system architecture, there are several changes in function of the application. First, the penetration of higher efficiency technologies have influenced in the system design and components. Second, the variety of systems and applications are increasing, and the system design is now done in a more integrative way considering the maximization of the energy produced and the reliability of the system (without forget the operation and maintenance activities of the plant).

After the key technological chapters on actual status of an n-type PV system and cost and sustainability analysis is presented in Chapter 5. This chapter compares the cost of ownership and the levelized cost of energy of the different solar cell and modules architectures considering different scenarios. Even if at current module prices, in high irradiance locations, extremely low LCOE can be achieved, a further cost reduction will open up more and more applications for PV generated electricity (for instance, green hydrogen, methane, or ammonia). Thus, PV-generated electricity will remain the cheapest energy source even when including the cost of storage in batteries or by other means. In addition, of course, also in lower irradiance regions, the LCOE of PV will become increasingly competitive compared to other methods of electricity generation widening the PV market. life cycle analysis (LCA) overview shows that the high efficiency of then-type technologies as well as their system performance (bifacial gain, degradation rate, etc.) offers a large potential to improve the environmental profile of PV system and make the PV energy "greener" by excellence.

A strong conclusion of this chapter deals with the impact of increased efficiency on LCOE and LCA analysis to improve cost but also sustainability. Today there is room for improvement to increase efficiency with boundary conditions related to reduce material use for example.

Chapter 6 shows the latest R&D results to increase efficiency based in two different concepts: single junction devices closing the gap to the silicon limit with nearly 27% efficiency and new silicon tandem architectures. In this sense, approaches based on passivated carrier-selective contacts with different materials are presented with the new requirements to overcome 27% efficiency. Thus, optimization concerns not by layer but also on device level are discussed with special focus on interface management. This approach is cross-linked with equipment suppliers though processing of the layer is extremely critical for moving forward.

Regarding tandem or multijunction integration on silicon, several approaches are presented based on different materials with compatible bandgap to achieve >30%. As mentioned, a lot of interest by the PV community is given to this approach in the last 3–4 years especially for the perovskite/silicon tandem architecture. Efficiencies are moving forward so quick, that during the book preparation, we need to update several tables and the latest record was published under the review phase by EPFL and CSEM and, for the first time, a silicon-based tandem solar cell exceeded 31% efficiency [4] showing the strong potential of this technology. This result was using n-type heterojunction solar cell. Of course, we are still at the R&D level, small areas, non-industrial processes. However, everything already learnt from PV mass production is easily transferable to this architecture, firmly believing that it will become a reality before 2030 [5].

To conclude, we can say that today the n-type PV community has entered the adult phase. The complete ecosystem needed for mass production is here and:

1. Several n-type technologies have demonstrate the lower LCOE and lower LCA.
2. Equipment manufacturers are here to make it possible with the different solar cell and module configurations.
3. PV module producers are manufacturing at the GW scale, and n-type is increasing rapidly market share as expected by the ITRPV [6].
4. There is a next step: efficiencies and performances are moving forward fast at the R&D level and supported by the industrial ecosystem to improve LCOE and LCA.

8.2 Impact of n-type PV

If we look back to the story of PV, every 3 years, the yearly installations are at least doubling. The milestone of 1 TW total PV installations was reached March 2022.

Moreover, this year 250 GW are still in construction and will be added. It means that in around 6 years (from 2028), we will enter a yearly 1 TW PV market. This numbers are amazing for the world and the planet, meaning that PV can contribute very fast to an energy system based on 100% renewables costly effective for 2050 as suggested by Breyer *et al.* [7].

Then PV production needs to follow this trend. The current production capacity is 0.5 TW. We need to double the PV production to feed the market needs.

An amazing similar trend is to look at funded European Projects based on n-type technologies allowing Europe to position premium technologies in the market again. This collaborative work is between countries, but also between all the segments needed to fill the value chain for PV: university, institutes, suppliers, equipment manufacturers, producers, etc.

In Figure 8.1, we can see a graph of majors n-type projects funded by European Commission in different program frameworks: FP7, H2020 and Horizon Europe. All these projects (and of course, others improving PV as a whole or contributing to global understanding) helped to:

1. Create the collaborative ecosystem in Europe.
2. Increase TRL level of the different n-type technologies

     One major example is the manufacturing of heterojunction technology. It started with the HETSI project in 2006 to demonstrate the TRL5. Then, HERCULES pushed the technology to TRL6, enhancing the capacities of Meyer Burger. Finally, AMPERE reached the pilot industrial phase at TRL7 around EGP production facilities. After AMPERE, Meyer Burger and EGP are the main producers of Heterojunction in Europe. In the European and complex

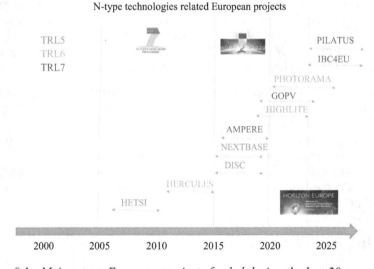

Figure 8.1   Main n-type European projects funded during the last 20 years in the different program frameworks with a logical TRL evolution

context of PV, we needed 15 years of collaborative work and funding to reach the market. Similar trend is observed with IBC-Heterojunction where HERCULES was at TRL4 and NExtBase at 5, and the new PILATUS project objective is to reach the TRL7 in 2–3 years. Alternatively, with IBC, where after the research phase, now IBC4EU is willing to push the n-type IBC to TRL7 creating the whole ecosystem for PV production.

3. Create different networks of PV European associative entities like Solar Power Europe (SPE), European Solar Manufacturing Council (ESMC), and European Technology and Innovation Platform for PV (ETIP-PV) which are pillars for European Commission to define the roadmaps, actions, strategy for PV in the whole value chain.

Finally yet importantly, in 2022, European Commission allocated the Innovation Fund to Enel to move up to 3 GW of solar cell and module hetero-junction production in Catania (Italy) [8]. The barrier of the European GW production will be then exceeded beginning of 2023. Meyer Burger is also planning to exceed 1 GW in Germany in 2023 [9]. Very recent news allocated another Innovation Fund from European Union for GW module production to REC Solar to build a 2-GW module factory in France [10].

8.3 What's coming next for PV and n-type

Even if in this book, we have focused on the technological effort to improve LCOE in PV systems, and how research has overcome all the challenges to succeed in the last decades, there is still work to do to enter the Terawatt era for photovoltaics and a new paradigm appear.

In this sense, efficiency, performance, and cost are not the only drivers for PV community as in previous decades. A new asset of words and challenges arise for PV. We can summarize them in three main families.

8.3.1 Environmental framework & circular economy of PV

The circular economy (CE), which is named in contrast to the current global economic model that treats materials in a linear take-make-waste manner, gives us the tools to tackle climate change and biodiversity loss together, while addressing important social needs. Renewable energy, and in particular PV, needs to fulfill CE approach, not only to be cost competitive, but also sustainable and aligned with United Nations Sustainable Development goals, especially *Goal 12: Sustainable production and consumption* [11]. The European Union has been one of the pioneers establish policies for CE, European Union's CE Action Plan (European Commission 2015, 2021).

The World Economic Forum defined CE as follows: *A circular economy is an industrial system that is restorative or regenerative by intention and design. It replaces the end-of-life concept with restoration, shifts towards the use of renewable energy, eliminates the use of toxic chemicals, which impair reuse and return to the biosphere, and aims for the elimination of waste through the superior design of materials, products, systems and business models.*

CE literature generally agrees on the classification of 10 CE strategies, labeled R0 to R9, to signify their stage of the life cycle: R0 = Refuse, R1 = Rethink, R2 = Reduce, R3 = Reuse, R4 = Repair, R5 = Refurbish, R6 = Remanufacture, R7 = Repurpose, R8 = Recycle and R9 = Recover (energy).

Today, the challenge created by the estimated growth of PV to 3 TW for 2030 is a concomitant increase in demand for materials to support the manufacture of PV technologies and reduction in environmental footprint. Creating a sustainable value chain and manufacturing capacity becomes mandatory to achieve such goal. Two CE strategies – refuse and reduce – are applied in the PV manufacturing phase. In the refuse strategy, materials that are environmentally toxic and pose hazards to human health are minimized or eliminated when manufacturing a PV system. These include, for instance, fluorinated backsheets, and lead-based solders.

Considering reduction, by far the largest mass fraction in a PV module is from bulk material: glass, Al, and Si. Luckily, demand for these bulk materials is not expected to cause market disruption or supply shortages [12]. The primary concern for photovoltaics is silver due to its scarcity and widespread use in essentially all current implementations of industrial silicon solar cell technologies [3]. In addition, there are significant concerns for the use of indium if the manufacturing capacity of SHJ and future tandem devices, and even the use of bismuth in the low-temperature interconnection approach. These two last materials are also considered in the EC CRM list [26].

Some studies show material consumption targets of 2 mg/W, 0.38 mg/W, and 1.8 mg/W for silver, indium, and bismuth, respectively, indicating significant material consumption reductions are required to meet the target production rate for sustainable multi-TW scale manufacturing in 2030 [12]. Continuous increase of efficiency and power of solar cells and modules will substantially reduce the consumption of these materials in terms of mg/W over time. n-Type technologies are moving forward in that direction but need a specific effort to reduce material intensity, especially low T° approaches like SHJ. Alternatives to indium are already under study and close to production. One route considers replacing indium-based TCOs by zinc-based TCOs (aluminum or boron doped) [13,14] on what is called In-free solar cells. The second route considers an extremely thin indium-based layer (< 30 nm) coupled with a dielectric and antireflective coating. Several options have been recently (SiN, SiO_x, TiO_x) with high efficiencies and promising results [15–18].

Finally yet importantly, components involved at the system level (Cu, Al, steel) are not yet blocking the TW deployment of PV in terms or raw material availability.

Another way to mitigate material supply challenges is to recover materials from products at the end of their lifetime through recycling, the most obvious end-of-pipe solution (and conforms best to linear economy thinking). Considerable research has focused on investigating the potential of this strategy and developing technological solution. Status of PV recycling shows three main challenges:

1. Lack of integrated process to recycle *all* materials in a c-Si PV module and preserve high-purity grade.
2. Cost for increased adoption of PV recycling (open and close loop).
3. Lack of robust projections of end-of-life market.

The pioneers of closed loop recycling in PV is First solar, with more than 90% module recycling material recovery rate [19].

Another example of CE approach is the reparability-by-design of PV modules improving their life span generating an environmental benefit by preventing premature and destructive recycling or landfilling of a significant volume of PV. In *repair*, functional issues and defects of a PV system are resolved and the PV system continues to be used. To date, repair has been found to be capable of addressing defects in a module's bypass diodes, encapsulant, junction box, backsheet, glass, and connectors [20]. In *reuse* technical guidelines for controlled and safe operations of PV systems are also proposed [21].

The emerging digital pathways for efficient PV CE strategy are summarized in Figure 8.2 and reported recently by Heath *et al*. [22].

Regarding n-type technologies today, as shown in Chapter 7, favorable LCA is shown, but that is not enough. Several new European Projects have started in the last few years to address sustainability: CIRCUSOL, RESILEX, PHOTORAMA, and a new paradigm has to be adopted where performance and cost are as important as sustainability. This road to completely circular economy approach will secure a sustainable and massive PV deployment. This is a must.

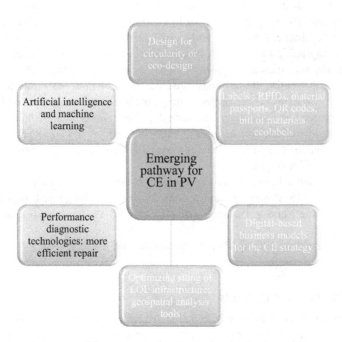

Figure 8.2 Summary of key aspects for PV technologies to achieve an efficient circular economy strategy

8.3.2   Social and market acceptance

The issue of social acceptance for PV remained largely neglected in the 1990s, because of a high level of general public support for renewable energy technologies. However, there is more than one aspect of social acceptance that arises in this decade with the massive deployment of PV worldwide [23].

Community acceptance refers to the specific acceptance of siting decisions and renewable energy projects by local stakeholders, particularly residents and local authorities [24]. Even if in Europe public opinion of energy developments is generally favorable for PV, many projects deal with local resistance or discontent, which in many cases is a significant barrier to development [25]. In the case of on-ground PV, the public acceptance of dedicated sites and their visual disturbance have been acknowledged as important sustainability criteria when these systems are not out of sight, and/or when the size of the project is big (large land areas covered) [24].

It is important to mention that social acceptance of a new infrastructure is possible when welfare decreasing-aspects of the project are balanced by welfare increasing aspects to leave all agent at least welfare neutral and indifferent to the completion of the project, or even better off and supportive. In contrast to traditional fossil fuels, the use of PV (and large-scale PV arrays in particular) makes new energy-oriented land use and landscape transformations visible because the energy generators are close to the places where people live.

Three factors are the most common barriers preventing further adoption of renewables and PV in particular: (1) low availability of information about the technology, (2) financial concerns and lack of business models and legal issues, and (3) sociodemographic factors (e.g., income level and educational level).

Therefore, activities aimed at disseminating information about PV technology should focus on tackling these issues as well as implement demonstrations that test innovative PV systems. These projects should include the development and conduct of feasibility studies. Moreover, additional policies need to be implemented to support these dissemination activities, including grants or loans. These policies could also promote a change in attitude about the financial and durability concerns related to PV technologies.

Thus, finding a balance among economic benefits, landscape protection, and people needs for local progress and preservation is a challenge that must be faced by energy planners and governments on different scales to establish a sustainable energy program.

Then, a need of linking PV, environment, and society arises, creating sustainable PV ecosystems.

8.3.3   Integration & landscape

Ignoring that PV will be one of the major drivers for energy transition is today a nonsense. There are some strong indications that c-Si photovoltaics could become the most important electricity source by 2040–2050 on a global scale. However, with the increase of both the number and size of installations, the attention to their impacts in terms of land-use and land-transformation is growing, as well as concerns about landscape preservation and possible losses of ecosystem services. The current design is generally straightforward and is aimed to the maximize energy generation given a certain land area. PV systems should be designed as an element of the landscape or the ecosystem they belong to, according to an "inclusive" design approach that does not focus only on the overall energy efficiency of the system, but extends to other additional ecological and landscape objectives.

The so-called X-IPV (integrated photovoltaics) was born to exploit as many surfaces as possible in configuration unlocking a huge potential for renewable electricity generation. It integrates the classical building approach (building integrated and applied photovoltaics BIPV, BAPV) but also the infrastructure and mobility one (vehicle-integrated PV, road-integrated PV, urban PV, etc.).

Thus, specific requirements arising from the applications, for instance, minimized weight per area or extremely high mechanical resistance can be meet by the choice of suitable materials, module and solar cell configuration. For example, maximum energy yield within a limited space is a requirement in mobility applications, which especially applies to vehicle-integrated PV modules. In addition, passenger cars and building-integrated PV applications place high aesthetic demands on integrated PV.

Besides, the integration of photovoltaics is often accompanied by an adaptation and optimization of the entire electrical system and intelligent energy management systems dedicated to the application.

Another relevant integration approach leads with the combination of PV with agricultural activity, the so-called Agrivoltaics or AgriPV. Through dual land use, agrivoltaics not only increases land efficiency but also the resilience of agriculture to the consequences of climate change preserving the global ecosystem water–energy–food. Agrivoltaics provides farmers with additional income and promotes the economic development of rural areas.

Similarly, the integration of PV on water, industrial sea shores, lakes, ponds, etc. (floating PV) also has a strong potential to exploit for PV deployment on large and not shadowed areas.

Both of them deal with different challenges, which affect the technology from the materials selection to the module architecture up to the global system design to fulfill the performance, cost, sustainability, aesthetics, reliability requirements to be socially and market accepted.

The n-type technologies presented offer a set of technologies which are flexible to be integrated in different configurations, tuning materials, bifaciality ratios, weight, and different module designs adapted to the final application.

The new paradigm for PV is to move to application-driven product design and moving away from the technology push era. Then, available technologies in the market cover a set of technologies varying the tradeoff between performance, cost, durability, and many others.

8.4   Conclusion

The goal of a great ending is to tie everything together, but that was already the goal of this book. A sum up of all lessons learned by the nPV community during the last years. The nPV workshop was one of the pillars for collaborative exchange and discussion. Now, the n-type book is also coming out as second pillar. The third pillar is the reality.

My impression is that many outstanding experts already say everything in the previous pages, so nothing is left to say at the end. Nevertheless, n-type technologies and PV in general have a lot to say in the next decades for the energy transition. Reducing carbon dioxide (CO_2) emissions is at the heart of the world's accelerating shift from climate-damaging fossil fuels towards clean, renewable forms of energy. The steady rise of PV energy generation forms a vital part of this global energy transformation. The global PV community plays a crucial role for the preservation of our planet and human wellbeing, and most importantly for our children's future.

References

[1]   Libal J and Kopecek R. *Bifacial Photovoltaics*: *Technology, Applications and Economics.* Stevenage, UK: The Institution of Engineering and Technology; 2018.

[2]   Yu J, Wang P, Chen K, *et al.* Improved bifacial properties of p-type passivated emitter and rear cell solar cells toward high mass production efficiency. *Phys. Status Solidi A* 2021;218:2100059.

[3]   Zhang Y, Kim M, Wang L, Verlinden P, and Hallam B. Design considerations for multi-terawatt scale manufacturing of existing and future photovoltaic technologies: challenges and opportunities related to silver, indium and bismuth consumption. *Energy Environ. Sci.* 2021;14:5587.

[4]   https://www.pv-magazine.com/2022/07/07/csem-epfl-achieve-31-25-efficiency-for-tandem-perovskite-silicon-solar-cell/

[5]   https://www.oxfordpv.com/

[6]   https://www.vdma.org/international-technology-roadmap-photovoltaic

[7]   Breyer C, Khalili S, Bogdanov D, *et al.* On the history and future of 100% renewable energy systems research, *IEEE Access.* 2022;10:78176–78218.

[8]   https://www.enel.com

[9]   https://www.meyerburger.com

[10]   https://www.pv-magazine.com/2022/07/13/french-hjt-solar-plant-batteryprojects-land-eu-funding

[11]   https://sdgs.un.org/goals

[12]   Valero A, Valero A, and Calvo G. Summary and critical review of the International Energy Agency's special report: The role of critical minerals in clean energy transitions. *Rev. Metal.* 2021;57(2):e197.

[13]   Cruz A, Ruske F, Eljarrat A, *et al.* Influence of silicon layers on the growth of ITO and AZO in silicon heterojunction solar cells. *IEEE J. Photovolt.* 2020;10(2):703–709, doi:10.1109/JPHOTOV.2019.2957665.

[14] Wu Z, Duan W, Lambertz A, *et al.* Low-resistivity p-type a-Si:H/AZO hole contact in high-efficiency silicon heterojunction solar cells. *Appl. Surf. Sci.* 2021;542:148749, doi:10.1016/j.apsusc.2020.148749.

[15] Boccard M, Antognini L, Cattin J, *et al.* Paths for maximal light incoupling and excellent electrical performances in silicon heterojunction solar cells. *Proceedings of the 2019 IEEE 46th Photovoltaic Specialists Conference (PVSC)*, Chicago, IL: IEEE, June 2019, pp. 2541–2545.

[16] Cruz A, Erfurt D, Wagner P, *et al.* Optoelectrical analysis of TCO+Silicon oxide double layers at the front and rear side of silicon heterojunction solar cells, *Sol. Energy Mater. Sol. Cells* 2022;236:111493.

[17] Du G, Bai Y, Huang J, *et al.* Surface passivation of ITO on heterojunction solar cells with enhanced cell performance and module reliability, *ECS J. Solid State Sci. Technol.* 2021;10(3):035008.

[18] Liu W, Zhang L, Yang X, *et al.* Damp-heat-stable, high-efficiency, industrial-size silicon heterojunction solar cells. *Joule* 2020;4:913–927.

[19] https://www.firstsolar.com/-/media/First-Solar/Sustainability-Documents/ FirstSolar_Sustainability-Report_2021.ashx

[20] Beaucarne G, Eder G, Jadot E, Voronko Y, and Mühleisen W. Repair and preventive maintenance of photovoltaic modules with degrading backsheets using flowable silicone sealant. *Prog. Photovolt. Res. Appl.* 2022;30: 1045–1053.

[21] Van der Heide A, Tous L, Wambach K, Poortmans J, Clyncke J, and Voroshazi E. Towards a successful re-use of decommissioned photovoltaic modules. *Prog. Photovolt. Res. Appl.* 2022;30:910–920.

[22] Heath GA, Ravikumar D, Hansen B, and Kupets E. A critical review of the circular economy for lithium-ion batteries and photovoltaic modules – status, challenges, and opportunities. *J. Air Waste Manag. Assoc.* 2022;72(6): 478–539.

[23] Peñaloza D, Mata É, Fransson N, *et al.* Social and market acceptance of photovoltaic panels and heat pumps in Europe: A literature review and survey. *Renew. Sustain. Energy Rev.* 2022;155:111867.

[24] Sconnamiglio A. 'Photovoltaic landscapes': Design and assessment. A critical review for a new transdisciplinary design vision. *Renew. Sustain. Energy Rev.* 2016; 55:629–661.

[25] Cohen JJ, Reich J, and Schmidthaler M. Re-focusing research efforts on the public acceptance of energy infrastructure: A critical review, *Energy* 2014: 76:4–9.

[26] https://ec.europa.eu/growth/sectors/raw-materials/areas-specific-interest/critical-raw-materials_en

Index